Physical Geography

Cuchlaine A. M. King

BASIL BLACKWELL · OXFORD

First published in 1980 by
Basil Blackwell, Publisher
5 Alfred Street
Oxford OX1 4HB

British Library Cataloguing in Publication Data

King, Cuchlaine Audrey Muriel
Physical geography.
1. Physical geography
I. Title
910'.02   GB54.5

ISBN 0-631-18950-5
ISBN 0-631-11081-X Pbk

Filmset and printed in Great Britain by
BAS Printers Limited, Over Wallop, Hampshire

Contents

Acknowledgements

The author and publisher are grateful to the following for permission to reproduce Plates: Aerofilms Ltd (22, 23, 32, 54, 58, 65); Janet and Colin Bord (4, 53); Canadian Government, Department of Mines and Technical Surveys (55, 56); J. Allen Cash (1, 26, 46, 51); Dundee University (70); Institute of Geological Sciences, NERC copyright reserved (7, 8, 13); Susan Griggs Agency (67—photograph: George Hall); Grant Heilman, Lilitz, Pennsylvania (cover, 2, 6, 10, 25, 45); Verna R. Johnston ©, National Audubon Society Collection, Photo Researchers Inc (19); Eric Kay (11, 12, 14, 16, 17, 24, 30, 31, 38, 48, 52, 64, 68, 69); Keystone Press Agency (5); Tony Morrison (43, 47); Natural History Photographic Agency (36—photographer: S. Krasemann, 37—photographer: J. Kroener); New Zealand High Commissioner (59, 66); Jo Philpot (34, 35, 60, 61, 62, 63); United States of America, National Parks Service (18). All other photographs were taken by the author.

The author and publisher are grateful to the following for permission to redraw and reproduce Figures: George Allen and Unwin (Publishers) Ltd and J. R. L. Allen (2.18a); American Association for the Advancement of Science and C. Emiliani (3.18b); American Geophysical Union and W. B. Langbein (3.16); Edward Arnold (Publishers) Ltd (3.8); Association of American Geographers and L. C. Peltier (3.17, A3.9a); Blackwell Scientific Publications Ltd (2.8, 3.10, 3.11); Cambridge University Press (A3.6); D. M. Churchill and *Quarternaria* (4.4c); L. F. Curtis (2.27); Duxbury Press (3.5); W. H. Freeman and Company Publishers (2.4, 2.6b, 4.3, A3.9b); Heffers Publications (1.3); the Controller of Her Majesty's Stationery Office (4.12, 4.13, 4.14); the same and the Hydrographer of the Navy (4.15); McGraw-Hill Book Company (2.3, 2.6a, 4.18); Methuen and Company Ltd (2.5, 2.7, 3.3, 4.16, 4.17, 4.22, 4.23, A4.9); G. F. Mitchell (3.21); *Nature* and Ian Douglas (3.16a); New Zealand Geographical Society and W. Packard (3.7); Prentice-Hall Inc (3.9, 4.6, 4.7, 4.8, 4.9b, 4.10b, 4.11); G. Østrem (3.6); Royal Geographical Society and *Geographical Journal* (2.13, 3.13); Society of Economic Paleontologists and Mineralogists and G. S. Visher (2.22); *Scientific American* (4.2, A4.1); Soil Survey of England and Wales (2.28); University of Chicago Press (4.24).

CHAPTER 1

Introduction: the scope of physical geography

Physical geography was studied in the nineteenth century before human and regional geography had developed very far. The latter aspects of geography were championed by Sir Halford Mackinder, whose influence was stronger than that of J. W. Judd, the main upholder of the physical side of the subject. Another development that was not beneficial to the growth of physical geography was the change of name in 1876 to *physiography*. Under that title the subject became more divorced from the familiar local studies and concentrated instead on astronomy and the solar system. This change has had its counterpart recently in the setting up of departments of environmental sciences in place of geography in some of the new universities. In some of them the course has also become more remote from the study of basic physical geography. As the title changed to physiography so the geomorphological aspect became stronger. The term *geomorphology* began to come into common use after 1897, when geography included geomorphology, geophysiology and anthropogeography. *Geophysiology* covered oceanography, climatology and biogeography, which together with geomorphology embrace the major aspects included in this book.

One of the most influential books on the subject of physical geography was published by T. H. Huxley in 1877, under the title *Physiography*. He started from the known and familiar, based on study in the field, and then worked to the less familiar. He began with observation and then moved on to classification, the systematic ordering of the observed facts. From the ordered and classified facts he could then lead on to deduce laws and relationships, whereby cause and effect were recognized.

The latter half of the nineteenth century, when physical geography was developing, was a time of great interest in the natural world and its workings. The general awakening of scientific curiosity partly developed out of the publication of Darwin's *On the Origin of Species* in 1859, but there were also many other problems that intrigued the scientists of the time, including such problems as those posed by the recognition of the Ice Age and the climatic changes that it implied. Science was being freed from the biblical strait-jacket concerning the length of geological time, whereby the whole history of the world had to be fitted into the period since the time of the Creation in 4004 BC, as proclaimed by Bishop Ussher in 1650. At this time also the extent of

Plate 1 Snow in Scotland on Rannoch Moor at the head of Glen Coe. Snowfall is greater in the uplands of Britain and lasts for considerable periods.

1

scientific knowledge was such that one person could still adequately cover several sciences, and this was one reason for the popularity of physical geography, which is by its nature a broad subject.

The nineteenth century 'natural science' has now become divided into the separate sciences of botany, geology, geophysics, physics, meteorology, chemistry and zoology and others, all of which in turn are becoming further subdivided. Physical geography still is, in some respects, a bridging subject or linking discipline, in that it retains contact with other sciences. The breadth of physical geography is still both its strength, and in some ways its weakness, because it must draw on the expertise of so many different sciences. Its breadth, however, also provides a valuable means of coordinating and relating a great range of separate fields of knowledge. The significance of the relationships between the different elements of the environment is becoming increasingly apparent and important.

The process of desertification is an example of the close links between the different branches of physical geography, and the need to understand them. In many areas of the world human interference has led to deterioration of the quality of the land. The areas suffering in this way often have marginal and difficult climates, with little reliable precipitation and extremes of temperature; and they may be of extreme relief, either too steep for stability, or so flat that flooding, if it occurs, will be extensive. Climate, meteorology and geomorphology are all involved in bringing about conditions of potential instability. These basic physical properties also determine the soil and vegetation—the biogeographical aspects of the land. The latter attributes are the most easily influenced by man. The introduction of new farming techniques, as in the dust bowl of America, or the overgrazing of animals, as in Otago, New Zealand, can cause disastrous results. Both overgrazing and deforestation for cultivation can lead to deterioration of the vegetation and

desertification. The process is often a vicious circle of positive feedback, i.e. as the land deteriorates so its capacity to support animals and crops decreases, and the process of decline accelerates (see page 135). In order to appreciate the problems of this type of land management it is necessary to study all the many closely linked aspects of physical geography and the relevant human factors. The links between the different aspects are particularly important.

Just as the scientific pioneers of the nineteenth century had to build up their concepts bit by bit, so each one of us in turn has to build up our own private store of knowledge both through formal education and through our own observations and thoughts. However, we have the great advantage that we can continue from the point at which others have left off. It is easy for us to take much of the early work of the pioneers for granted and not to appreciate the great leaps of imagination required to formulate concepts which now seem obvious and self-evident. For example, we have grown up with the idea that rivers carve their own valleys by erosion. This was in fact a point of controversy during the early nineteenth century and was not generally accepted until later in the century, although Hutton and Playfair, at the end of the eighteenth century, and later Lyell, had appreciated the powerful part played by rivers in creating the valleys in which they flow. Thus, what seems obvious to us was not at all so to earlier generations. It is useful to develop the ability to appreciate the interrelations of the natural world environment. One of the aims of this book is to include as many practical exercises as possible that are concerned with these interrelations.

For the purposes of this book the field of physical geography will be considered in seven main subdivisions. Most of these subdivisions are related to each other in a complex way. One difficulty is to establish a starting point when all the aspects of the subject are so interwoven in this way in the environment.

Geomorphology is concerned with the form of the earth's surface, and the processes that create and modify it. The form cannot be understood without a consideration of the materials of which it is composed. Geomorphology developed from physiography, which in turn grew out of the nineteenth century physical geography, as discussed by Huxley in his book of 1877. The development of geomorphology was strongly influenced by W. M. Davis, whose geomorphological work took over the field of physiography and restricted it to the study of landforms. Davis attempted to explain landforms in terms of three main variables: structure, process and stage. The time element gave rise to his important concept of the cycle of erosion.

The three essential elements in geomorphology are morphology, structure (material) and process. The term *structure* can be taken to include the material of the earth's surface. It includes both the nature of the solid rocks and their structure, as well as the superficial materials, such as coastal sediments, terrace deposits, alluvium and glacial till. The inclusion of morphology in the trilogy should perhaps be justified, as morphology is the aspect that is being studied and explained. It is included to stress the importance of the bonds between it and the other two elements. Morphology can be used to study both process and material. The omission of stage does not signify the irrelevance of time in landform development. Time is, however, implicit in process, which is a dynamic element and hence necessarily occurs over time. Its operation varies according to the scale of study as pointed out by Schumm and Lichty.

From the point of view of physical geography, one of the most important materials of the earth's surface is the thin layer of soil, the subject matter of the science of pedology.

Pedology (from Greek *pedion* = ground) is concerned with soil, which is derived from the break-up of the solid rocks by weathering, a process that is intimately related to the climatic conditions. Thus there is a very important link between geomorphology, pedology and climatology. Soil depends essentially on the rock type and climate, but the nature of the relief and past geomorphological processes also play an important part. These three aspects of the environment, geomorphology, climate and soil together play a vital role in the field of biogeography.

Biogeography is the science that considers the non-human living elements of the environment. Plants are the basis of life, because only they, with the aid of sunlight and the nutrients in the soil or water, can synthesize living matter from the inorganic chemical elements. Plants create the foundation of all life, including human life. They provide the sustenance for the animals, who have filled almost every available niche on the earth's surface, in the air, under the ground and within the oceans. Animals in their turn provide food for each other and for man, and eventually, when they excrete and when they die, help to recycle the nutrients that go to maintain the fertility of the soil and water, thus aiding the growth of plants. The natural vegetation and the animals that it supports depend on the soil and climate as well as the relief and altitude. One of their most vital needs is for water.

Hydrology is concerned with the study of water in all its forms. It is of rapidly increasing importance and is closely linked to all other aspects of human life. Water that falls as precipitation on land may pass through the hydrological cycle in a number of ways. Some runs off directly into rivers and soon reaches the sea again, some infiltrates into the earth to become ground water, working its way slowly through the rocks to feed the streams in dry spells. Other water may fall in the solid state, as snow. Some of this may not melt at once but be kept in 'cold storage' as ice in glaciers and ice sheets and so remain out of the cycle for a long time, up to hundreds of thousands of years in the case of the Antarctic ice sheet, but much less in most glaciers. Nevertheless the water held in cold storage does exert an important influence on the land and the ocean through its influence on sea level, because when the water is locked up as ice sea level falls, rising again as the ice melts and returns to the ocean. The existence of ice sheets in high latitudes affects the circulations of the atmosphere and oceans, and thus modifies the climate.

Meteorology is concerned with the circulation of the atmosphere and the resulting weather. Hydrology and meteorology are closely linked because the processes that go on in the atmosphere carry the water from its great reservoir of the oceans to the land, distilling it from salt water to the fresh water that falls on land as precipitation. The processes leading to precipitation are amongst the most important in meteorology, and precipitation is the starting point of the hydrological cycle on land. The science of meteorology seeks to understand the operation of the atmosphere as a complex three-dimensional dynamic machine covering as one unit the whole of the earth's curved surface. This aim, requiring widespread observations throughout the atmosphere and all over the earth, has become more feasible with the introduction of satellites to observe the whole earth from above. Essentially it is a matter of taking observations over a wide area at one time, a type of observation called *synoptic*. Observations of temperature, humidity and other elements of the weather provide the data for studying the physical working of the atmosphere. The weather pattern is continually changing since the system is a dynamic one, and for practical purposes it is necessary to consider variation of weather systems over a short period to provide daily weather forecasts. Changes of weather must also be studied over both longer periods of time and larger areas than those associated with day-to-day forecasting of variations in the weather. The long-term, world-wide study is called climatology,

another important branch of physical geography.

Climatology can be fully understood only if it is based on a sound knowledge of meteorology. Climate is sometimes referred to as average weather. However, mean values often hide very considerable variations, especially in a marginal climate such as that of the British Isles. This so-called temperate group of islands can experience prolonged drought, heat waves, gales and torrential downpours, all in a few months. The average values of temperature and precipitation, nevertheless, are normally moderate, and compared to some climates, relatively reliable.

Climate is always varying to a greater or lesser degree over a range of time scales. These variations have given rise to ice ages and interglacial periods, as well as briefer, smaller fluctuations. The changes over time are an important aspect of climatology, but the spatial changes that provide the different climatic zones over the earth's surface are also very significant in the field of physical geography. Spatial variations can be related to other aspects of physical geography, including pedology, biogeography and hydrology, as already mentioned. The contrasts between maritime and continental climates on a global scale indicate the importance of the distribution of land and sea on climatology, showing that the oceans play an important part both in meteorology and climatology. Their study is important, but it is often neglected in physical geography.

Oceanography is the scientific study of the oceans from all points of view and it is a very large and rapidly growing subject. Its importance in many spheres of human activity is becoming increasingly apparent. The oceans interact intimately with nearly all the other aspects of physical geography. They provide essential evidence on which the recent theory of global tectonics and sea floor spreading is based. It is evident that, from the structural point of view, the earth must be considered as one whole unit: the oceans, continents and islands all being part of this unit. Similarly the oceans and atmosphere are inextricably coupled at their interface, where sea and air meet. The circulations of these two fluid bodies are dependent on each other, and together they create the world's weather and climate. Life on land was derived from that in the sea, and the biogeography of the oceans is in some respects equally important to that of the land.

The hydrological cycle can be considered as the dynamic system that provides a link between the many and varied aspects of physical geography. A number of its stages will be explored in more detail in subsequent chapters on a variety of scales.

1.1 Scale in physical geography

Views of the world or parts of it seen through a microscope or from a satellite in space are very different, owing to the very diverse sizes of the fields of view. Features revealed at various scales of viewing must be investigated by different methods because the problems raised vary partly according to the size of the area under consideration. Thus, scale is an important aspect of physical geography. The characteristics of the close-up view of a small area will be considered first, as these are easier to appreciate at first hand. It is usually best to start with the familiar, small scene and to move on from it to larger areas, which raise quite different problems, particularly when they are set in a different environment. Methods of study that can be used in one's own familiar surroundings can usually be used in other similar sized areas, with suitable modifications; but as the size of the area under study increases so the technique of study and the problems raised change.

Just as the size of the study can vary from the small local area one can walk around, so the period of time studied can range from a brief spell of a few minutes to the whole period of the earth's existence as a planet, a matter of 4,000 to 5,000 million years. Different principles apply over different time spans. In general it is best to study small areas over shorter time spans, within one's own experience. The longer periods of geological time, on the other hand, are more appropriate to studies that involve the earth as a whole.

The tempo of events differs in the same way as the actual time spans vary. Some events take place quickly, others over a longer time. It is necessary, therefore, to consider both the actual span involved and also the tempo of processes, i.e. the rates at which they act. It must also be appreciated that some act continuously while others are episodic, occurring only at intervals. The extent of both space and time must be continually borne in mind in physical geography. Different scales of study are the basis of the arrangement of this book. They are the small scale or local study, the medium scale, which is regional or continental in size, and the global scale.

The small scale environment for the purposes of this book is considered to cover an area that can be studied on large scale maps, such as 1:50,000, 1:25,000 or larger. More important, the area can be studied directly on the ground by field-work, thus providing first hand knowledge and the opportunity to discover its characteristics for oneself. Relationships studied in the field can then be applied to other similar areas or problems. Field methods are the most suitable techniques for studying the small scale environment in physical geography, although aerial photographs and maps may also be useful. For some purposes natural units provide good subjects for study on the small scale. For example a first-order drainage basin, a stream valley, a single corrie, a small bay on the coast, or two hillsides with different aspects may be compared in terms of their microclimates, biogeographies or geomorphologies.

At the medium scale of study the area becomes too large for an individual to explore personally throughout. This regional scale must normally be studied on maps or from other published sources, although parts can be visited. The British Isles provide a manageable sized area from this point of view. Other such areas include, for instance, New Zealand, Baffin Island or one of the larger American states. Areas of these dimensions have much greater varieties in their natural environments and present different problems. The Alps, the North Sea or a major drainage basin provide another type of medium scale region. These regions have natural homogeneity and boundaries.

In medium scale regions the structure, geology and landforms, as well as the climate, vegetation and soils, must be considered over a longer time span in order to understand regional differences within the area, and to appreciate the stages by which it has achieved its present character. At a rather larger scale a complete continent or ocean becomes the object of study. The approach must again become more generalized as the variety of climate, vegetation and relief, for example, increases; size is determined by the pattern of land and sea, and a longer time span must also be considered.

On a still larger scale the whole earth becomes the unit of study. Again individual involvement in its study is reduced because the area is much larger. The problems are also of a different type, and in seeking solutions the earth must be considered as one unit; examples include the global circulation of the atmosphere and the oceans, and the theories of the new global tectonics. The spherical character of the earth is an important element of studies on the global scale. The earth is a curved, three-dimensional surface of rock (the lithosphere), covered by a thin shell of water over 70 per cent of its surface (the oceans). A thinner, vital layer of soil covers part of the remainder, and everything is contained in an envelope of air (the atmosphere). Within these outer layers the life of the biosphere exists.

For some purposes it is useful to consider the earth as part of the solar system, which is the largest scale view of the earth relevant to physical geography. On this scale the earth becomes a small planet, one of the family of planets that circulate around the sun, the nearest and most important star to the earth. The sun is vital to life on earth for several reasons. Firstly, it supplies energy in the form of heat which maintains the circulation of the atmosphere and the oceans. Secondly, it provides the light and heat, and indirectly the water, without which nothing living can grow. We also depend on its past activity in the form of fossil fuels, coal and oil, without which modern society could not exist as it does at present.

1.2 Systems

An important concept in physical geography was referred to in passing in the last paragraph. The earth was described as a member of the solar system, particularly when it is studied on the largest scale. This reference to the solar system implies that the sun and the planets revolving round it make up an organization of individual members which are mutually dependent. Together they are often said to make up a whole that is greater than the sum of the individual parts. A system is defined by Chorley and Kennedy (1971) as a structured set of objects and/or attributes. The objects, which are sometimes referred to as variables, are related to each other so that they operate together as one complex whole. Some of the variables are dependent on others, while some may be independent. The human body is a good example of a system. It consists of parts which cannot function in isolation, but when they are put together in the right way they function as a unified whole, which is more than the sum of the separate parts by themselves. The interdependence of the separate parts of the system is one of its essential characteristics, and a property by which it can be recognized and defined.

Just as the human body consists of arms, legs and other parts, so each arm or leg consists of different elements, the bones, nerves and skin for example, each of which is essential to the whole limb. Similarly the earth is one of the planets circling and depending on the sun and itself also consisting of a complex hierarchy of systems. There are many sets of systems that are interlinked and which operate on different scales. At each of the scales already enumerated there are different types of systems, each of which forms an element of the next one up the scale. The nested pattern of systems is sometimes referred to by the terms *subsystems*, *systems* and *supersystems*, indicating the nesting hierarchy of their mutual relationships. Physical geography can be studied as a series of systems, each of which operates as a whole, but which itself consists of many smaller units, as well as forming part of a larger system. The interlocking of the different systems is dynamic, as they are functional units.

The way in which a system functions provides a means of identifying different types. Three different types are recognized: the isolated, the closed and the open systems. The isolated system is one which functions as a completely independent unit, in that there is no transference of energy or mass across its boundaries. Systems of this type are rare in nature, owing to the interdependence and hierarchical nature of most systems. In fact the only true example is that of the solar system, which is independent as far as the criteria specified are concerned. Virtually no energy or mass crosses its boundaries. The solar system, however, is itself part of a larger system, the galaxy, of which the sun is one component star. Even the solar system is not strictly an isolated system as light energy from other stars reaches it, but this is not essential to its operation as a functional unit.

A closed system has boundaries through which mass cannot pass, but energy can pass into and out of the system. The earth as a unit provides the best example of a closed system; it receives energy from the sun in the form of heat and light, but mass transference is confined within the earth, with only very minor exceptions, such as meteorites and artificial satellites. There are also other closed systems within the major earth system. The hydrological cycle, for instance, forms an important closed system, within which water circulates in any one of its three states: it can be solid, in the form of snow, hail or ice; liquid, as fresh or salt water; or a gas, the water vapour in the air. On a smaller scale an isolated pocket beach can be considered in some respects as a closed system—beach material cannot enter or leave the system, although energy in the form of waves and wind can influence the movement of material within the bay.

Most natural systems are open in type, in that both energy and mass can enter and leave the system. A drainage basin provides an example of an open system. Matter and energy in the form of precipitation, heat and products of weathering enter the streams, while water and material, including both organic debris and inorganic sediment can leave the system through the stream that flows out at its lower end. Many open systems become adjusted so that the amount of material entering balances that which leaves, creating a steady state of dynamic equilibrium, in which the system is self-regulating. The idea of self-regulation is an important one in open systems, because it allows the steady state to develop. This is only changed by some extreme or unusual event, which may result in the passing of a threshold and the introduction of new controls. The system must then adjust to a new dynamic steady state. A slope, for instance, may be adjusted to the vegetation and rainfall and the amount of weathered material passing slowly across it until one day a sudden torrential storm exceeds the stability threshold. The slope then collapses in a major landslide, and a new set of conditions must be established with a new threshold, until the vegetation, soil and gradient become readjusted to a new steady state condition. The mutual adjustment between the different parts of the system introduces another important concept that applies to all types of system. This is feedback.

Feedback means that as one variable affects a second variable, the second one in turn causes

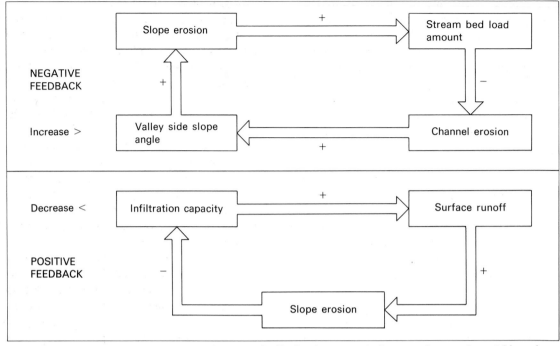

Figure 1.1 Diagram to illustrate negative and positive feedback. The + sign indicates an increase in variable at the head of the arrow and the − sign a decrease.

a change in the first variable. Feedback can be direct, only influencing the two variables, or more often indirect, through the modification of other variables. It can be negative or positive. In negative feedback the change in the second variable causes a change in the first such that it tends to return to its former state. This results in the self-regulation of the system, and a dynamic steady state is produced. In positive feedback, on the other hand, the change in the second variable causes the first one to change still further in the direction of the initial change. This leads to the self-generation of change and an increase in its magnitude. The self-generation of positive feedback must, however, eventually bring its own destruction, when a threshold is exceeded.

An example of negative feedback is provided when slope gradient increases, leading to an increase in slope erosion, which in turn leads to an increase in stream bed load; this change in its turn leads to a reduction of channel erosion, which allows the valley slope

angle to decrease, resulting in reduced slope erosion. The system thus returns to its former state of steady equilibrium by negative feedback, as illustrated in Figure 1.1.

Glacial erosion, on the other hand, operates to a certain extent at least by positive feedback. It is intensified when the ice flows compressively and as it thickens. As a glacier erodes into its bed to form a hollow, compressive flow and ice thickening are induced, and this in turn leads to further erosion of the hollow. Thus hollows in the floors of glacial valleys tend to become exaggerated as they induce conditions that enhance glacial erosion. Eventually, however, a situation is reached beyond which the ice can no longer deepen the hollow further.

The examples given so far have mainly been drawn from inorganic systems of geomorphology. One of the most important systems, however, is that which involves the whole of the living environment. This complex system is usually referred to as the 'ecosystem'. It can be defined as the mutual interdependence of the plants and animals of a living community and their environment. In a natural ecosystem the living organisms become adjusted to their environment in a mutually satisfactory way, which results in a steady state of self-regulation and stability. The balance is usually precarious and liable to disturbance, if any of the delicate thresholds are overstepped. Thus human interference in the natural ecosystem may have serious consequences, if its natural working is not adequately understood. One illustration of this is the disturbance of the prey–predator relationship, which is a natural mechanism developed both for the benefit of the predator and its prey. The weak and ill of the prey are eliminated to the benefit of the species as a whole, and to the obvious advantage of the predator species.

The natural cycle of lemming numbers in the tundra environment is followed by a similar cycle in numbers of its predators, the arctic owls, foxes and wolves. The same effect may be induced by human interference—when one pest is eliminated by artificial means a second may be initiated because of the removal of the predators of the second species. Many animals have an inbuilt capacity to adjust their numbers to the prevailing environmental conditions. For example, grey squirrel numbers are related to the climate, which in turn affects their food supply.

Study of the ecosystem involves all fields of physical geography as living organisms cannot be isolated from their environment. Each plant and animal has its own particular niche in the ecosystem and its own requirements, as well as often providing for the needs of other organisms. The whole is a complex, living, dynamic system, through which matter passes and changes, but is not lost. Most of the processes involved repeat themselves on a regular basis; plants grow, mature, seed and decay to produce new nutrients, on which the seeds can germinate to continue the sequence of life, death and rebirth. The sequence introduces another important aspect of physical geography that can be studied in many different ways: the *cycle* concept.

The circulation of matter in a closed system is often cyclic in character, and one of the most important cycles so far mentioned is the hydrological cycle. The cycle concept is a necessary corollary to the law of the conservation of matter, because if matter cannot be destroyed it must be recycled if the earth is to maintain its essential characteristics. Cycles are imposed on the earth by its membership of the solar system—providing the natural cycles of the year, with its seasonal contrasts, and the day, with its changes between light and darkness. The effects of these cycles in many different fields of physical geography will be examined in subsequent chapters.

Another well-known cycle, which has recently been questioned, is the geographical cycle or the cycle of erosion, first suggested by W. M. Davis towards the end of the last century. This cyclic concept may have grown out of the early work of a pioneer of geomorphology, James Hutton of Edinburgh, who was a medical man. Through his knowledge of the circulation of blood in the body he came, by analogy, to appreciate the circulation of matter in nature. He saw that erosion is followed by sedimentation; and in time the sediments are raised up to make new land; the cycle of erosion is restarted and more sediments are derived from the destruction of the old. Davis' cycle of erosion has been criticized lately because of its oversimplification of reality. However, the idea should be considered in terms of varying scale in both space and time. Over a large area and in the long term there is probably justification in thinking in cyclic terms—over a long period of time, tens of millions of years, a land surface can be worn down to an area of low relief. In the shorter term and within a smaller area, such as a first-order drainage basin, it may be more useful to think in terms of open systems and a steady state of dynamic equilibrium.

However, dynamic equilibrium is not confined to short-term processes, but also applies to much longer-term processes and cycles. It plays a part in the orogenic cycle of mountain building, the creation and destruction of ocean basins, and the major ice ages, which operate over periods of hundreds of millions of years on the global scale. These long-term, large-area cycles and processes can be viewed in terms of systems, just as well as the short-term, small-area cycles and processes can. Thus the systems approach provides a useful means of linking processes, morphology and materials at all scales of enquiry in the study of physical geography.

One of the advantages of considering physical geography in terms of systems is that a system stresses the interrelationships between the units that form the system. Throughout this book one of the major points to be emphasized is the dependence of variables upon each other and the relationships between them. Many of the practical exercises are based on the analysis of such relationships, and various statistical methods of establishing relationships are used to add rigour to the analysis.

The ecosystem stresses the interdependence of living organisms on each other and their environment. The field of physical geography embraces the whole environment. The subject is an immense one, so that in a short book it is essential to be selective. The aim has been to exemplify a considerable range of situations, and for this purpose the division of the book into different scales of enquiry provides a means of considering a wide range of approaches and situations. Within each scale of study reference is made, where possible, to all fields of physical geography, but some are easier to study at one scale than another, so again the exercises only provide a glimpse of the great variety of material that could be included. Rather than covering a wider field in what would of necessity be a very superficial way, a smaller number of examples are treated in more detail to indicate methods of analysis. If an appreciation of the complexity of the natural environment is gained then the book will have served its main purpose.

Physical geography is a field of study of considerable importance. The wide-ranging interest of the geographer is particularly suited to such an extensive range of knowledge. The importance of physical geography grows as man's ability to interfere with the natural order increases with the advance of technology, and as the increase in population makes ever greater demands on our natural resources of food and raw materials, many of which are not inexhaustible or renewable. One must know how nature works before one can safely modify it. There is still much to be learnt concerning all aspects of physical geography, and it is thus a constantly growing subject. The more people who know something about it, the better will be our long-term chances of survival.

1.3 Position and place

Geography is the study of the earth's surface. One of the first essentials is to define position with respect to the earth's surface. Thus the first exercises are concerned with methods of locating places on the earth's surface by means of latitude and longitude observations, which can only be made by reference to bodies outside the earth, i.e. the sun or stars. It is also necessary to consider some other aspects of the earth's relation to the sun such as the equation of time. This is because longitude can only be defined by reference to the time differences of astronomical events between two places.

It has been known for a very long time that the earth is roughly spherical in shape and that it rotates about its polar axis once a day. Thus the position of any one place can be described by reference to its distance from the pole or equator. This gives its latitude as an angular value. In order to establish the size of the spherical earth the early Greek astronomers measured the latitudes of two places on the same meridian by reference to the elevation of the sun at noon. The angular measurement could then be referred to a distance measurement by surveying the distance between the two points. Thus a reasonably accurate estimate of the size of the earth was obtained.

The science of *geodesy* is concerned with the accurate measurement of the figure of the earth, which is described as a geoid, i.e. earth-shaped. Very accurate survey observations using the stars are needed to provide a detailed and precise value for measurements of the earth's minor and major axes and to determine the extent to which it departs from the true spherical form. Accurate measurements of elevation are also required to provide a complete picture of the figure of the earth. These observations are complicated by variations in gravity which affect the isostatic element and the vertical plumb line used to measure height.

Latitude and longitude

The reference system of latitude and longitude has been devised to divide the earth's surface and to allow the positions of places to be described uniquely. Latitude defines position relative to the equator or pole, and longitude defines position relative to some assumed origin, the prime meridian. A meridian is defined as the shortest distance along the circumference between two poles. Meridians all pass through both poles so they are examples of great circles, which are circles which have their centres at the centre of the globe. Of all the lines of latitude only the equator is a great circle, that is, the largest circle that can be drawn around the globe.

Each place has its own local meridian or line of longitude, measured in degrees east or west of the prime meridian, normally the Greenwich meridian, i.e. the meridian which passes through the Royal Observatory at Greenwich. Longitude can vary from 180 degrees to the east to 180 degrees to the west of Greenwich. In terms of length 1 minute of longitude or latitude around the equator is 1 nautical mile, which is 6,080 feet. There are 60 seconds of arc in each minute of arc so that one second of arc is 101·33 feet or 30·9 metres. The lengths of degrees of longitude decrease in size as one moves away from the equator since the

lines of latitude become small circles. Their lengths can be calculated by multiplying the earth's radius, R, by the cosine of the latitude. The whole length of a parallel or line of latitude is $2\pi R \cos$ (latitude).

Exercise 1.1. Latitudes

Calculate the length of your own parallel of latitude and its distance from the equator along the meridian, and from the nearest pole.

One of the main uses to which latitude and longitude can be put is in fixing position in surveying for map making and for navigation—ships and aircraft need to know where they are and they fix their positions by measuring latitude and longitude.

In order to fix positions relative to the earth's surface it is necessary to refer to points outside the earth. These are the heavenly bodies, primarily the sun and stars; the former is easier to use, but the latter are more accurate. In the early days of navigation sailors had little problem in determining their latitude, but longitude was much more difficult to measure, because it requires a means of reference to the prime meridian, which involves time keeping. The development of the chronometer was an essential step in the measuring of longitude, and this was first achieved about the time of Captain Cook's explorations.

The measurements of the positions of heavenly bodies must be made relative to the earth, and the points used for this purpose are the horizon, which is easily identified at sea, and the zenith, which is the point immediately overhead. These two reference points are 90 degrees of arc apart.

The astronomical triangle. The astronomical triangle, shown diagrammatically in Figure 1.2, provides the framework into which observations can be fitted. It is a spherical triangle, its sides as well as its angles being given in degrees or radians. The sum of its

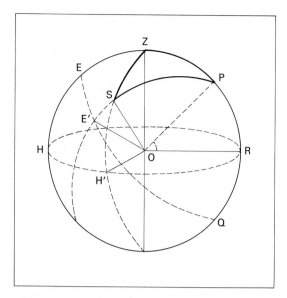

Figure 1.2 Diagram of the spherical astronomical triangle. Z = zenith, S = star or sun, P = pole, EQ = equator, HR = horizon, angle ZPS = hour angle, angle PZS = azimuth, SP = co-declination, angle E′OS = declination, SZ = co-altitude, angle H′OS = altitude, ZP = co-latitude, angle POR = latitude. The outer circle represents the meridian.

a great circle; it represents the distance between the zenith and the star or sun. If this angular distance is subtracted from 90 degrees, the resulting value gives the elevation of the star or sun above the horizon. This is the value that can most easily be measured at sea, where the horizon is easy to observe. It can also be measured from the air by means of artificial horizons. The distance given by PS is called the polar distance, and it represents the distance between the star or sun and the pole. In the case of the sun this value, which is called the declination, varies with the apparent movement of the sun north and south of the equator with the passage of the seasons. The declination is the amount by which the sun is north or south of the equator. At the equinoxes the declination is zero; it reaches $23\frac{1}{2}$ degrees north in the northern summer at the solstice, when PS is $67\frac{1}{2}$ degrees, and $23\frac{1}{2}$ degrees south at the southern solstice, when PS would be $113\frac{1}{2}$ degrees, measured from the north pole. The value of the declination can be looked up in tables. The distance PZ is the angular distance between the zenith and the pole, and when subtracted from 90 degrees, it is the distance between the equator and the zenith. It is a measure of latitude, so that PZ is called the colatitude. Its value is needed to calculate latitude by observation of the sun or star.

The point Z in the triangle represents the position of the point under consideration. The angle at P gives the difference of longitude between the sun or star and the point of observation, and this is a measure of longitude. It is called the hour angle in the spherical triangle.

As with plane triangles, if two sides and the included angle of a spherical triangle are known, then the third side or other angles can be calculated. Similarly, if all the sides are known, the angles can be found. In a plane triangle trigonometry gives the required values. The sine formula

$$\frac{a}{\sin A} = \frac{b}{\sin B} = \frac{c}{\sin C}$$

angles can exceed 180 degrees and its sides are parts of great circles. In fact all its angles can be right angles, making a total of 270 degrees when its sides are also 90 degrees each. The three apices of the astronomical triangle represent the zenith, Z, the pole, P, and the sun or star, S. The points can be thought of as transferred from the sky to the curved surface of the earth, and can thus be marked on a sphere representing the earth. A great circle drawn at 90 degrees from the pole represents the equator, EQ, and another at 90 degrees from the zenith represents the horizon, HR.

The three sides of the triangle can then be identified with reference to these points and circles. Two of the sides are parts of meridians, these are the sides PS and PZ, both of which pass through the pole, through which all meridians pass. The third side SZ is also part of

12

or

$$a = \frac{b \sin A}{\sin B}$$

or

$$\sin A = \frac{a \sin B}{b}$$

and the cosine formula

$$\cos A = \frac{b^2 + c^2 - a^2}{2bc}$$

are the most useful.

The spherical triangle equivalents of these formulae are used to solve the spherical triangle to calculate latitude and longitude. The sine equation is used in the form

$$\sin HA = \frac{\sin zd \ \sin AZ}{\sin pd} .$$

and the equivalent of the cosine equation is

cos zd = cos (colatitude) cos pd + sin (colatitude) sin pd cos HA.

Thus HA can be established if zd and AZ are measured; pd can be looked up in tables.

Measuring latitude. When the sun is on the meridian of the observer the spherical triangle can be simplified since the angle at P (the hour angle HA) is zero, and all the points P, Z and S lie on one great circle, the local meridian. This occurs at local noon when the sun reaches its maximum altitude. The maximum elevation of the sun above the horizon at noon gives the colatitude when corrected for the declination. This simple way of measuring latitude was used by early navigators and is still used. The early navigators had an instrument that enabled them to measure the elevation of the sun above the horizon fairly simply. It was called a quadrant, being in the form of a quadrant of a circle with a sighting device to observe the sun's elevation. The quadrant was the forerunner of the sextant, which is the instrument now normally used at sea to measure the sun's elevation using the sea horizon as the plane from which the elevation is measured by a system of mirrors. At the

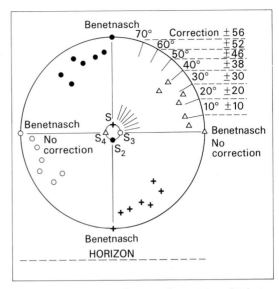

Figure 1.3 Diagram to illustrate the position of Polaris, S, in relation to the star Benetnasch, the tail of the Great Bear constellation, and the north pole. (After F. Debenham, 1942)

equinoxes no correction for declination is needed, but at other times the correction must be applied.

There is another simple way of measuring latitude using the stars. The Pole Star, Polaris, is almost above the north pole and its elevation above the horizon gives a fair measure of latitude. It would be exact if Polaris were situated exactly above the pole, but in fact Polaris moves around the pole at about 1 degree from it so a correction is necessary to obtain an accurate latitude. A simple way of making the correction is to use the position of the Great Bear's tail, the star Benetnasch, as shown in Figure 1.3.

Exercise 1.2. Latitude measurement

Solve the following problems:
(1) A night watchman found that he could just see the Pole Star over a wall 16·8 m high north of him. Sometimes he had to move as far as 22·6 m away from the wall to see it and sometimes as close as 21 m. What was his latitude, allowing 1·5 m for the height of his eye?

(2) A highly observant Frenchman noticed that when the water in his circular well was half as far from the top as the top was wide, the shadow of the well top reached half way across the surface of the water at noon on 21 March. Deduce his latitude.

(3) Steve Williams and Nick Evans lived at the southwest and northeast corners respectively of one of those sections of the Western States where all the roads run either due north and south or due east and west. When Steve went to visit Nick he always went north first and then east. Nick, the mutt, used to go south first and then west, when visiting Steve. Why was he a mutt? Give a quantitative answer.

Measuring longitude. Longitude is more difficult to measure as it requires some means of relating one's position to that of Greenwich or another prime meridian. This can be achieved by the use of the hour angle and a measure of time is essential. The chronometer is set on Greenwich time so that when it indicates noon the sun is at its maximum elevation on the Greenwich meridian. The time at which the sun reaches its maximum elevation at the position to be determined then provides a measure of the distance of longitude between the two places. The sun moves 15 degrees each hour to complete its circuit of 360 degrees in 24 hours, so that time can be converted into angular measurement on this basis.

There is no need to wait for the sun to reach the local meridian to obtain the longitude, although the calculations are simpler at this time. At other times of the day the elevation of the sun and the azimuth, the angle AZ in the spherical triangle, must be observed. The declination of the sun can be looked up. The astronomical triangle can then be solved for the hour angle, HA, which will give the longitude if the Greenwich time of the observation is noted. These calculations require the use of trigonometrical tables, or they can be done graphically using a stereographic projection on which angles are preserved correctly.

The date line, where the date changes, is fixed along the meridian antipodal to that of Greenwich, but it is not straight to allow for certain islands to be on the same side. As one travels eastwards by plane it is necessary to put one's watch forward continually, until the date line is crossed, when it is retarded 24 hours. Similarly in travelling west one's journey time is shortened when adjusted to local time. Thus, leaving London at 11.00 hours GMT one arrives at New York at 13.00 hours local time after travelling for 7 hours, as New York is 5 hours behind London. With supersonic travel it is possible to arrive before one left, according to local time. In these days of fast air travel time changes are significant. There are other complications concerning time that should be mentioned briefly and which must be taken into account in accurate survey observations.

Exercise 1.3. Latitude and longitude

(1) An airman starts from Wellington, New Zealand, longitude 175°E, at 11.30 local time on the morning of 6 August, and arrives at Tahiti, longitude 149°W, a distance of 3,840 km, at 11.30 local time on the morning of 6 August. What was the average speed of the plane?

(2) Robinson Crusoe found that the shadow of his 12·3 m flagstaff was sometimes as much as 10 m to the north and sometimes 1·7 m to the south at noon. He also remembered that an eclipse which took place at noon for him was due at Greenwich at 16.04. What island was he on?

(3) A London boy scout tries to find true north by the sun on 4 November. He marks the shadow of a stick at 09.00 and 15.00. How far is his meridian (taken as midway between the marks) in error?

The equation of time. Time is measured by clocks that theoretically move evenly and regularly, but the movement of the earth around the sun is not so regular. Noon is

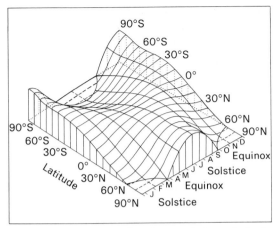

Figure 1.4 The pattern of insolation relative to latitude and season over the whole globe, assuming no atmosphere, which explains the high values at the poles in summer.

marked by the passage of the sun or other heavenly body across the local meridian. Thus one day is the interval between two such passages. Time can be measured by the stars and as they are so far away time obtained this way is more accurate, but the sidereal day is 4 minutes shorter than the solar day, and thus would be inconvenient for regular use. The lunar day is 50 minutes longer than the solar day, the significance of which is considered further in Chapter 4.

However, we live by the sun, which determines night and day, and so our clocks are adjusted to solar time. The day, as measured by the sun's transits across the meridian, is not uniform because the earth does not go round the sun in an exactly circular orbit. The earth's orbit is elliptical and the position of the sun relative to the ellipse is not central. There is also an effect due to the obliquity of the ecliptic. The rate of revolution round the sun changes sign twice a year relative to the mean or standard time kept by regular clocks. The effect of the obliquity of the ecliptic is to change the sign four times a year. The difference between mean time, or regular clock time, and sun time is called the equation of time. It is found by adding the two

effects and reaches its maximum value of $+16$ minutes 20 seconds in early November, and its minimum of -14 minutes 21 seconds in mid-February. The $+$ and $-$ signs mean respectively, one must add to or subtract from mean time to obtain sun time.

The equation of time means that the longest evening of the year occurs in mid-December, while mornings do not start getting longer until into the first week of January. Table 1.1 gives values for the declination of the sun and the equation of time throughout the year.

Exercise 1.4. Equation of time

Plot values for the equation of time and for the declination of the sun throughout the year. Plot times of sunrise and sunset for the periods covering the solstices and explain why the longest evening occurs before the shortest day and the shortest morning occurs after the shortest day. What is the pattern at the northern hemisphere summer solstice?

The declination of the sun is relevant to a consideration of the seasons.

The seasons. The seasons are the result of the obliquity of the ecliptic. The sun is overhead at different latitudes as the earth moves in its orbit around the sun through the year. The tropics mark the latitudes between which the sun is overhead twice a year. It is overhead at the tropic of Cancer on 21 June, the summer solstice of the northern hemisphere, and at the tropic of Capricorn on 21 December, the summer solstice of the southern hemisphere. At the equator the sun is overhead at the equinoxes on 21 March and 21 September. The Arctic and Antarctic circles mark the latitudes to the north and south of which for at least one night each year the sun never sets. At the north pole the winter night lasts for 176 days and the summer period of continuous daylight lasts for 189 days. The inequality of the periods is due to refraction which makes the sun appear above the horizon when it is not.

15

Table 1.1 *Sun's declination and the equation of time*

January

Declination ° '	Equation of time min s
1 23 03S	− 3 21
6 22 34	− 5 40
11 21 54	− 7 47
16 21 03	− 9 39
21 20 02	−11 14
26 18 52	−12 31
31 17 33	−13 28

February

Declination ° '	Equation of time min s
1 17 16S	−13 27
6 15 48	−14 10
11 14 13	−14 21
16 12 32	−14 14
21 10 46	−13 48
26 8 56	−13 07

March

Declination ° '	Equation of time min s
1 7 49S	−12 36
6 5 54	−11 33
11 3 57	−10 19
16 1 59	− 8 56
21 0 00	− 7 28
26 1 58N	− 5 58
31 3 55	− 4 27

April

Declination ° '	Equation of time min s
1 4 18N	−4 09
6 6 13	−2 40
11 8 05	−1 16
16 9 54	+0 02
21 11 39	+1 10
26 13 19	+2 07

May

Declination ° '	Equation of time min s
1 14 53N	+2 52
6 16 22	+3 24
11 17 43	+3 42
16 18 57	+3 46
21 20 03	+3 36
26 21 01	+3 12
31 21 50	+2 25

June

Declination ° '	Equation of time min s
1 21 58N	+2 27
6 22 35	+1 38
11 23 03	+0 42
16 23 20	−0 20
21 23 27	−1 25
26 23 23	−2 30

July

Declination ° '	Equation of time min s
1 23 09N	−3 32
6 22 45	−4 27
11 22 12	−5 13
16 21 28	−5 49
21 20 36	−6 13
26 19 35	−6 22
31 18 26	−6 17

August

Declination ° '	Equation of time min s
1 18 11N	−6 15
6 16 53	−5 51
11 15 28	−5 12
16 13 57	−4 19
21 12 20	−3 14
26 10 38	−1 57
31 8 52	−0 30

September

Declination ° '	Equation of time min s
1 8 31N	−0 11
6 6 41	+1 26
11 4 48	+3 09
16 2 53	+4 55
21 0 57	+6 40
26 1 00S	+8 25

October

Declination ° '	Equation of time min s
1 2 57S	+10 05
6 4 53	+11 39
11 6 47	+13 03
16 8 39	+14 15
21 10 29	+15 12
26 12 14	+15 54
31 13 55	+16 17

November

Declination ° '	Equation of time min s
1 14 14S	+16 20
6 15 48	+16 20
11 17 15	+16 00
16 18 35	+15 18
21 19 47	+14 15
26 20 50	+12 51

December

Declination ° '	Equation of time min s
1 21 43S	+11 10
6 22 25	+ 9 12
11 22 57	+ 7 01
16 23 17	+ 4 39
21 23 26	+ 2 11
26 23 23	− 0 18
31 23 09	− 2 45

+ (−) in front of equation of time means add to (subtract from) mean time.

The movement of the sun north and south of the equator with the seasons also affects the pattern of solar energy received over the earth's surface owing to variations in the altitude of the sun and hence the amount of radiation received. The amount of solar energy received is much greater when the sun is overhead than when it is at a low angle. The pattern of insolation varies with latitude and season as shown in Figure 1.4.

Table 1.2 Pattern of insolation outside the atmosphere in cal $cm^{-2} d^{-1}$

Date	90°N	70	50	30	0	30	50	70	90°S
Dec. 22	0	0	181	480	869	1,073	1,089	1,114	1,185
Feb. 4	0	25	298	586	905	1,003	937	809	834
Mar. 21	0	316	593	799	923	799	593	316	0
May 6	796	722	894	958	863	560	285	24	0
June 22	1,110	1,043	1,020	1,005	814	450	170	0	0

Exercise 1.5. Solar radiation

Plot the figures given in Table 1.2 to illustrate the variation in solar radiation, sum the totals for each latitude and explain the variation with latitude.

The excess radiation received in the low latitudes gives rise to the dynamic systems discussed in Chapter 4, whereby energy is distributed more evenly over the surface of the earth. Nevertheless there are still marked contrasts between the different zones: the polar, temperate and tropical areas. One of the greatest contrasts between the different areas results from their varied seasonal phenomena. These differences affect all aspects of physical geography, but perhaps most of all the biogeography, as will be considered in the following chapters.

References and further reading

*Chorley, R. J. and Kennedy, A. B. (1971) *Physical geography: a systems approach.* Prentice-Hall, London.

Darwin, C. (1859) *On the origin of species.*

Davis, W. M. (1899) The geographical cycle. *Geographical Journal* **14**, 254.

*Debenham, F. (1942) *Astrographics.* Heffer, Cambridge.

*Ebdon, D. (1977) *Statistics in geography: a practical approach.* Blackwell, Oxford.

Hutton, J. (1795) *Theory of the earth with proofs and illustrations.* Edinburgh.

Huxley, T. H. (1877) *Physiography.* Macmillan, London.

Judd, J. W. (1911) *The student's Lyell.* 2nd Edn. Murray, London.

Judd, J. W. (1911) *The coming of evolution.* Cambridge University Press, Cambridge.

Mackinder, H. J. (1887) On the scope and methods of geography. *Proceedings of the Royal Geographical Society* NS9, 141–60.

Stoddart, D. R. (1975) That Victorian, Huxley's physiography and its impact on geography. *Transactions of the Institute of British Geographers* **66**, 17–40.

Publications recommended for further reading are marked by asterisks.

CHAPTER 2

Local scale studies

2.1 Introduction

The best way to learn about the natural environment is to study it in the field. This can be done in sufficient detail to be rewarding only if the area is small. For this reason the small scale study is a useful beginning, and having gained experience at first hand of some of the many interactions that can be observed and measured in the field, it is easier to appreciate these relationships in other environments at the same scale, and also the rather different relationships that apply over larger areas.

The first aspects to be considered are the weather and climate. Although the climate of Britain is classed as temperate and maritime, the mean values of temperature and precipitation hide a great range, both over space and through time. Many interesting aspects of the weather and climate can be investigated at the local level, and microclimatology is an important subject in its own right. There is a close relationship between climate, vegetation, soil and land use, which brings in the human element. The urban climate provides a

Plate 2 Genesee River gorge, New York State. The Genesee River is cutting through the glaciated Allegheny plateau. The river now has a smaller discharge than that which cut the gorge. Note the thinly bedded, horizontal strata.

field well worthy of study, because large cities considerably modify the weather in their vicinity.

Many of the relationships between rocks, soil, vegetation and climate can be explained through hydrology. Figure 2.1 illustrates the hydrological cycle in a drainage basin on the local scale. The rocks and superficial material determine whether the rainfall or snowfall can penetrate into the ground, while the nature of the soil depends closely on the water available, which is in turn a function of the weather and climate. The rainfall pattern is very closely linked to the relief; there is, therefore, an important feedback relationship involved in these interrelationships—relief affects rainfall, which in turn affects the relief through erosion.

One basic aspect of a small area is its relief or morphology. This is dependent upon the material, both solid rock and superficial cover, of which it is composed, and the processes that have acted on this material. The three aspects of the landscape—morphology, material and process—are closely interrelated, but for convenience they will be dealt with separately. The solid material involves the rocks, and hence involves the geology, while the weathered mantle or superficial material forms the basis of the soil. Thus pedology follows closely on from the geology and geomorphology, and in turn leads to a consideration of the vegetation. There is very little natural vegetation left in the British Isles. Nevertheless the vegetation does bear a close relationship to the soil, and through this to the rocks; it is also affected by the climate.

2.2 Meteorology

Introduction

At the local scale meteorology provides a considerable range of opportunities for both field-work and practical analysis of the field data. The emphasis is on the effects of relief, aspect and ground conditions on the weather elements. It is possible to study interesting examples of microclimatology, such as variations of temperature within a metre of the ground, under different conditions. At a rather larger scale the influence of forests and urban areas on the local weather and climate can be examined. The influence of relief on many aspects of weather provides valuable opportunities for field-work and data that can be related to other aspects of physical geography, such as valley-side asymmetry, in which feedback relationships can be studied. Some examples of the relationships between meteorological and hydrological variables are also analysed. Local winds can also be related to relief, and the distribution of land and sea. At this scale it is possible to record the intensity of short-term variations in weather elements, including diurnal temperature fluctuations, and the pattern of particular precipitation episodes. Thus both space and time elements allow field observations to be made usefully, providing data for subsequent analysis. Some examples of the many possibilities are discussed in this section.

Precipitation is one of the most important weather inputs into the small-area system, so its measurement must be considered. Rainfall is measured by means of a rain gauge at one point over a period of time. However, there are problems when rainfall is related to runoff.

Normally rain gauges are only measured at daily or at best hourly intervals. They are also restricted to particular places, whereas rain falls intermittently, with varying intensity both over space and time. The problem of temporal variation can be overcome by use of a continuously recording rain gauge. The problem of recording accurately over a whole drainage basin is more difficult to solve.

The difficulty varies with the type of rainfall. Frontal rainfall is usually fairly steady in time and covers a considerable area, although it may vary consistently with altitude and exposure in a hilly drainage basin, and extreme contrasts can occur on either side of a high watershed. It is likely to be fairly uniform over a small first-order drainage basin of the type under consideration.

Convectional rainfall characteristic of shower and thunderstorm activity is much more variable both spatially and temporally. It is usually shorter in duration, but more intense. This type of rainfall can lead to flash floods in small basins and it provides more difficult problems in recording, especially where the rain gauge density is low.

The rainfall over the entire catchment area must be estimated to relate to the runoff at the outlet of the basin. There are different methods of obtaining the total precipitation from spot observations.

One method is to interpolate isohyets, i.e. contour lines of equal rainfall, across the basin. This can be done simply by eye, or more accurately by numerical methods. Trend surface analysis, for example, fits a plane or more complex surface through the data points in such a way that the squared distances between the data points and the plane are reduced to a minimum. When the isohyets have been inserted on a map of the drainage basin the areas between them can be calculated, multiplied by the mean value between the isohyets and summed to give the total precipitation.

Another commonly used method is by Thiessen polygons. The sites of the rainfall gauges are plotted on the map of the basin and

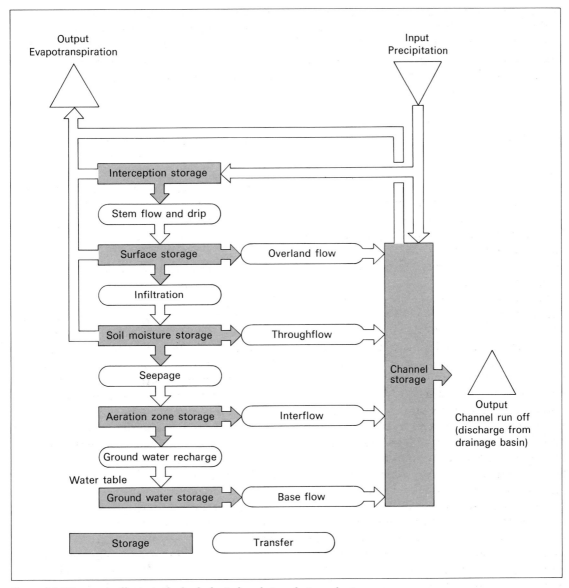

Figure 2.1 Diagram to illustrate the hydrological cycle in a drainage basin.

connected by straight lines. The lines are then bisected and further lines are drawn at right angles to the original lines to meet to form a polygonal pattern. The areas of the polygons are measured and calculated as fractions of the total area. These fractions are then multiplied by the rainfall at the gauge in that polygon and the total is summed to give the whole precipitation over the catchment.

Exercise 2.1. Thiessen polygons

Figure 2.2 shows a pattern of rainfall observations in a small basin. Calculate the total precipitation by means of Thiessen polygons.

The problems of measuring precipitation in the form of snow are quite different. One advantage is that snowfall can be recorded

21

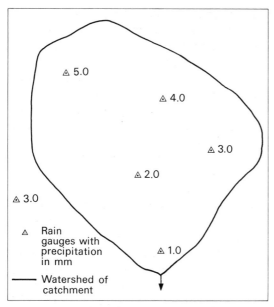

△ 5.0

△ 4.0

△ 3.0

△ 2.0

△ 3.0

△ Rain
gauges with
precipitation
in mm

△ 1.0

—— Watershed of
catchment

Figure 2.2 Hypothetical pattern of rainfall for one storm in a small catchment to calculate the total basin precipitation by means of Thiessen polygons.

throughout the basin by simple depth measurement sampling. Drifting, however, presents one difficulty, especially when the snowfall is accompanied by strong winds which make the depth variable. The process of solidification and packing must also be considered, as well as subsequent melting before observations are made.

Snowfall can provide a substantial element to ground water under some conditions, particularly as temperatures and evaporation associated with the winter period are low. Much melt water can seep into the ground when thawing is slow. On the other hand rapid thawing can lead to major floods, such as those in eastern England in 1947, when a long cold and snowy spell was suddenly followed by mild, wet weather, causing a rapid thaw when the land was already waterlogged or frozen. (See Plate 3.)

Temperature is another important element that can be related to relief, rainfall, vegetation and soil on a wide range of scales. The relationship with relief depends on the lapse

rate with height, a topic that will be considered in the next chapter in more detail. Rainfall and temperature are related through the moisture holding capacity of the air, which depends on temperature: the warmer the air the more moisture it can contain before it is saturated, and hence the heavier the rainfall is likely to be. At one extreme is the heavy equatorial rain and at the other, the cold deserts of the Arctic tundra. The relations between soil, vegetation and temperature are intimate at all scales. In this section some of the relationships that can be recorded in the field and that apply over a small area will be discussed.

Temperature is a quantity that varies in both space and time; it is also continuous. It can be recorded at points in both space and time with thermometers, or continuously at one place over a period of time with a thermograph. Dry and wet bulb thermometers are both usually used. The former gives the temperature of the ambient air, and the latter provides a value for the moisture content of the air. The relative humidity is the ratio of the amount of moisture the air holds to the saturated value. Useful field observations can be made by sampling temperature in a variety of ways, including readings taken over both space and time.

A vertical sampling plan provides valuable information on the change of temperature with height. Thermometers can be placed in the soil at depths up to 1 m, on the ground surface and in a standard meteorological screen at 1 m above the ground, as well as higher up where suitable equipment is available. The ground temperatures are particularly useful when they are associated with different soil types or during a period in which freezing takes place.

Temperature changes across the freezing point can be related to the formation of ice in the ground, frost heaving and similar geomorphological processes. The penetration of heat or cold into the ground can be related to the type of material, the site exposure and the ground moisture. Such observations must be continued over a reasonable period of time, such as a year, preferably at daily intervals, as there is a considerable time lag in heat penetration into the ground.

Variations of temperature with time can be measured at a standard level relative to the ground in different places. A useful project is to record the air temperature at various sites on either slope of a small valley, preferably having northern and southern exposures. The measurements can be repeated at different times of the day and in different weather conditions. On still, quiet anticyclonic days the measurements should reveal considerable contrasts between the two sides of the valley. A contrast between the valley bottom and upper slopes should also emerge, and differences between seasons should become apparent. Plotting and analysing these variations in temperature provide useful insights into micrometeorological processes.

The contrasts both in space and time will be greater under anticyclonic conditions than when it is windy and cloudy. Contrasts between hill slopes tend to be greater in winter than summer, owing to the low altitude of the sun at midday in winter, particularly in higher latitudes. Contrasts over time may be greater in summer when the altitude of the sun varies most between day and night, although in midsummer the short nights of higher latitudes may allow insufficient time for maximum night cooling compared with day temperatures. When temperature changes are at their maximum between day and night, often near the equinoxes, when nights are fairly long, temperatures may cool below the dew point. The air then becomes saturated, and mist or fog can form under anticyclonic conditions.

Plate 3 Snow in northeast England on the North Yorkshire Moors above the Vale of Pickering. The snow-covered moors contrast with the bare steep slopes of the Newtondale meltwater channel that cuts across the Moors from north to south, emptying into the Vale of Pickering, at one time a glacial lake.

By taking observations at regular intervals during the day and night it is possible to study the change of temperature with time at various positions on the slopes. The lowest temperatures are usually recorded early in the morning and the highest about mid-afternoon. The fall in temperature is usually more rapid in the evening than the rise in the morning. There are also significant differences in the changes of temperature on opposite slopes of the valley where these face north and south. On north and west facing slopes the temperature rises more slowly in the morning than on south and east facing slopes. The difference will depend both on the slope angle and the time of year in high latitudes. West facing slopes, however, have the advantage of higher temperatures of the afternoon, while those facing south benefit from the higher elevation of the sun at midday, when they receive maximum insolation in the northern hemisphere.

Observations of local temperature variations on slopes of different aspects are especially valuable in attempting to explain the asymmetry of slope gradients. The relationship between microclimate and geomorphological processes, such as soil creep, is complex. Many variables, including latitude, rock type, temperature fluctuations, precipitation type and distribution over time, as well as the effect of the river at the base of the slope, affect this relationship and can be recorded in the field. The variables listed control vegetation to a large degree, and this factor in turn strongly influences the geomorphological processes. Studies of this type integrate the whole range of physical geography and are especially rewarding if they can be carried out in the field in suitable small drainage basins.

Temperature records taken at different positions relative to the valley floor over time also illustrate important climatological principles. The range of values recorded on the upper slopes of a valley is usually considerably smaller than that recorded in the valley bottom. The reason for this contrast is that cold air is denser than warm air, and

Plate 4 A close up view of hoarfrost, resulting from condensation with temperatures below freezing point in winter during night time cooling.

hence tends to gravitate to the valley bottom, especially on a still, long night. In winter when temperature falls below freezing, frost hollows can develop. There is a very well known example at Rickmansworth on the southern flanks of the Chiltern Hills, where temperatures are much lower than the surrounding higher ground and late frosts occur long after they do in adjacent areas. Interesting relationships can be observed between land use and microclimatic phenomena, such as the siting of orchards on warmer slopes, and avoiding valley bottoms for frost-susceptible crops. (See Plate 4.)

The important principle demonstrated by the occurrence of cold air at low elevations under still conditions is that air density depends on temperature, and that cold air is heavier than warm air. Thus air will rise if it is heated. It can also be forced up mechanically as when air is moving towards a mountain barrier in its path. Air is also forced upwards when cold, dense air undercuts warmer air at a front. The heating of air by the ground causes it to rise; as it does so it cools, thus reducing its capacity to hold moisture. In time the condensation point is reached and the air becomes saturated. Clouds may form and under some conditions precipitation may also occur.

Changes of temperature lead to movement of the air, such as downslope gravity flow that occurs on calm nights. These movements give rise to local winds, which can be recorded by means of a simple hand-held anemometer— the number of rotations of the cups is counted over a specific interval to give the wind speed. The drainage of cold air under gravity into the valley bottom at night gives rise to a downslope wind, called *katabatic*, which will blow most strongly in the early morning as the lower levels reach their minimum temperature. The reverse wind blows upslope during the day as the upper slopes warm up and the heated air rises, creating an upslope air movement from the valley floor. These winds are called *anabatic* winds. They will only develop fully when the general air movement is slack under anticyclonic conditions, and they are generally light.

Other microclimatic wind effects are concerned with different types of land use. Shelter belts are often planted to protect vulnerable areas from excessive wind. Recording the wind velocity in the lee of, within and upwind of shelter belts provides useful data relating biography to microclimatology. Indeed the forest microclimate provides much scope for detailed field-work, covering wind, insolation, temperature and moisture measurements. Conditions in different types of woodland and at different seasons provide opportunities for a wide range of studies. For example, the effect of deciduous woodland on wind speed in winter can be compared with that in summer.

Another environment that has been shown to create its own microclimate is the urban one. A considerable amount of work has been done on this topic, a whole book having been written on the climate of London (T. J. Chandler, 1965). Buildings influence urban climate in three main ways: (1) they affect temperature, (2) they modify the composition of the atmosphere, and (3) they alter the surface configuration and hence the wind pattern. One of the major effects of towns is on the temperature. The burning of fuel, absorption of heat by bricks and concrete to be released during the night, and the reflection of heat towards the ground from a layer of atmospheric pollution also adds to urban temperatures. Thus one of the most obvious features of the urban climate is the heat island effect that develops, especially during the night when winds are light. Differences of 1 to 2 °C between central London and the surrounding countryside have often been recorded. Although mean wind speed may be reduced over urban areas, the winds are often gusty as they are funnelled between tall buildings.

Wind in the coastal environment is important from several points of view. It can either enhance or inhibit the action of the waves and thus their effect on the beach. An onshore wind blows water towards the shore on the surface and this creates a seaward-flowing bottom current, enhancing the destructive effect of steep waves. An offshore wind, on the other hand, strengthens landward movement on the bottom and reduces the height of waves advancing shoreward, thus increasing their constructive effect. When the wind blows alongshore it will help to generate short waves that refract less than longer ones. Short oblique waves can move much material along the shore and thus exert an important effect on coastal geomorphology.

Another more strictly meteorological aspect of wind in the coastal zone is the creation of land and sea breezes. They are the result of the different responses of land and water to insolation. The land heats up quickly during the day, particularly in summer and when the pressure gradient is slack. The heated air tends to rise as its density is reduced. Water, on the other hand, absorbs much more heat through a deeper layer so the sea remains cool during the

25

day. Air is drawn from the sea to flow over the heated land during the late morning and early afternoon, when the sea breeze reaches its maximum velocity. At night the land cools much more quickly than the sea, the air over it becomes denser and sinks, causing a land breeze to develop as the sea is now warmer than the land. Observations of land and sea breezes can readily be made by means of a hand anemometer. It is of interest to trace the inland penetration of the breeze, as well as the time of onset of the sea breeze, the time it reaches its maximum intensity and when it is replaced by a land breeze during the night.

The creation of the alternating air flow illustrates fundamental facts concerning the heating of land and water, and is basic to an understanding of the contrasts of oceanic and continental climates, and the much larger, but in some respects, similar, meteorological phenomena, such as monsoons. Land and sea breezes operate most effectively in relatively low latitudes or in summer in higher latitudes, when the differential heating between land and water reaches a maximum. In low latitudes the Coriolis effect is weak and sea breezes can penetrate up to 80 km inland. The breeze may reach a velocity of 6 to 30 km/h under these conditions. The land breeze is always lighter and rarely exceeds 8 km/h.

Specific exercises have not been given in this section, but many suggestions for useful observations have been mentioned. They can provide data of the type analysed in the following section and a valuable practical experience of small scale local variations in the main weather elements.

Analysis of local meteorology

Analysis of local weather recordings can provide much information of practical importance. It is of value to relate morphological aspects to meteorological ones, in terms of relief–weather relationships. These relationships can be either over space at one time, or a comparison can be made of con-

ditions over time at one or more places. The time intervals used can also vary from detailed changes over a diurnal cycle to variations over longer periods of time; or the occurrence of particular conditions, such as frosts or snowfall, can be examined.

Examples of temperature variations across a valley at different times of the day have already been noted. The difference between the sides of the valley can be tested for significance, and the effect of exposure between the sides of the valley shows that the south and west facing slopes tend to be warmer than the opposing ones in a temperate northern hemisphere situation. The changes can be plotted to show both the change in space across the valley and in time at any one point on the valley transect.

Variations of temperature at different sites over longer periods of time provide useful data for analysis. Some sites are noted for their extreme conditions, such as the frost hollow at Rickmansworth north of London, already mentioned. This frost hollow lies in a narrow valley into which cold air can drain readily. Records show that the average minimum temperature is below freezing from mid-November to early April, being similar to records from Braemar in the Scottish Highlands. The temperature is often 8 to 11 °C lower than that at Kensington only 27 km away. Frost has also been recorded in each of the summer months, and very high ranges occur in this restricted dry valley.

Two sites near Durham provide suitable data for analysis. One set of readings was taken at Ushaw, which is situated on a ridge at 181 m, with land falling off in three directions. Another site 5·6 km to the east lies in a low position close to the River Wear at 49 m, where the river is incised in a steep-sided valley with a floor about 1·6 km wide. On one occasion in February the minimum night temperature was −7·2 °C at Ushaw while it fell to −15·5 °C at Houghall in the valley bottom. The value at Durham observatory was −10·0 °C, this station being sited on a ridge at 102 m.

Exercise 2.2a. Temperature effects

Table 2.1a gives the average daily maxima, the average daily minima and the extreme minima for each month at the two stations. Plot these values on a graph to illustrate the significance visually. Use the *t*-test applied to the differences between the paired values to assess the statistical significance of the differences.

The results for the maximum values are given in Table 2.1a, but the minima and extreme minima are more interesting. The results illustrate the importance of site in local temperatures, showing that the drainage of cold air to the valley bottom on still nights is very important. The maximum values represent day temperatures and at this time the differences are less. The lower, more sheltered sites are consistently warmer by a few degrees. This could be due to the lower wind strength in the valley bottom as opposed to the open ridge site.

At coastal sites the contrast between Scarborough and Bridlington provides an instance of the effect of proximity to the sea. The Scarborough values, given in Table 2.1b, were taken at a point only 200 m from the sea at 36 m elevation, while the Bridlington station, although only 23 m high was about 1·5 km from the sea at the foot of the Wolds, where air drainage can take place effectively. The moderating influence of the sea can be shown by analysing the figures in a similar way. On a local scale the ameliorating effect of the sea can only penetrate a short distance inland, but on a much larger-area basis the maritime influence is important over the whole of the British Isles.

Exercise 2.2b

Another interesting example is given in Table 2.1c, and concerns the effect of soil on local temperature values. The figures give the difference for each month for the average daily minima and extreme minima for Lynford and Cambridge. Lynford is situated on a dry sandy soil in the Breckland, whereas Cambridge is typical of the East Anglian countryside, with its damper, heavier soil. The figures show that the dry sandy soil cools much more than the Cambridge soil, as the sites are comparable from the relief point of view. Analyse the figures as before. They refer to a series of dry, cool anticyclonic days in early spring. The results show that although the light, dry sandy soil tends to warm up quicker during the day, giving higher maxima, it cools down much more quickly to reach lower temperatures as shown by the minimum values.

These examples illustrate the importance of local effects on the temperature, including relief, marine influence and soil type.

The effect of relief on precipitation can also be studied. Because of its marginal position the British Isles is particularly liable to a wide range of snowfall in different situations. The average number of days with snow lying each year on low ground varies from less than 5 in the southwest peninsula to over 30 in northern England and over 35 in northeast Scotland.

Variations in rainfall can be considered in relation to relief and exposure. Figures for a lowland station can be compared with those for an upland station. The mean precipitation recorded in the Brecon Beacon area of South Wales, for example, with westerly conditions, exceeds 14 mm, while the comparable figure for the western peninsula of Pembrokeshire is less than 4 mm. The precipitation gradient becomes very steep as the mountains are approached, showing the effectiveness of orographic precipitation.

The precipitation pattern can be analysed by relating the amount of precipitation to the type of weather. The daily weather map provides a means of evaluating the type of weather. Eight directional types can be recognized from the pattern of the isobars, with the addition of cyclonic and anticyclonic types when the isobars are curved, a final unclassified type allows for difficult situations that cannot be fitted into any of the others. The effects of site and elevation can be studied

Table 2.1 Temperatures (°C).

(a) Ushaw and Houghall, County Durham

Month	Maxima		Minima		Extreme minima		
Jan.	5·77	6·67	+ 0·3	− 0·6	−4·9	−8·4	The first value refers
Feb.	5·88	7·06	0·5	0·0	−4·6	−7·4	to Ushaw and the
Mar.	8·50	9·61	1·8	+ 0·9	−3·2	−6·5	second to Houghall
Apr.	10·44	11·39	3·2	2·7	−1·6	−4·6	
May	13·61	14·39	5·7	5·0	+0·2	−3·1	
Jun.	17·22	18·17	8·4	7·4	+3·6	+0·7	
Jul.	19·22	20·33	11·0	10·5	+6·9	+3·9	
Aug.	19·17	20·05	10·6	10·0	+6·5	+2·9	
Sep.	16·17	17·22	8·4	7·2	+3·5	−0·2	
Oct.	11·78	13·16	5·4	4·4	−0·4	−3·0	
Nov.	8·22	9·33	2·9	2·0	−1·8	−5·2	
Dec.	5·77	6·61	1·1	+ 0·1	−3·8	−6·8	

(b) Coastal stations

	Scarborough	Bridlington
Jan.	0·9	3·2
Feb.	1·2	1·9
Mar.	1·3	2·2
Apr.	1·3	1·8
May	1·2	1·9
Jun.	1·3	3·7
Jul.	1·4	2·4
Aug.	1·4	3·3
Sep.	1·8	3·5
Oct.	2·3	2·8
Nov.	1·6	2·9
Dec.	1·3	3·2

(c) Soil types

	Lynford (sand)	Cambridge
Jan.	1·2	3·3
Feb.	1·4	3·4
Mar.	1·4	3·2
Apr.	1·5	3·8
May	1·7	3·4
Jun.	1·7	4·2
Jul.	1·7	3·7
Aug.	2·1	4·2
Sep.	1·6	3·4
Oct.	1·3	3·3
Nov.	1·1	2·6
Dec.	1·1	3·0

The values refer to differences in temperature in °C, with Bridlington below Scarborough and Lynford below Cambridge. The first set of values refers to the average daily minima and the second to average extreme minima.

Year	1·4	2·8		1·5	3·4

by comparing the values for the 11 different weather types for Dale Fort on the Pembrokeshire coast and Glyntawe in the mountains.

Exercise 2.3. Precipitation analysis

Figures for Dale Fort and Glyntawe are given in Table 2.2 for January and July. Analyse them in the same way as the temperature figures for both stations at two seasons for the

Actual temperatures on different soil types, in 1941 at Lynford and Cambridge.

	Maxima		Minima	
Date	Lynford	Cambridge	Lynford	Cambridge
May 4	15·0	13·9°C	−9·4	−2·2°C
May 8	10·0	8·9	−7·8	−1·1
May 9	11·7	9·4	−7·2	−1·7
May 10	12·2	11·1	−8·3	−1·7
May 11	14·4	16·1	−9·4	−2·8
May 16	15·6	15·6	−4·4	−0·6
May 18	20·6	20·0	−2·8	+3·3

Table 2.2 Precipitation (in mm) for different weather types for Dale Fort and Glyntawe in Wales

	Dale Fort		Glyntawe	
	Jan.	Jul.	Jan.	Jul.
West	3·4	1·8	15·0	6·8
Northwest	0·6	0·4	2·0	2·5
North	1·2	1·0	1·6	1·3
East	1·0	2·2	1·7	9·2
Northeast	0·0	0·2	0·2	0·8
Southeast	5·0	0·0	3·7	3·5
South	3·5	5·5	7·3	5·3
Southwest	5·2	2·7	17·5	15·0
Anticyclonic	1·0	0·5	0·5	1·2
Cyclonic	3·6	3·5	12·5	9·4
Unclassified	6·5	3·4	10·0	5·0

11 different weather types. The relationship between weather type and precipitation can be plotted to show the effect of wind direction for the eight directions. The local effects of relief and site help to explain the contrasts between the two stations which are about 110 km apart. The low values for precipitation with north to east weather types compare with the high values for southwest and west types at these stations with a southwesterly exposure. The importance of wind direction relative to frontal and orographic rainfall can be assessed.

2.3 Hydrology

Introduction

The local scale hydrological processes can be recorded in the field over a relatively short time scale. These processes include the input of precipitation, which was mentioned in the last section, and the runoff in the form of stream flow. Discharge can be relatively easily recorded in the field in small streams, and it is particularly valuable to relate discharge to sediment load. The relationship provides an important link between hydrology and geomorphology, and is also frequently affec-

ted by both natural biogeography and human intervention. The water that does not run off or is lost by evapotranspiration becomes ground water. The behaviour of ground water can be related to relief and geology; some examples of this aspect are considered, such as artesian conditions. On a local scale it is of interest to relate discharge of a small catchment to its surface characteristics. The land use affects the runoff and again urban areas can be differentiated from natural catchments by their runoff characteristics.

Field measurements

The aspects of hydrology that can be studied in the field include the recording of stream discharge and its variation both in the downstream direction at any one time, and its variation at any one point over time. The measurement of spring flow and the groundwater level are also important locally. Precipitation that neither runs off immediately to the stream nor enters the ground-water system is lost through evaporation and/or transpiration. This aspect should also be measured where possible. The relationship between these variables can be expressed by the water balance equation in the form:

$$Q = P - E \pm \Delta S$$

where Q is the stream discharge, P is the precipitation, E is the evapotranspiration and ΔS is the change in storage, which is the volume of ground water. (See Plate 2.)

Stream flow is one of the quantities that can be measured in the field. It can be obtained by simple or more sophisticated means. The simplest method is to record the time it takes a floating object to cover a measured distance down the stream. The surface velocity is not, however, the average velocity. The mean velocity occurs approximately at a depth of 0·6 of the total depth. The flow is quicker on the surface, falling off with depth to the bottom. The surface velocity should be multiplied by 0·8 to give an estimate of the mean velocity. It is also necessary to know the cross-sectional area of the stream where the observations are made. This can be obtained by measuring the depth at regular positions across the channel and drawing out a cross section, the area of which can be calculated. A more accurate value can be obtained by using a current meter. It should be held at a depth of 0·6 of the stream depth at various positions across the stream to make allowance for changes of velocity near the banks. A more accurate value will be obtained if the means of readings taken at depths of 0·2 and 0·8 are averaged.

Water engineers provide more permanent gauging stations by building special weirs across the river bed. The weirs are often V-shaped and have a known cross section. All that is required is to measure the level of the water at the weir once the flow velocity has been related to the different levels. The relationship between the water level at the gauge and the discharge as recorded by current meter at this cross section is called the rating curve. Once the relationship has been established it is simple to record the height of the stream to obtain the variation of discharge over time. A recording gauge provides a continuous measure of discharge. There is a considerable advantage in a continuous record as some streams vary rapidly in level. This applies for example to streams from melting glaciers in summer, where there is often a diurnal variation, with low flow in the morning building up to high flow in the afternoon.

The relationship between the type of river bed and the nature of flow is also important. In some small mountain streams the water appears to be flowing very fast over large boulders and small falls, and pot-holes complicate the bed flow. Careful observations will reveal that although the water is flowing fast in some parts there are eddies and backwaters. The mean flow velocity is not necessarily very high, although the slope may be steep. On lower ground, where the stream is flowing over a smooth sandy or muddy bed, its flow is

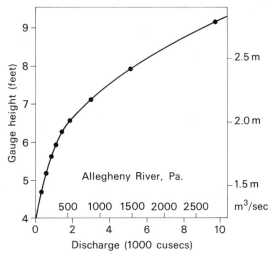

Figure 2.3 A rating curve for the Allegheny River, Pennsylvania. (After M. Morisawa, 1968)

much more uniform. Despite the lower longitudinal slope the mean velocity is often greater. Thus in many streams the mean velocity increases in a downstream direction despite a decrease in gradient in this direction. The basic reason for this result is the reduction in friction downvalley as the stream bed becomes smoother, owing to the reduction in sediment size. The stream bed also increases in size and thus has a more efficient hydraulic radius.

The rating curve shows that stream velocity increases as the discharge increases at any one place on the river, usually called at-a-station velocity to differentiate it from changes in a downstream direction. Figure 2.3 illustrates a rating curve and Figure 2.4 indicates the characteristics of changes at-a-station and downstream with variations in discharge. Figure 2.4 distinguishes changes that take place at one point on the river as the discharge rises, from those that take place downstream at any one discharge at one time. The dashed lines A–C and B–D refer respectively to small and large channels, while the solid lines A–B and C–D refer respectively to changes downstream at small and large discharges. The graph showing width indicates that the variations are greater downstream than at-a-

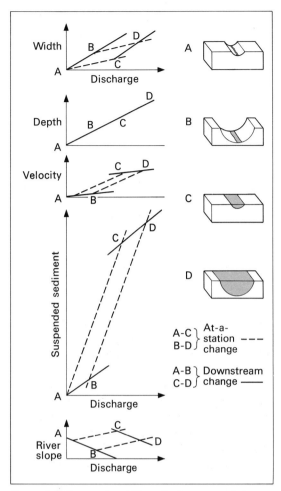

Figure 2.4 Changes in width, velocity, suspended sediment and river gradient at-a-station and downstream with increasing discharge. (After L. B. Leopold, M. G. Wolman and J. P. Miller, 1964)

station, and the reverse applies to the velocity and suspended sediment load, when the greater changes occur at-a-station as discharge increases.

As with so many other field measurements, the longer the available records are the more valuable they become. This applies to records of river discharge because the flow in rivers is very variable. The longer-term variations will be considered in the next chapter, but useful information may be gained from detailed short-term records. These are the records of individual storms and their effects on stream

discharge. In small catchments the runoff usually takes place rapidly, and the relationship between an individual storm precipitation and stream runoff is called a flood or storm hydrograph. Figure 2.5 illustrates the hourly rainfall and the discharge, which is shown as a continuous curve. The rainfall covers a 3 hour period and there is a considerable delay before the peak of water discharge in the stream. The steep curve up to the peak is called the rising limb and the gradual lowering of the water level is called the falling limb, which may not return the low level of before the storm for a long time, owing to prolonged outflow of ground water.

Exercise 2.4. Discharge analysis

Table 2.3 gives a set of data derived from observations made by Gregory and Walling (1968) in five small catchments in south Devon. The peak discharges for the basins are given and a value adjusted to the area of the basin for comparative purposes. The percentage of each basin covered by woodland, enclosed and unenclosed land is also given. The discharge is given for two separate periods of heavy rain in November 1967. The exercise aims to study the effect of land use on the discharge for the two storms. Plot the percentage of woodland against the peak discharge corrected for the drainage basin area, using different symbols for the two storms. Examine the results and test the significance of the correlations by regression and correlation analysis. Assess the effect of woodland and enclosed land on the discharge.

The individual characteristics of the storm hydrographs vary from stream to stream with the pattern of precipitation in time and space, and in relation to the shape of the basin. The height of the flood will depend on several factors, one of the most important being the amount and concentration of the rainfall. Intense thunder rain when 12 to 15 cm can fall in an hour or two causes very high discharges

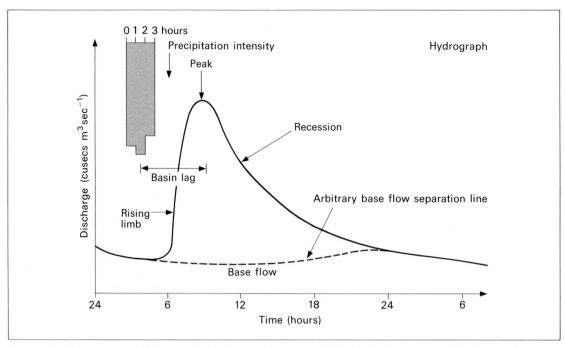

Figure 2.5 A hydrograph to illustrate the relationship between rainfall and discharge in a small catchment for a storm lasting 3 hours. (After R. J. Chorley, (Ed.), 1969)

Table 2.3 Devon catchment data from Gregory and Walling (1968)

Basin	A	B	C	D	E
Discharge m³s⁻¹ (unit area)⁻¹					
1 Nov. storm	0·77	0·93	0·46	1·27	1·00
4 Nov. storm	0·35	0·81	0·43	0·73	0·56
Land use (%)					
Woodland	1·1	5·7	77·0	11·0	39·5
Unenclosed	92·2	20·8	2·1	19·7	11·7
Enclosed	6·7	73·5	20·9	69·3	48·8

and flash floods occur in small upland catchments. The 1952 Lynmouth disaster in north Devon illustrates this type of flash flood. The small East and West Lynn streams carried as much water as the Thames at high flow for a short time. The intensity of the rain is important as there is a limit to infiltration depending on many variables, including vegetation, soil type, rock permeability, relief and others. (See Plate 5.)

The antecedent rain is another significant variable, because if the ground is already saturated less will be able to infiltrate. Other variables are evaporation and transpiration, which vary seasonally, both increasing in summer. This effect is not so marked during a short, intense storm, when relative humidity will be high.

Storms of the intensity that result in flash floods are relatively rare, and it is useful to assess the frequency with which they may be expected to occur. In general it is true that the more extreme an event the more rarely it occurs. The recurrence interval, illustrated in Figure 2.6a and b, is the period over which it is statistically likely that an event of a given magnitude will occur. A 5-year flood, for example, is likely to occur once in any period of 5 years. It will not necessarily happen at 5-year intervals. A 100-year flood will be much higher but is only likely to occur once in a

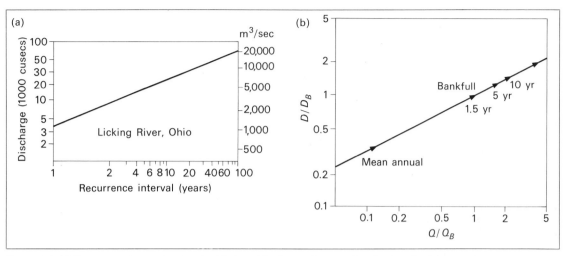

Figure 2.6 (a) Recurrence interval using double logarithmic graph paper for plotting discharge against recurrence interval in years for the R. Licking, Ohio. (After M. Morisawa, 1968) (b) Relationship between bankfull discharge and depth for the mean annual flow, bankfull discharge, 5-year and 10-year flood. (After L. B. Leopold, M. G. Wolman and J. P. Miller, 1964)

Plate 5 Debris left by the floods of the Lyn Valley of north Devon in 1952. During the course of the flood the stream changed its channel and caused much damage. Its ability to carry a heavy load is apparent in the large debris left as the flood subsided.

period of 100 years. This idea is important in planning flood protection schemes and water supply constructions, such as dams.

Near the lower end of the scale is the bankfull discharge, which is the amount of water that will fill the stream to the top of its banks before flooding occurs. This discharge occurs in British rivers about 0·6 per cent of the flow duration curve, a value that has a return period of about 6 months. A bankfull discharge would be expected on about 2·2 days in the year. Different values apply to rivers in the United States owing to differences in precipitation regime and other variables. The importance of the bankfull discharge is that it is thought that the river adjusts its bed and banks to carry this volume of flow most effectively. A series of values relate the bankfull discharge to various parameters of the river bed, including the following: width $= 1 \cdot 65 \, Q_b^{1/2}$, where Q_b is the bankfull discharge, depth $= 0 \cdot 545 \, Q_b^{1/3}$, area $= 0 \cdot 9 \, Q_b^{5/6}$ and mean velocity $= 1 \cdot 112 \, Q_b^{1/6}$.

In plan form relationships have also been established between bankfull discharge and the dimensions of meanders, including their wavelengths and channel widths. The width of the channel adjusts to discharge at bankfull most readily. Thus former discharge can be estimated from the size of old channels.

Only a very small proportion of the rain that falls actually directly enters the stream, most of the water reaching the stream must come from the banks or bed. Unless the rain is torrential it is rare for water to flow over the surface, especially where vegetation is thick. Sheet flow can occur, however, in very flat country in semi-arid areas which are liable to extreme precipitation at rare intervals. The whole ground surface may be covered by a shallow sheet of flowing water. In more temperate climates with more frequent and less intense rain, the drops are absorbed by the soil and move gradually downslope. The water that passes more or less directly to the stream is called throughflow. When the rain is very heavy rills can form and these carry soil to the stream in the process of rain or rill wash.

Where rocks are permeable a considerable proportion of the precipitation will penetrate below the soil layer to become ground water. It then moves slowly through the rocks eventually seeping into a stream or emerging as a spring. Occasionally the structure is suitable for water to flow out naturally from the ground under pressure. This situation is called *artesian* after a well in Artois, France. The fountains of Trafalgar Square in London used to flow naturally under artesian pressure until the water table was lowered too far.

Figure 2.7 illustrates a type of artesian structure. There must be a collecting area in permeable rock at a higher level than the same rock where it is overlain by an impermeable bed. In the London area the chalk of the North Downs and Chilterns provides the aquifer, which is overlain in the London Basin syncline by impermeable London clay. If a well is drilled into the chalk beneath the clay in the syncline the water will rise naturally under artesian pressure to the level of the lowest chalk outcrop. Although the artesian structures are usually large their effects can be examined locally.

Water supply in the areas of hard, impermeable rock is derived largely from surface sources. Stream flow is important and it is maintained during dry spells by ground-water flow. This is called *base flow*, and it reaches the streams through springs and seepage along the river beds. Precipitation and temperature both play an important part in ground-water hydrology. Precipitation provides the moisture that can penetrate into the ground, while temperature is an important control on evapotranspiration. In areas of permeable rocks where the surplus water can penetrate into the ground there is a direct relationship between the level of the ground water table, precipitation and evapotranspiration through the percolation rate. Field observations on these variables have provided results that illustrate the processes involved.

The movement of the water table in an area of permeable rocks, such as Bunter sandstone and chalk, is significant because the rocks

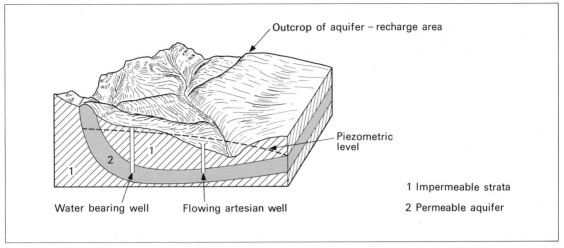

Figure 2.7 Block diagram to illustrate the formation of an artesian well. (After R. J. Chorley, (Ed.), 1969)

provide important aquifers for water supply from underground sources. Water can only be successfully extracted when the wells from which it is withdrawn extend below the water table. Thus the movement of the water table is very important. It separates the permanently saturated zone from that through which the water is passing. Usually in a permeable rock the boundary between the saturated and unsaturated rock is sharp. Its surface is related to the head or pressure of percolating ground water, and this in turn depends in part on the relief. The water table tends to follow the relief in a subdued fashion, being highest under the interfluves and falling towards the valleys.

The water year starts about September or October when the autumn rains penetrate through to replenish the ground water. The water table begins to rise at this stage. The variation of the level of the water table can be obtained from measurements taken in wells, provided the level is not disturbed artificially by water extraction. Wells that are pumped for water supply show an artificial lowering of the water table in a cone-shaped area around the well. This is called the *cone of depletion* and in chalk it normally has a radius of about 1·6 km. This distance sets the minimum spacing of wells.

Figure 2.8 illustrates the movement of the water table throughout the year at a site in a chalk area of eastern England. One of the notable points concerning the movement is its asymmetry. From a maximum in February there is a slow fall in level throughout the summer until about mid-December, when there is a very rapid rise, which decelerates during January, slowing down and reversing in February. The main variables on which the pattern depends are precipitation and evapotranspiration, which together determine the percolation in the absence of runoff. The full height of the bars on the graph above the horizontal line indicate precipitation, which is split into two parts. (a) is the loss by evapotranspiration and (b) is the percolation, thus (a)+(b) = precipitation. The part of the bar below the horizontal line (c) is the loss from soil moisture.

The amount of precipitation in summer is not enough to balance the increased evapotranspiration when temperatures are high and cause much more effective evaporation. The growing plants also need more water, thus increasing transpiration also. In October the rainfall is just sufficient to make up for the loss by evapotranspiration, which is beginning to fall off as the season advances into autumn. By November the deficiency in soil moisture in the upper layers above the water table is being

35

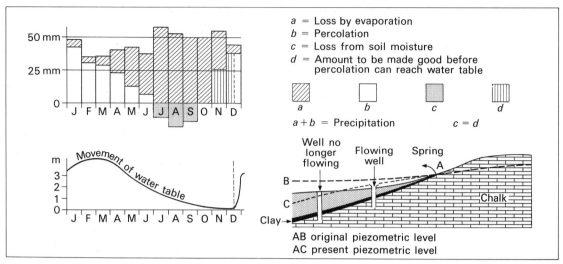

Figure 2.8 The movement of the water table in relation to precipitation, evaporation and percolation in the absence of surface runoff. The lower diagram illustrates the effect of lowering the piezometric level on the flow of an artesian well. (After W. B. R. King, 1957)

made good, as percolation starts to fill the pores in the dry ground above the water table. It is not until mid-December, however, that the ground water can reach the water table, and this results in the sudden rise that takes place about this time. The date of the rise in the water table is determined by the point on the graph where $c = d$, which is the loss of soil moisture to be made good before percolation can reach the water table. In a wet season c is small and hence the rise occurs early, while in a dry season it may be delayed.

The pattern of evapotranspiration has a markedly seasonal character, with high values in summer and low ones in winter, paralleling the movement of the water table to a considerable extent. The rainfall distribution, on the other hand, is fairly even throughout the year. The pattern of the balance between evapotranspiration and precipitation varies over the British Isles. The southeast usually has a water deficit during the warmer part of the year, while the northwest normally has a surplus throughout the year. These contrasts are significant from the point of view of water supply from underground sources. The best aquifers in the chalk and Mesozoic sandstones outcrop in the driest, southeast part of the

country where water deficits are much more widespread.

Exercise 2.5. Evaporation and precipitation

The figures in Table 2.4 give the evaporation and precipitation monthly for Harrogate, Yorkshire, and Selbourne/Petersfield, Hampshire, for a series of years. The pattern for evaporation shows little annual variation, but the precipitation varies considerably. The long-term average evaporation is 43·2 cm at Petersfield and 49·7 cm at Harrogate for 1908–1942. The precipitation pattern is very different in the two areas and varies both throughout the year and from year to year. Use the values provided to calculate the water deficiency and surplus throughout the year. Plot the data to provide a visual impression of the contrasts between the two variables.

A very different situation is shown by the figures for San Antonio in Texas. The figures show that in every month there is a water deficit, which reaches its maximum, despite higher precipitation, in the summer months. The annual pattern of evaporation is such that

36

Table 2.4 Rainfall and evaporation values (in cm) for Harrogate, Yorkshire, Petersfield and Selbourne, Hampshire, and San Antonio, Texas

Rainfall (the first value refers to Harrogate and the second to Selbourne)							Evaporation (the first value refers to Harrogate and the second to Petersfield)					
Month	1946		1947		1948		Month	1946		1947		1948
Jan.	9·17	10·85	7·04	10·01	17·70	19·56	Jan.	0·84	0·33	0·58	0·00	0·94	0·05
Feb.	8·76	8·33	8·10	5·87	4·67	6·02	Feb.	1·02	0·41	0·64	0·51	1·02	0·41
Mar.	3·45	3·45	16·46	16·51	3·48	3·84	Mar.	1·57	1·85	0·58	0·64	2·64	3·33
Apr.	4·52	6·76	8·64	8·31	6·35	5·23	Apr.	4·88	5·00	5·51	4·42	5·05	5·64
May	7·57	10·54	10·26	4·39	2·49	7·44	May	7·21	7·67	7·42	6·22	7·32	9·27
Jun.	5·77	9·17	9·63	6·27	9·17	9·22	Jun.	7·09	6·25	6·96	9·32	7·24	7·47
Jul.	9·74	5·79	8·94	3·76	2·79	4·47	Jul.	7·92	9·88	8·46	9·09	7·19	7·72
Aug.	11·23	15·44	2·24	1·47	10·90	14·53	Aug.	6·27	5·94	7·95	10·72	5·46	5·82
Sep.	14·86	16·08	4·98	4·09	4·57	7·32	Sep.	3·94	2·74	6·02	5·41	4·19	4·09
Oct.	4·85	3·61	2·79	3·25	2·57	7·52	Oct.	2·87	2·74	2·21	2·69	3·12	2·49
Nov.	15·72	19·43	7·49	5·21	2·82	5·23	Nov.	2·24	1·04	2·34	0·56	1·32	0·66
Dec.	6·15	10·87	6·27	9·45	7·65	16·00	Dec.	0·99	0·51	0·71	0·25	0·97	0·00
Year	101·52	120·32	92·84	78·59	75·16	106·38	Year	46·84	44·73	49·38	49·83	46·53	46·94

San Antonio, Texas												
	Jan.	Feb.	Mar.	Apr.	May	Jun.	Jul.	Aug.	Sep.	Oct.	Nov.	Dec.	Year
Evaporation	6·1	7·6	11·4	14·0	16·8	20·1	23·1	23·1	17·3	12·7	7·9	6·1	166·1
Precipitation	3·6	4·1	4·3	7·9	8·1	6·1	5·3	5·8	7·6	5·6	4·6	4·1	67·1

the water loss is much greater during the summer when temperatures are very high.

A comparison of these values provides a good indication of the equability of the British climate and the importance of maritime influence on the country. The summer temperatures are relatively low, the air is moister, with higher relative humidities, and hence evaporation is much lower. Precipitation in Texas is, however, similar to that in lowland Britain. The balance between precipitation and evaporation provides a useful link between meteorology and hydrology.

Ground water is particularly valuable as a source of water supply because once the water is underground below the level of the capillary layer it does not suffer from evaporation and is available for withdrawal from wells and springs in suitable rocks. For this reason modern water supply methods include ground-water recharge schemes. Under this system the surplus water of the winter season that would normally flow wastefully to the sea is extracted from the rivers by pumping and allowed to percolate through permeable aquifers to replenish and raise the water table. It is then available for pumping from wells in the summer season of deficit. Percolation also purifies the water as it passes down through the pores of the permeable strata. This method is adopted from the water flowing down the Thames, which is pumped back into the chalk of the Chiltern Hills.

Hydrographs

When the catchment area consists of impermeable rocks the excess precipitation will run off and be added to the stream flow. It can then be recorded at a gauging station. The relationship between the rainfall and runoff can be analysed by the construction of a hydrograph. The flood hydrograph represents the runoff from one particular wet spell, and can be constructed for a small drainage basin. The method of calculating the amount of precipitation over the catchment area by means of Thiessen polygons, for example, has already been mentioned. The first part of the precipitation, amounting to at least 1 mm, will be lost by interception by vegetation, then the soil must be saturated and surface hollows filled before overland flow can begin. Usually 2 to 5 mm are needed to saturate the soil, depending very much on the previous conditions, the nature of the soil and the vegetation. Eventually, however, if the precipitation intensity is sufficient, water will reach the stream by overland flow and runoff will begin to augment the base flow.

The easiest way of recording the flood is to establish a rating curve, whereby the discharge of the stream can be obtained from a record of the depth of water or stage at a particular point on the stream where the relationships have previously been established. It is worth noting that the rising limb of the flood hydrograph, when the water level is increasing, differs from the falling limb. The former is steeper when discharge is plotted on the vertical axis and stage on the horizontal one. For any recorded level the discharge is higher on the rising stage than the falling one. This is because the surface slope of the water and the velocity are both greater during the rising stage.

The hydrograph shows the relationship between stage and time, or discharge and time. The short-term flood hydrograph is constructed for one particular heavy rainfall event. The longer-term hydrographs refer to monthly values and provide the river regime, an aspect that covers a longer time span and greater area, and hence will be considered in the next chapter.

The flood hydrograph is useful in studying the small drainage basin and short-lived

storm. A sample hydrograph is illustrated in Figure 2.9. It consists of the following components: a = surface runoff, b = interflow, c = ground-water flow, and d = channel precipitation. The interflow is the precipitation that seeps through the soil laterally into the stream a short distance below the surface. The ground-water flow provides the base flow. Channel precipitation is usually only very small, and its duration depends directly on that of the precipitation. The interflow starts soon after the precipitation and gradually increases to a maximum, falling off rapidly to zero when the rain stops. The major contribution is the surface runoff, which also increases to a sharp peak and then falls off rapidly when the precipitation ceases. The duration of the rising limb, the crest flow and the falling limb can be approximately measured by drawing a horizontal line from the start of the rising limb to where it cuts the hydrograph again, although if the amount of base flow varies considerably, the estimate will not be accurate. The change in base flow is usually fairly slow, except under special conditions.

The amount of runoff can be calculated by measuring the area above the base flow on the hydrograph. This is made easier if it is divided into hourly columns as the rainfall is usually plotted in this way. The percentage runoff can vary with many variables, but it can never exceed 100 per cent. Some storms can produce 80 per cent runoff if they occur in impermeable basins in which the soil and vegetation are already saturated from previous precipitation and the evaporation is low, due to low temperatures and high relative humidities. On the other hand the figure can fall as low as 20 per cent in basins with permeable rocks, dry absorbing soil and high temperatures and evaporation rates. The annual figure for a drainage basin in the English Midlands is about 40 per cent.

The variables that affect the relationship between runoff and rainfall include static and dynamic ones. The former include the nature of the bedrock, especially its permeability,

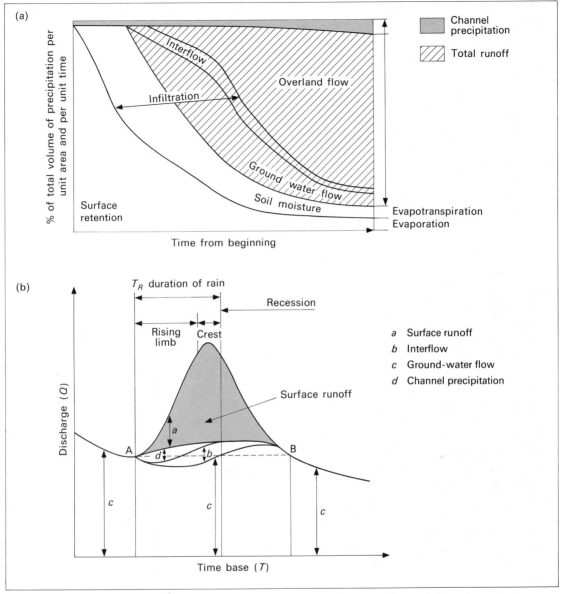

Figure 2.9 A flood hydrograph to show surface runoff, interflow, ground-water flow and channel precipitation through time relative to one storm. The upper graph (a) shows the percentage of total precipitation against time, and the lower one (b) discharge against time.

and the nature of the soil and vegetation, which influence the immediate throughflow, evapotranspiration and interception. These variables may be the most important ones in a short-lived storm which produces only enough water to replenish the soil deficit and does not penetrate to the water table. Another static variable is the shape of the drainage basin. Some shapes tend to concentrate the water more effectively than others, thus producing sharper hydrograph peaks.

The dynamic variables include four important ones: (a) the total rainfall, (b) the intensity of rainfall, (c) the antecedent rainfall, and (d)

39

changes in loss rates. The total rainfall (a) affects the total volume of the flood hydrograph. The greater the rainfall amount the greater the flood volume and usually the higher the percentage of runoff to rainfall. The intensity of rainfall (b) affects the runoff through the capacity of the soil to absorb water, the infiltration rate. The greater the intensity the smaller the proportion that can infiltrate, so that a lower intensity makes for a proportionately smaller runoff, as more can percolate into the ground, at least until the upper layers are saturated. The antecedent rainfall (c) is important in this connection. If it has been high the ground will tend to be saturated and so runoff will increase in proportion. If the ground is dry much of the water will be absorbed, unless it has been so dry that the surface develops a hard crust that must be dampened before infiltration takes place. However, under these conditions cracks often occur, and these readily absorb water. Previous snowfall can usually absorb much water, but if this falls as rain it may melt the snow thus adding very greatly to the flood water in some instances. The floods of 1947 when warm rain fell on snow, causing rapid thawing, provide an example. Frost can render a normally permeable layer impermeable, and hence increase the runoff.

Changes in the rate of loss of water, (d), are involved in variations in evaporation, interception and transpiration with the seasonal change in vegetation. Deciduous trees, for instance, intercept much more rainfall in summer than in winter, when they are bare, and they also transpire more in the summer season, thus considerably reducing the runoff in summer compared with winter. Air temperatures and wind speed also influence the loss rate to a considerable extent. The effects of vegetation are very important in this respect; the barer the soil and the poorer the quality of the vegetation, the more rapid will be the runoff, as soon as the rain starts. There is a close relationship between the quality of the vegetation and the runoff. Rapid runoff can produce gullying and erosion of the land

surface. This relationship has a positive feedback effect, in that once the surface soil has been washed away the vegetation becomes further depleted and the runoff further intensified as a result. In this way soil erosion is a self-generating process. The importance of maintaining an effective vegetation cover is amply demonstrated and is indeed the main aim of soil protection methods.

The Thames provides an example of a valley in which soil erosion is not severe, and this can be at least partly explained by the low runoff to rainfall percentages. Table 2.5 provides figures for rainfall and runoff for the Thames at Teddington for two separate years and the mean for 4 years. Evaporation at Camden Square in London is also given.

Exercise 2.6. Runoff–rainfall ratios

Use the values given in Table 2.5 to calculate the percentage of runoff to rainfall for the Thames at Teddington for the two very different years, 1934 and 1936. Compare the values with the mean for 4 years. Plot the values on graphs to show the contrast between the two years, 1936 being considerably wetter than 1934, apart from the wet December of 1934. The significance of the high summer evaporation is shown clearly, particularly for the 1936 values. This effect can be demonstrated by plotting runoff–rainfall percentage against evaporation, labelling the months on the graph. The values are highest at the beginning of the year, falling to very low values for the summer period, and increasing again at a slower rate during the autumn and early winter. The mean values for the four years should be used.

The data illustrate several important points. In some instances the runoff exceeds the rainfall. This occurs when the previous month has been very wet, as in January 1934, because in a drainage basin the size of the Thames at Teddington there is a lag between precipitation and runoff. This is particularly

Table 2.5 Rainfall and runoff at Teddington and evaporation at Camden Square, London (Values in mm)

Month	1934		1936		Four year mean	
	Rainfall	Runoff	Rainfall	Runoff	Rainfall	Runoff
Jan.	55·60	10·16	108·97	76·96	59·44	33·78
Feb.	6·35	6·86	55·12	52·07	56·39	28·19
Mar.	59·18	11·94	48·00	38·10	49·53	34·80
Apr.	53·09	9·40	43·94	29·97	56·13	21·59
May	18·03	6·35	18·80	16·26	29·97	13·46
Jun.	34·29	3·81	81·79	13·21	67·56	9·91
Jul.	40·13	3·30	109·73	13·72	53·54	7·87
Aug.	56·13	3·56	15·24	9·91	37·34	5·84
Sep.	52·58	3·56	84·58	9·91	80·26	6·60
Oct.	42·42	4·06	43·69	9·14	58·93	10·16
Nov.	51·56	5·59	82·04	25·40	73·15	22·86
Dec.	167·64	29·46	70·10	32·00	84·07	27·18
Year	793·75	68·33	829·56	406·91	Year Oct. to Sep.	
	1934–5		1936–7			

	Camden Square											
Month	Jan.	Feb.	Mar.	Apr.	May	Jun.	Jul.	Aug.	Sep.	Oct.	Nov.	Dec.
Evaporation	2·54	5·08	17·78	38·10	60·96	63·50	76·20	58·42	35·56	15·24	7·62	2·54

likely to occur in winter when some of the precipitation may fall as snow, and stay on the ground for a while before melting and running off. The low percentage of runoff to rainfall in January 1934, compared with February is the result of very dry weather in the last few months of 1933, when only 3·75 cm of rain fell in November and December together. The percentage also tends to be lower in the early winter than the late winter and spring because the precipitation in autumn and early winter goes mainly to replenish the depleted ground-water supplies, as already mentioned.

The exceptionally high runoff in early 1936 followed a very wet autumn and winter in 1935–36. During the last four months of 1935 the precipitation amounted to 43 cm, and it continued high in January 1936. The values show the great variability characteristic of conditions in the British Isles, which are marginal from so many points of view. The heavy precipitation of June and July 1936, was not, however, accompanied by the equivalent percentage of runoff, although the flow remained high in comparison with drier years. The evaporation values provide an explanation for the different percentages of runoff between the summer and winter seasons. Evaporation shows less annual variation, but seasonal changes are considerable. The marked variations in precipitation in different years may mask the normal relationship that holds when mean values for precipitation are used. This type of analysis illustrates important relationships between different aspects of the hydrological cycle and weather conditions.

Sediment load, vegetation and discharge

The last two sections have dealt with some

relationships between various elements of the hydrological cycle and atmospheric conditions. In this section other elements of physical geography are introduced: vegetation and geomorphological response in the form of sediment yield. Variations in sediment yield have an effect on the nature of stream processes, as well as other effects, including soil erosion, flood plain aggradation and flooding.

Measurements have been made in small catchments in various parts of the world to establish the relationship between sediment yield and river discharge. This relationship enables the rate of erosion in the catchment to be calculated from records of discharge. It is possible to calculate modern rates of erosion from these observations. However, care must be exercised if the modern rates of erosion are extrapolated over a longer period of time, because it can be shown that the character of the vegetation plays an important part in the relationship. Modern farming techniques have seriously altered the natural vegetation over much of the world, and caused considerable changes in the relationship between sediment yield and discharge. In this instance man is an important geomorphological agent. Farming is not the only factor which alters the sediment yield–discharge relationships; the effects of urbanization may be even more important, especially during construction phases.

The first stage in an analysis of this type is to observe the stream flow and its suspended sediment load. By establishing a rating curve and recording the variations in stream level on a continuous gauge the annual suspended sediment yield can be obtained. This value is then expressed as sediment yield per unit area of the drainage basin.

The values derived by Douglas (1967) for a number of drainage basins in Australia provide a relationship that can be analysed by regression analysis in the form

$$\log_{10} Sy = 0{\cdot}527 \log_{10} Qy + 0{\cdot}0668$$

where Sy is the sediment yield in $m^3 \ km^{-2} \ yr^{-1}$

and Qy is the runoff in mm. The catchments chosen for the analysis were as far as possible unaffected by human interference. It is significant that the basins showing the highest sediment yield to discharge ratio are those in which human interference in the form of agriculture has been greatest.

Relief was also considered in the analysis. In general it can be shown that where other conditions are equal, the higher, steeper basins produce the greater sediment yield. However, when the streams flowed from well vegetated upper steeper reaches onto lower, less well vegetated slopes they showed a marked increase in sediment yield downstream. Vegetation therefore more than counteracts the effect of relief.

Precipitation intensity was shown to correlate even more closely with the sediment yield than the discharge. The significance level was raised from the 0·01 to the 0·001 level by using a modified measure of precipitation intensity, given by p^2/P, where p is the maximum mean monthly precipitation and P is the mean annual precipitation. The correlation coefficient was increased from 0·499 to 0·687. This highly significant relationship can be explained through the effect of climate on vegetation. In areas where the rainfall intensity is high and seasonal in character the sediment yield is greater. This is because in these areas the vegetation is less luxuriant, and the heavier rainfall can have an increasingly important effect in removal of surface soil, thus further weakening the vegetation. When the rainfall occurs seasonally with high intensity a dense vegetation cannot become established. (See Plate 6.)

The lowest sediment yield will be from areas where rainfall is considerable, but well distributed and less heavy, allowing an effective forest cover to grow. However, it is under these conditions that human interference may be most marked. Agriculture in the humid tropics is often associated with a greatly increased sediment yield. A value of 2,500 $m^3 \ km^{-2} \ yr^{-1}$ was recorded in the island of Java, where steep slopes are cultivated. The

Plate 6 The Grand Canyon of the Colorado, Arizona. The river Colorado has incised its bed into the uplifted, horizontally lying strata. Note the cliffs formed by the resistant beds and the slopes on the less resistant. The semi-arid environment leads to bare slopes.

disturbance of tropical soils is particularly dangerous as they are liable to rapid deterioration, followed by severe erosion and very high sediment yields. For example, a catchment with 94 per cent of its area under natural vegetation had a sediment yield of $21 \cdot 1$ m^3 km^{-2} yr^{-1} compared with $103 \cdot 1$ m^3 km^2 yr^{-1} where only 64 per cent of the catchment was under natural forest.

The figures given in Table 2.6 illustrate the characteristics of three catchments in northeast Queensland, Australia. The upper and lower reaches of the rivers are given separately. Despite their steepness the upper reaches produce considerably less sediment than the lower ones, where agricultural development is more intensive. This is true even in those upper reaches where there has been some agricultural activity which has increased the sediment yield. Despite good husbandry, such as contour ploughing and

similar measures, under the extremes of the tropical climate erosion is severe in the lower reaches of the valleys.

There is a strong contrast between conditions of low flow and when occasional intense storms occur. Under storm conditions gullies are initiated and rapidly increase in size. Man can be shown to be a very important geomorphological agent under these conditions, especially in the more extreme environments.

A study of the relationships between sediment yield and mean annual precipitation illustrates some of the variables involved and the effect of climate, through vegetation, on geomorphological process. The sediment yield data given in Table 2.7 were obtained from stream gauging stations and observations in reservoirs for a considerable number of small to medium sized catchments in the United States, totalling 257 observations in all.

Table 2.6 Queensland catchment data (Douglas 1967)

Catchment	Barron		Davies		Millstream	
	Upper	Lower	Upper	Lower	Upper	Lower
Area (km²)	11·91	225·33	14·17	105·67	19·81	91·95
Mean annual rain (mm)	1,778	1,562	1,397	1,270	2,032	1,900
Catchment relief (m)	274	587	718	924	274	412
Sediment load 1963–4 (m³ km⁻² yr⁻¹)	5·65	13·60	2·02	3·75	6·15	12·25

Exercise 2.7. Sediment yield and precipitation

Plot the figures given to illustrate the relationship between effective precipitation and sediment yield. Effective precipitation is used rather than the actual amount to make allowance for variations in temperature and with it, evaporation. The effective precipitation is defined as that required to produce a known volume of runoff, a relationship that has been established from many observations. Calculate the correlation coefficient and regression equation for the two sets of data separately. Take the independent variable, X, as the average effective precipitation, and the dependent variable, Y, as the sediment yield. Draw the regression line on the graph. Both lines show a similar pattern with a peak of maximum sediment yield where the effective precipitation is about 30 cm. The sediment yield decreases as effective precipitation increases over the range of data available.

There are two main influences determining the relationship between sediment yield and effective precipitation, and these can be expressed as $S \propto R/V$, where R is the annual runoff and V is the vegetation. The runoff is a function of precipitation, which can be given as $S = 10P^{2\cdot3}$, where P is the effective precipitation and S is the sediment yield. The effect of vegetation is given by $S = 1 + 0\cdot0007P^{2\cdot3}$. Because V is inverse to R their effects oppose one another, i.e. the greater the runoff the greater the sediment yield will be, but on the other hand the denser the vegetation the less the runoff will be. Thus the

Table 2.7 Sediment yield in relation to effective precipitation (Langbein and Schumm 1958)

Average effective precipitation (mm)	Sediment yield (tonne km⁻²)	Number of records at sediment stations
203·2	259	9
317·5	301	17
444·5	212	18
609·6	212	20
889·0	154	15
1,270·0	85	15
Reservoir data		
219·9	541	31
254·0	456	38
279·4	579	12
482·6	436	18
698·5	552	10
901·7	305	20
990·6	216	11
1,143·0	182	18
1,854·2	170	5

curve relating sediment yield to effective precipitation reaches a peak at low precipitation value, where the vegetation will be sparse and ineffective. The lusher vegetation more than compensates for the greater precipitation when this exceeds 30 cm.

It is possible to quantify the relationship between vegetation and annual precipitation in terms of the weight of vegetation. The type can also be specified, and it can be shown that forest is more effective than grassland in lowering sediment yield and hence erosion, which is directly related to sediment yield. The analysis illustrates some of the complex relations between geomorphological process, climate and vegetation, which are important from the human point of view as they affect soil fertility and food supply.

2.4 Geomorphology

Introduction

Geomorphology at the local scale is concerned with the study of form, material and process in the field, which provides material for analysis. Geomorphology at this scale can be approached from a number of directions. Field observations provide direct input to support theoretical, model and mathematical approaches. In this section morphology is considered first, followed by a consideration of material; both solid rock and sediment are included. Then processes are examined by various examples of possible field techniques. The first part of the section is concerned with fluvial and slope processes and the second with coastal processes. Wherever possible processes are related to morphology and material.

Observations of both two- and three-dimensional situations are exemplified. The three-dimensional aspects are particularly important in dealing with geological strata and their representation on geological maps. On the other hand, the analysis of two-dimensional profiles of rivers, slopes and beaches can provide much valuable data. The more complex deterministic mathematical models, using differential calculus, are not considered, although this approach will probably eventually provide a fuller understanding of the relationships between form, material and processes, which is one of the main aims of geomorphology. Until the basic forms, materials and processes have been accurately recorded, and their changes studied, this ultimate goal cannot be achieved.

A study of the physical geography of a small drainage basin provides an opportunity for using many useful techniques. The form of the basin must first be accurately recorded and its area delimited. With the help of a large scale map and/or vertical aerial photographs, the watershed of the basin can usually be plotted with reasonable accuracy. The map contours, however, are rarely adequate to provide a sufficiently detailed and geomorphologically valid picture of the relief.

Morphology

The technique of morphological mapping is specially designed to bring out the significant features of the relief, while slope profiles provide essential information in studying the relationships between morphology, material and process. The method of morphological mapping is based on the recognition and plotting of breaks and changes of slope in the landscape. Breaks of slope are abrupt changes of gradient, and are convex when the slope becomes steeper downhill, and concave when it becomes flatter. As indicated on Figure 2.10 the change of slope is indicated by a V-shaped symbol on the bounding line. It is always placed on the steeper side of the break of slope. The bounding line is dashed for a change of slope and a full line is used for a sharper break of slope. Cliffs and microchanges can be shown by the special symbols indicated. A cliff is defined as a bare rock slope steeper than 40 degrees. The method of morphological mapping applies well in areas where there are sharp changes of gradient, but is less easy to apply in gently rolling country, where it is difficult to establish the exact position of a change of slope.

Exercise 2.8. Morphological mapping

Construct a morphological map of a suitable, accessible area, using a 1:10,000 or 1:25,000 scale base map. Good vertical aerial photo-

45

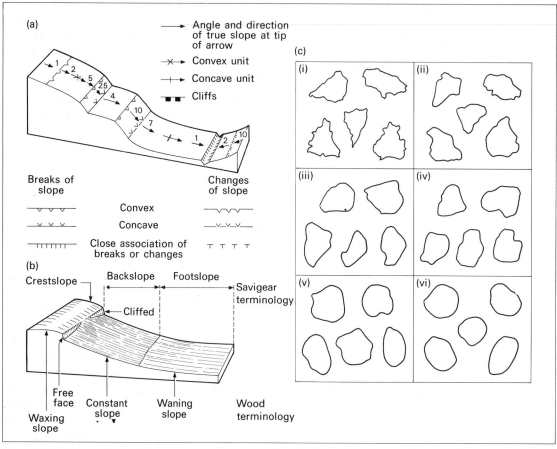

Figure 2.10 (a) Symbols used for morphological mapping. (b) The slope terminology of Wood and Savigear. (c) Stone roundness diagram in 6 classes.

graphs should also be consulted if available.

To make the map more quantitative the angle of slope can be measured in the field and entered in the appropriate unit on the map. The units can also be designated as having convex or concave curvature by suitable symbols, as indicated on Figure 2.10. The mapping of the breaks and changes of slope provides a valid base to study other aspects of the geomorphology of the basin. Drainage lines can be added and in some instances the geology can be related to the form of the land surface. This applies, for example, in country where different rock types provide clear evidence of their outcrop in the relief. The limestone scars and gritstone edges of the Pennines provide good examples, while the alternation of clays, sandstone and grits, as in parts of northeast Yorkshire, provide clear dip and scarp slopes that show up well on a morphological map. However, the danger of mapping geology from relief and then arguing that relief is controlled by geology must be avoided. Outcrops, on which dip and strike can be measured, should provide confirmation of the geology wherever possible. (See p. 157.)

The survey of a slope on the ground provides a much more detailed picture of the character than the morphological map, but it also involves problems. One of the first is the problem of selecting the slope to be surveyed. The whole area, which can be shown in its totality on a morphological map, consists of

46

slopes, so there is theoretically an infinite number of possible positions for the slope surveys. Some of these will be ruled out for various reasons, such as growing crops or animals that must not be disturbed, while some may be built over or impassable for other reasons. However, assuming that the field area is open moorland with access to the whole area, then the problem of selecting the slope must be faced. This introduces the problem of sampling.

If the slope survey results are to apply to the whole drainage basin under consideration then the rules of statistical sampling must be followed. The sample selected must be a random one if results of sample measurements are to be used to infer the characteristics of the total population of slopes in the basin. There are various ways of selecting a random sample of slopes to survey. One method is to use random numbers to select points along the drainage axis of the basin from which to start the surveys. An alternative method is to use a systematic sample. A single point is randomly selected near the mouth or top of the basin and profiles are then surveyed at regular intervals up or down the valley. Another possibility is to use a stratified random sample. A given number of profiles are selected randomly from each area covered by a certain rock type for example. This method has the advantage that statistical analyses can then be carried out to determine the effect of rock type on slope. In some circumstances it may be more meaningful to select the slopes on geomorphological grounds in order to provide typical data.

The profiles must run along the line of steepest slope at right angles to the contours. It is often useful to survey slopes on opposite sides of the valley from the same point in the valley. This allows a statistical study of the differences resulting from variations in aspect and other variables on either side of the valley to be made. An example will be considered later.

Exercise 2.9. Slope sampling

The drainage basin shown in Figure 2.11

provides a situation for practising the selection of different types of sample. Using random numbers, where appropriate, determine the positions of 12 slope surveys: (a) on a random basis; (b) as a systematic sample; and (c) as a stratified random sample. In practise the number of slopes in the sample should be based in part on the variability of the slopes. The more uniform the slopes are, the smaller is the number needed to represent the total population reliably.

Exercise 2.10. Slope survey

When the position of the slope survey has been selected then it must be surveyed. There are two main ways of surveying the slope, one of which needs two people, while the other can be carried out by one person. The equipment needed for the first method is a clinometer and a tape or chain to measure distances. An Abney level can be read with a vernier scale to an accuracy of 10 seconds of arc. The Suunto clinometer is even easier to use, but can only be read to $\frac{1}{2}$ degree. The starting point of the survey must be marked by some fixed object and the line of survey established along the maximum slope by fixing a marker on the line. The tape or chain can then be run out along the line until the first break of slope or the end of the tape, usually 30 m, whichever is reached first. The angle of slope is then read, making sure that a point at eye height is sighted or a correction for eye height made. It is possible to check the reading by sighting both up and down the slope each leg of the survey.

The second method of survey makes use of the slope pantometer, as designed by Pitty (1971). This simple instrument enables steps of equal length to be measured along the slope. It consists of four lengths of wood hinged together so that the angle of slope can be recorded on a simple protractor fixed to one piece of wood to which a spirit bubble is attached. When two of the slats are horizontal and the other two vertical the spirit bubble must be central and the protractor must read

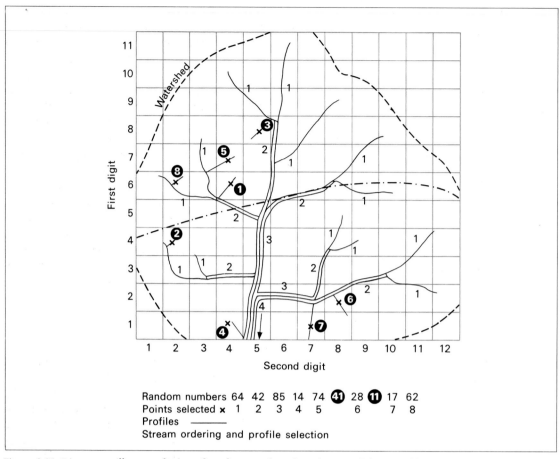

Figure 2.11 Diagram to illustrate the use of random numbers for selection of slope profiles. Stream orders are indicated by the number of lines and figures. The watershed is shown by a dashed line, and the dash-dot line indicates the outcrop of the junction between two strata.

zero. When the instrument is placed on the slope the bubble is centred, and the angle of slope can then be read off the protractor. The length of step is usually between a metre and two metres.

When the field-work has been completed the results must be plotted as a profile. It is usual to use a small vertical exaggeration of scale unless the slope is very steep. The most distinctive sections are: (1) the straight parts, called rectilinear slopes; (2) the steep parts with bare rock outcropping when the angle exceeds 45 degrees, called the free face; (3) the concave parts, called the waning slope, and usually found near the base; and (4) the convex parts, called the waxing slope, and

often occurring near the top of the slope. Any one slope need not include all four elements, which vary in proportion from slope to slope. These different elements on the slope are thought to indicate a relationship between form and process. As part of Exercise 2.10 identify the elements represented on the profiles shown in Figure 2.12 and suggest which processes have been important in their formation.

Analysis of form—valley asymmetry

The study of valley asymmetry must involve both geomorphological aspects and micro-

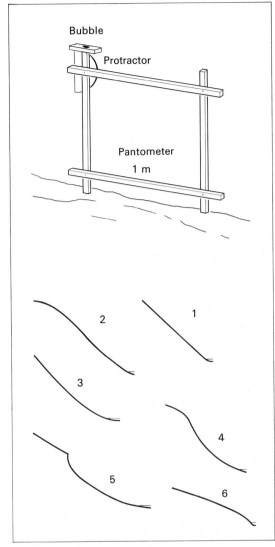

Figure 2.12 (a) Diagram of a slope pantometer as designed by A. F. Pitty. (b) Examples of slope profiles combining different elements.

meteorological ones as they are closely interrelated. There are many possible reasons for opposing slopes of small valleys to differ in slope. Some of these are not associated with meteorological variations, although many are, either directly or indirectly. Asymmetry must first be demonstrated and the next step is to attempt to explain it where it can be demonstrated. An easy way of testing the asymmetry of slopes is to use the sign test or *t*-

test, the first being applicable to paired values on a nominal scale, and the second to a set of values of slope gradient given in degrees or some other numerical form.

The use of the sign test to establish the statistical probability that the sides of small valleys differ significantly in gradient is illustrated by values recorded in the Chiltern Hills. There are a number of dry valleys that run down the dip slope of the chalk near Chesham. These valleys trend approximately northwest to southeast parallel to the dip of the chalk, so that there is no structural reason why the valley sides should differ in steepness. A slope map can be constructed for these valleys, as shown in Figure 2.13, which divides the slopes into five categories of gradient. (See Plate 7.)

Exercise 2.11. Slope analysis sign test

Take a random starting point at the head of the valley and note the steepness of the opposing valley slopes at intervals of 500 m down the valley. Tabulate the series of gradient categories in paired values and put +, − or 0 according to their relative steepness. Count the pluses and minuses. Check the significance of the number in Siegel's tables. Three valleys 4 to 5 km long have been sampled in this way to provide a total of 33 paired values. These are listed in Table 2.8. The valleys that have a steeper southwest facing slope are recorded as + and the equal slopes as 0. There are 14 zeros, 16 pluses and 3 minuses. In this instance the probability that the result could have occurred by chance is about 0·004, which means that there is a statistically valid probability that the southwest facing slopes are indeed steeper than those facing northeast.

When the asymmetry has been established it must then be explained. One of the important variables that must be recorded when the surveys are being made is the orientation of the valley and its dimensions. It has been shown that moderate sized valleys are more

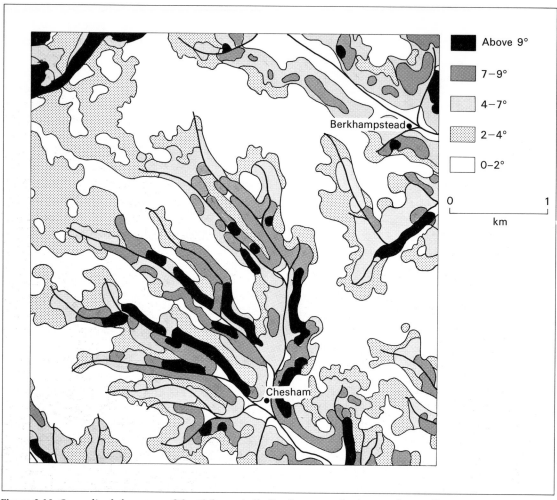

■	Above 9°
▨	7–9°
▤	4–7°
▦	2–4°
□	0–2°

Figure 2.13 Generalized slope map of the Chiltern chalk dip slope near Chesham to show valley asymmetry. (After C. D. Ollier and A. J. Thomasson, 1957)

likely to show asymmetry than either very large or very small ones. Valleys more than 50 m deep in the chalk of southern England do not show asymmetry nor do ones less than 8 m deep, while those between 20 and 30 m deep show the maximum asymmetry. The reason for recording the orientation is because different processes work most effectively under specific conditions, and these are likely to depend on the exposure to insolation, wind and other climatic variables. The orientation can also be related to the geological structure, particularly the strike and dip. Slopes are likely to be steeper when they oppose the dip where other conditions are not important. The contrast between dip and scarp slopes confirms this.

The relationship between processes and slope angle is more complex and varies from place to place. Different workers have found that in some areas slopes facing south or southwest are steeper than those facing the opposite direction, whereas in other areas the reverse applies. Where the orientations of the valleys do not conform to a simple pattern it may be useful to plot the results for all orientations. This can be done by recording the maximum angle or representative angle,

Plate 7 The dry valley shows asymmetry in its slopes, that to the left being steeper than that on the right. The valley is cut in the Cheviot lavas at Wooler Dean, Northumberland.

Table 2.8 Slopes in the Chiltern Hills near Chesham sampled every 0·4 km

Aspect

Southwest	3	3	3	4	4	4	4	4	4	3	3	3	3	3	5	5	5	5	4	5	5	5	5	5	2	3	4
Northeast	2	3	2	4	3	4	4	4	2	3	3	4	4	3	3	3	4	4	3	5	4	4	4	3	2	3	3

Southwest	3	3	4	5	5	3
Northeast	3	4	4	4	4	3

Slope gradient groups

1 = 0 — 2 degrees
2 = 3 — 4 degrees
3 = 5 — 7 degrees
4 = 8 — 9 degrees
5 = more than 9 degrees

such as the mean slope angle, for each 10 or 20 degree orientation, giving 36 or 18 groups of data. The results can then be plotted on a circular graph, which shows visually the pattern of slope asymmetry with respect to orientation. The groups can also be arranged according to other variables, such as height, rock type or some other variable. Fewer orientation classes may need to be used if there are not many observations. If a chi square test is used there should be at least five expected values in each class.

In considering the cause of asymmetry in terms of process a number of variables are important. Asymmetry is sometimes associated with a periglacial climate, but it can also occur in other areas. The Chiltern Hills valley asymmetry has been ascribed by Ollier and Thomasson (1957) to more efficient creep on the southwest facing slopes. This causes more material to move towards the valley floor on this side of the valley, pushing the stream, which would have been flowing under periglacial conditions, towards the opposite hillside, once material can accumulate on the valley floor. In the early stages of development, however, when all the material can be evacuated by the stream, the process of creating the asymmetry is essentially erosional. The gentler slope on which creep is less efficient acts as a slip-off slope, opposing the steeper one on which creep is more efficient. Downcutting deepens the valley as the stream migrates down the less steep slope, thus enhancing the asymmetry with time. The contrasts in soil type on either side of the Chiltern valleys tend to confirm this type of development. There is a flinty, loam and clay soil on the gentler slope, while chalky head and bare chalk occur on the steeper slope where creep has been more effective. The soils are also reflected in the land use, the heavier flinty clays and loams being associated with woodland. Thus both morphology, geomorphological process, soil and vegetation can be used to demonstrate the possible cause of asymmetry in this area. The processes, however, are probably not operating so effectively

at present as they were when the area was under a periglacial climate, and valley deepening could take place.

Each area must be treated as a specific problem because it is not possible to generalize concerning the causes of slope asymmetry. It has been shown, for example, in some parts of Canada that slope asymmetry depends more on the presence or absence of a stream at the foot of the slope than on other variables. Slopes that are being actively undercut by a stream tend to be steeper than those without a stream at their base. Such a situation is, therefore, not necessarily related to valley orientation. In order to test the situation a two-way analysis of variance test can be used.

Exercise 2.12. Slope asymmetry

Table 2.9 presents data on slope angles in relation to two variables, orientation and stream presence or absence. Work out a simple two-way analysis of variance test to establish whether either the orientation or stream presence affects the slope.

Material

Another aspect of the physical environment that must be examined in the field is the nature of the material. This includes both the solid rocks and the unconsolidated material that often covers them. The solid rocks must be

Table 2.9 Analysis of variance data for mean slope angles for four orientations and two stream situations

	Orientation				Total
	North	East	South	West	
No stream at base	5	6	8	9	28
Stream at base	8	7	12	11	38
Totals	13	13	20	20	66

considered both from the point of view of their composition and of their structure. The mapping of rocks in the field is the task of the geologist, but physical geographers should be able to appreciate the qualities of different rocks and to recognize them in the field. Solid rocks are mapped by geologists normally according to their age, but the type of rock is more important to the physical geographer.

Rocks are divided into three main groups (igneous, metamorphic and sedimentary) according to their modes of formation.

Igneous rocks are those solidifying directly from molten magma derived from inside the earth. They are further subdivided according to the mode of formation and related cooling rate, which is related to their crystalline structure. Their mineralological content provides a second criterion for classification, giving the groups shown in Table 2.10.

Table 2.10 Classification of igneous rocks

Chemical type Crystal type	Acid	Intermediate	Basic	Ultrabasic
Volcanic—fine grained Hypabyssal	rhyolite quartz porphyry	andesite porphyry	basalt dolerite	
Plutonic—coarse grained	granite	granodiorite	gabbro	dunite

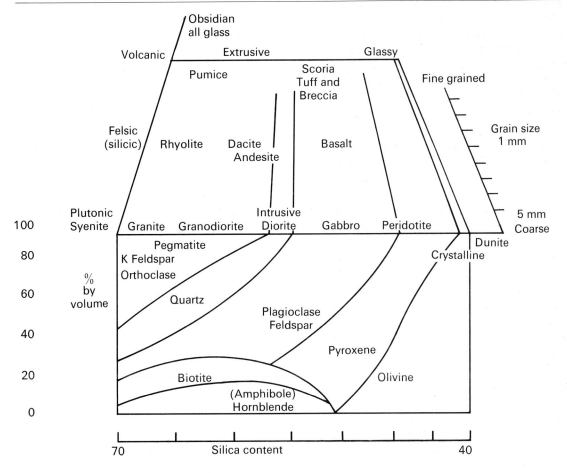

Plate 8 The volcanic material ejected beneath the sea cools quickly and sometimes it forms well defined rounded forms called pillow lava. This is an example from Anglesey. The concentric layering of the lava is visible in some of the pillows.

The plutonic rocks are cooled deep in the earth's crust and are only exposed on the surface after long continued denudation; they include the granites, which are acid rocks, and the denser, darker gabbros, which are basic. All the plutonic rocks tend to be coarsely crystalline, owing to slow cooling. They usually form massive rocks, their planes of weakness being mainly joints, some of which are vertical; horizontal joints may result from pressure release, for example by melting of thick ice.

At the other end of the igneous rock scale are the volcanic rocks that cool quickly in contact with air or water, as they are ejected by volcanoes. They are fine grained or glassy, and the most widespread and common is basalt. This is a basic rock that flows freely and so forms extensive nearly horizontal sheets. These are often cut by near vertical joints sometimes with a polygonal pattern. The rock is tough and resistant, often forming plateaux. The acid equivalent is an ignimbrite; in this case the material is ejected as incandescent ash that fuses to form a massive resistant material. Rhyolitic pumice is another acid volcanic material, which may also occur in glassy form.

The intermediate rocks are finely crystalline, having cooled at a moderate rate after injection to form nearly horizontal sills or near vertical dykes. These hypabyssal rocks often differ in strength from the rocks into which they were intruded, and so they may stand out as scarps or ridges, or less commonly, be eroded into negative features. The Whin sill of northern England is a good example of a resistant dolerite sill, along part of which the Romans built Hadrian's wall in Northumberland. (See Plate 8, 9.)

Another group of usually massive rocks, often with little structural control, are the metamorphic rocks that have undergone changes due to heat and/or pressure. They have been altered to form, for example, schists, gneisses and quartzites. These rocks occur in the older parts of the crust, such as the northwest Highlands of Scotland, and parts of north Wales, where another group of metamorphic rocks, the slates, are important both scenically and economically. Slates are formed from mudstones or shales by pressure. They are hardened and develop cleavage planes allowing them to split into thin sheets, hence their value for roofing material. Their homogeneity results in smooth hillslopes, such as those of Skiddaw in the Lake District. On the other hand the gneisses and schists of the Scottish Highlands form knobbly country, especially where glaciers have etched out the weaknesses. (See Plate 28, p. 123.)

The third group of rocks are the sedimentary strata that have formed from the destruction of earlier rocks. Weathering, erosion, transport and finally deposition have all taken place in their formation. The size of the particles provides a first criterion of classification. (See Plate 10.)

The finest material is clay, which will only settle in very quiet conditions. Thus most clays are laid down in deep, still water, but there are some shallow areas where clay can accumulate, for example in a salt marsh sheltered from the action of waves. From the point of view of landforms the main property of clay is its usually unresistant nature, which means that it often forms lowlands. It is also impermeable, which means that water cannot penetrate through it. Clay country is often waterlogged as a result. It does not provide stable foundations for building on sloping surfaces. Shales and mudstones are slightly coarser than clays and have been rather more consolidated, thus forming rather harder material, which is also usually impermeable.

The coarser material that forms sandstones is often largely derived from the weathering of granite, and consists mainly of quartz grains, a constituent of granite and gneiss. These grains are resistant to weathering and thus sand-

Plate 9 Thick basalt flows in southeast Iceland showing well developed columnar jointing, giving rise to the polygonal blocks seen in the stream bed below the waterfall. Many of the blocks and columns are hexagonal.

Plate 10 The middle falls of the Genesee River in New York State. The water is falling over the thinly bedded, horizontal strata, in which the bedding planes form the most conspicuous features.

stones are usually not very fertile. Because the grains are much larger they will transmit water readily, unless very well cemented, and thus form permeable rocks. Sandstone areas tend to be dry. When the rocks are strongly cemented, however, they are often resistant to erosion and can form scarps, where the structure is suitable, for example the green-sand scarps of the Weald.

Conglomerates are coarser still, formed of rounded pebble-sized material set in a matrix. Breccias, consisting of angular pebble-sized particles, are sometimes fossilized screes. Both rock types must have been laid down in vigorous environments, such as a mountain stream or scree. The clay type rocks are called

argillaceous, and the sandy ones arenaceous, while a third group consists of limey rocks, which are called calcareous, as they consist mainly of calcium carbonate. (See Plate 11.)

The calcareous rocks are mainly organic in origin, while another important form of organic material provides the carbonaceous rocks, or coals. Coal originated as peat deposits which accumulated in swampy conditions and have since consolidated. Limestones and chalk, the main calcareous rocks, were deposited in clear, warm seas, being the consolidated or crystallized remains of many small marine organisms that secrete calcium carbonate shells.

The Yoredale rocks of the central part of the

Plate 11 Malham Cove, North Yorkshire. The 100 m high abandoned waterfall of Malham Cove is formed in the Great Scar Limestone of the Lower Carboniferous. Limestone pavements have developed on the limestone exposed at the top of the Cove, and the view shows the clints and grykes of the pavement in the foreground, exploiting the joints in the limestone.

carboniferous period outcrop in the Yorkshire Dales and provide a very good example of different types of sedimentary rocks. The sequence consists of a repetition of strata starting with a limestone, followed by shales, sandstones and often ending with a thin coal seam. This sequence indicates first a quiet, calm sea with little sediment reaching it from the land, thus implying low relief inland, followed by an increase of relief allowing fine material to reach the sea and settle to form the shale. The sands accumulated under estuarine conditions and higher relief inland allowed coarser material to be carried to the sea. The coal seams indicate drying out of the area to

swampy conditions as the level built up and the sea was excluded. The next cycle starts with a sudden transgression of the sea, and a cutting off of the inland sediments, allowing limestone to accumulate once more. This sequence of rocks, which is called a cyclothem, illustrates well the relationship between erosion inland, which supplies much of the sediment, and deposition offshore, forming the sedimentary sequence. The repetition of the sequence is probably due to the gradual build up of stress on the sea floor by addition of sediment, leading to subsidence when the earth's crust is depressed, while at the same time weight is removed from the land

Plate 12 Giggleswick Scar in north Yorkshire lies along the line of the South Craven fault marking the junction between the block and basin zones of the southwest part of the Askrigg block on the central Pennines. The Carboniferous limestone on the left is faulted against the Millstone Grit on the south, right side of the fault in the view, with the downthrow to the south. The less resistant beds on the south have been eroded to form the lower ground along the fault-line scarp.

through erosion, allowing uplift. An examination of the rocks on the earth's surface can tell much, both of the past history of the area and the processes related to erosion and deposition. It is also an essential preliminary step in the understanding of the modern environment, its relief, soil, vegetation and animal life. (See Plate 12.)

The qualities of the rocks that are most important from the geomorphological point of view include their material and its resistance to weathering and erosion. The material and its cementation, jointing and bedding characteristics exert an important influence on slope processes, largely through their effect on the passage of water through the rock. Limestone

is well-known from this point of view. It is liable to solutional weathering, especially along the joint and bedding planes—water readily disappears underground, giving rise to dry valleys and cave formation below the surface. Chalk has similar characteristics and often has dry valleys, though with less well developed underground drainage systems. Sandstones also allow the ready transfer of water underground, unless they are strongly cemented, but clays and shales are so fine grained that water cannot readily penetrate. The capacity of the rock to transmit water influences the slope processes and is also important from the point of view of water supply. Permeable rocks provide good water-

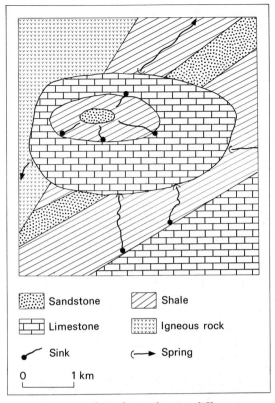

Figure 2.14 Hypothetical map showing different types of strata, with igneous rocks faulted against older sedimentary strata and these are overlain by unconformable younger strata. The positions of springs and sink holes are indicated.

bearing beds, called aquifers, because water can penetrate easily into them and can also be withdrawn from wells and springs. Clays and shales may contain considerable amounts of water, but it cannot be extracted readily because it is firmly held within the very small pores between the rock particles.

Exercise 2.13. Rock identification

On the evidence of the rock characteristics mentioned already identify the limestone and shales on the map shown in Figure 2.14, and explain.

Sedimentary rocks can usually be recognized by their bedding, as each stratum was laid down on the one below, giving their stratified nature. The pattern of the bedding can be mapped by recording the dip and strike of the strata. The dip and strike on the bedding plane represent the equivalent of the hachure and contour on the land surface. The dip is the maximum angle of slope of the bedding plane, while the strike is always at right angles to the dip. It is a straight line if, but only if, the bed is dipping in one constant direction. The strike lines are equally spaced if the dip is constant both in direction and amount. Dip and scarp slopes form a conspicuous feature where rocks of alternating resistance slope gently in one direction. Such a pattern occurs widely in southern and eastern Britain, scarps are commonly formed of sandstone, limestone or chalk, with clays and shales forming the intervening vales, often outcropping on the scarp slope beneath the more resistant cap rock. (See Plates 49, 51.)

Exercise 2.14. Geological map analysis

Figure 2.15 provides examples of simplified geological maps for analysis.

Map 1: give the strike direction and the direction and amount of dip. Give the order of deposition of the beds, and the thickness of beds Y, P and W. Draw a section through the isolated outcrop of bed Y. The section should be drawn true to scale to give the correct dip.

Map 2: describe the structure and give the strike and dip of beds B, G, R, W and Y. Measure the thickness of as many beds as possible. Draw a section to illustrate the structure from southwest to northeast across the map.

Map 3: calculate the depth of the coal seam at A, B, C and D. Give the depth of the mine shaft needed to reach the coal from the information given on the map.

Geological maps are sometimes issued in two series, one called the solid geology, which consists of the rocks just considered, and the other called the drift geology. The latter map

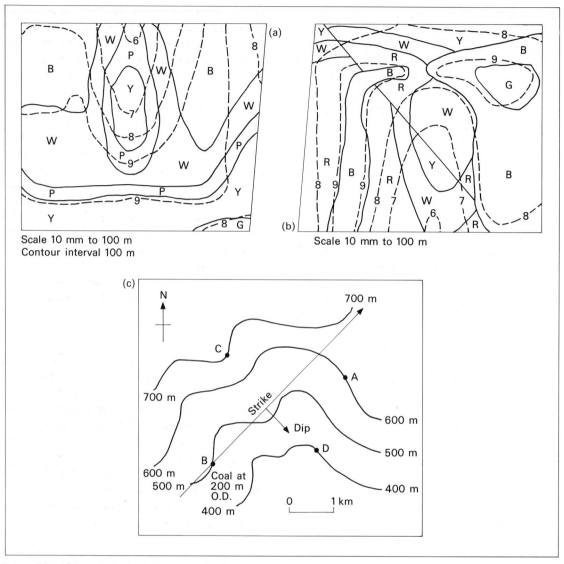

Figure 2.15 (a) Map 1. Geological map on which outcrops are shown by solid lines and rocks distinguished by letter. Contours at 100 m intervals are shown by dashed lines. The beds dip uniformly in direction and angle. (b) Map 2. Geological map with symbols as map 1. The beds dip uniformly. (c) Map 3. Contours, strike and dip in uniformly dipping strata are shown.

shows the distribution and type of superficial deposits, which in glaciated areas usually consists in part at least of glacial deposits. Till, which used to be referred to as boulder clay, is the deposit laid down directly by the ice and is recognizable by its unstratified nature. Fluvioglacial gravels and sands, and sometimes lake clays, can be recognized by their

stratified bedding as they were laid down by glacial melt waters. They can be either ice contact deposits or proglacial sediments. Some of these deposits, such as kames and eskers, which are ice contact deposits, can be recognized both by their morphology and their material.

Other superficial deposits shown on drift

maps include river terrace deposits, raised beaches along the coast, peat and warp in low lying marshy areas or high moorlands. The peats of the uplands, formed where lack of drainage and heavy rainfall cause excessive moisture, and lowland peats, usually the result of poor drainage, often contain pollen grains, because of their preservative properties. These pollen grains can provide valuable evidence of past vegetation, allowing past climatic conditions to be assessed. A great deal can be learnt by studying superficial deposits in detail. Two examples of the type of measurements that can be made in the field and their analysis will be considered.

Stone shape and material size both provide useful information in appropriate conditions. Material size is particularly relevant to a study of the nature of fluvioglacial deposits for instance. If the material is coarse and gravelly it must have been laid down in a vigorous environment of fast flowing water, especially if it is well sorted. On the other hand fine material can only accumulate in quiet conditions, such as a glacial lake. These lake sediments often show interesting structures and sometimes a rhythmic pattern occurs, where a fine layer alternates with a coarser one. The two layers are called rhythmites or varves, and each couplet represents the seasonal sedimentation in a lake adjacent to the ice. Because the deposits are seasonal they provide a useful time scale and a chronology of glacial retreat can be established by analysing them.

The measurement of stone shape can be done in many different ways as there is no simple method of accurately describing an irregular pebble. One of the simplest is to match it against a set of outlines of increasing angularity as shown in Figure 2.10. This provides an ordinal scale of pebble roundness. The Cailleux roundness provides a more precise measure. It is obtained by measuring the maximum dimension of the pebble and the radius, r, of the sharpest corner in the principal plane. The roundness is given by $2r/l$, where l is the maximum length, and

multiplying the result by 1,000. This value provides a measure of abrasion. Care must be taken to measure pebbles that have the same lithology and roughly the same size.

Exercise 2.15. Stone roundness

Collect a sample of 50 stones from a suitable deposit, such as the bed or banks of a stream. Measure their long axes, which should be between about 2 and 4 cm. Measure the long axes and minimum radii of curvature, using a series of arcs of known radius to fit the sharpest corner. Count the number of pebbles in each R class $(R = (2r/l) \times 1,000)$ in steps of 50 and plot a histogram. The mean and standard deviation of the roundness value, R, for the sample of 50 should be calculated so that the sample can be compared with other samples.

Studies of this type allow the efficiency of various types of erosion to be assessed. For example, samples may be taken at different points in a river, or from successive moraines in a glaciated valley. Different processes can also be distinguished by this means. In the Arctic environment of Baffin Island stones sampled for size and roundness from different depositional features differentiated between moraines, solifluction, eskers, kames, deltas and raised beaches. Some of the results are given in Table 2.11. The results can be explained by reference to the processes forming the different features.

A study of rock type also provides useful information; for example, it can indicate the source of ice and its flow direction by studying glacial erratics, derived from sources outside the areas where they were found. Similarly, stone type can provide useful data in studying river capture in river terrace deposits.

Till fabric analysis is another method that has been used to study ice flow and processes of glacial deposition. The method is based on the measurement of stone orientation and dip in suitable deposits. The directions of the long axes of elongated stones can be measured with a compass to the nearest 5 degrees, while the dip can be recorded with a clinometer.

Exercise 2.16. Till fabric analysis

Till fabrics must be measured in the field where suitable exposures of till or other deposits are available. The results can be plotted, as indicated on Figure 2.16, on circular graph paper either as a two-dimensional or three-dimensional pattern. The former does not show the dip of the pebbles. In a three-dimensional pattern the lower hemisphere is shown with the dip indicated by the distance from the outside edge of the graph, horizontal pebbles being plotted on the edge, with increasing dip towards the centre. For the two-dimensional plot the observations are divided into 20 degree frequency groups and plotted at the intermediate values, 10, 30 degrees, etc. There is a relatively simple method of analysing the two-dimensional pattern to obtain a preferred orientation and a measure of the statistical significance of the results. This is called the Raleigh test and the method of working the test is exemplified in Table 2.12. The frequencies in each 20 degree class are multiplied by the cosine and the sine of twice the angle and summed. The preferred orientation is found by the equation $\tan 2\theta =$

Table 2.11 Stone length and roundness data from Baffin Island

	Length measurements (mm)	
Deposit	Mean	Standard deviation
Moraines	386	251
Ice-contact	440	247
Lag deposits	589	328
Eskers	237	138
Deltas	121	59
Beaches	79	28

Differentiation between landforms using length measurements—t-test

	Difference between means	Degrees of freedom	Significance level
Lag deposits and ice-contact	149	17	NS
Eskers and moraines	149	30	0·1
Eskers and ice-contact	203	29	0·02
Deltas and ice-contact	319	31	0·001
Deltas and eskers	114	44	0·01
Beaches and deltas	42	31	NS
Moraines and ice-contact	54	17	NS

Differentiation between landforms using roundness measurements—t-test

	Means		Difference between means	t value	d.f.	Significance level
Beaches and deltas	398	334	64	3·32	22	0·01
Deltas and eskers	334	332	2	0·14	19	NS
Deltas and kames	334	238	96	5·65	16	0·001
Eskers and kames	332	238	94	6·86	13	0·001
Moraines and solifluction	138	185	47	1·68	7	NS
Moraines and kames	138	238	100	5·44	10	0·001
Moraines and deltas	138	334	190	10·90	16	0·001
Moraines and eskers	138	332	194	12·60	13	0·001

A/B, where $A = \Sigma f \sin \theta$ and $B = \Sigma f \cos \theta$. The result must be divided by 2 to obtain θ and 10 degrees subtracted. The statistical significance of the distribution is found from $R = \sqrt{(A^2 + B^2)}$ and $L = (R/n) \times 100$. The null hypothesis is that the pattern shows no preferred orientation, and if the L value exceeds that given in tables then the null hypothesis can be rejected. The distribution covers 360 degrees so allowance must be made using the signs of A and B to determine the right quadrant. The example given in Table 2.12 shows a highly significant result, with $L\% = 35$. Calculate the values of the preferred orientation and R and $L\%$ for the other values given in Table 2.12, and plot the circular graphs of the frequencies.

The results may be difficult to interpret because the position of the stone depends on many things, such as its shape, size and angularity, as well as the way the ice or other process was moving it. Nevertheless it is generally agreed that in most circumstances the till fabric shows a preferred orientation in the direction of ice movement.

Processes

The study of material provides evidence of the process depositing it, but it is also possible to study process directly in a small drainage basin. One of the difficulties is that most of the processes act very slowly, and observations only become really valuable when they are carried out over a very long time span of at least several years. Another difficulty is that processes do not necessarily act continuously, or at the same rate, thus short-term measurements may not be representative of the longer-term periods under which the slow processes operate.

The substitution of space for time is one way of making long-term study possible. This is called the ergodic principle. An example of the principle applies to the development of slopes below a rapidly retreating waterfall. In

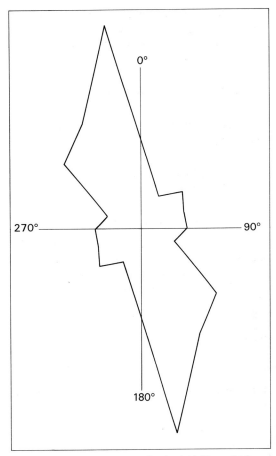

Figure 2.16 Two-dimensional spherical diagram of a till fabric with 20° divisions and a sample of 50 stones. The length of the vector is proportional to the frequency of stones.

the Yorkshire Dales, where the Yoredale rocks outcrop there are many waterfalls. Some of them can be dated fairly accurately as they have formed as a result of glacial interference and so date from deglaciation about 15,000 years ago. The waterfalls retreat by the undercutting of the softer shales beneath the harder, horizontally lying limestones and sandstones. Beside the waterfall a vertically standing cliff in the hard rock is left. This is slowly modified by slope processes as the waterfall retreats. The slope nearest the fall shows the youngest profile, with a vertical face above a steep slope on the shales, but as the free face is traced downstream it is gradually reduced in height and eventually eliminated, leaving a constant slope. Further from the fall a waning slope develops at the base of the slope as weathering reduces the material size and so the angle at which it can remain stable. At the top of the slope rounding produces a waxing slope. Thus, surveys in space in this type of situation provide a time sequence of slope development, illustrating the ergodic principle.

Some processes that operate on slopes occur fairly steadily and slowly, while others are larger scale, more violent and intermittent in operation. Creep and rain wash are examples of the former type, while landsliding and slumping exemplify the latter type. A sudden major adjustment is triggered by the passing of a threshold, when the strength of the material is reduced below the stress imposed upon it. This can happen when excessive rain reduces the strength of a clayey material so that it can no longer support the weight of overlying material. The rain provides the trigger in this instance.

Sometimes the movement is predictable: for example the seasonal movement of soil on solifluction slopes when snow melt and thawing ground provide the necessary trigger in the form of excess pore water pressure.

Table 2.12 Till fabric data and example of Raleigh test

Raleigh test of significance

θ	f	$\sin 2\theta$	$f \sin 2\theta$	$\cos 2\theta$	$f \cos 2\theta$
0–20	6	0·000	0·000	+1·000	+ 6·000
20–40	4	+0·643	+ 2·572	+0·766	+ 3·064
40–60	6	+0·985	+ 5·910	+0·174	+ 1·044
60–80	5	+0·866	+ 4·330	−0·500	− 2·500
80–100	5	+0·342	+ 1·710	−0·942	− 4·710
100–120	4	−0·342	− 1·368	−0·942	− 3·768
120–140	11	−0·866	− 9·526	−0·500	− 5·500
140–160	13	−0·985	−12·805	+0·174	+ 2·262
160–180	22	−0·643	−14·146	+0·766	+16·852
			+14·522		+29·222
			−37·845		−16·478
		A =	−23·323	B =	+12·744

$\tan 2\theta = A/B = -1\cdot83 = 61°\,21'$ in 4th quadrant

$2\theta = 360 - 61\frac{1}{3}° = 298\frac{2}{3}°$

$\theta = 149\frac{1}{3}° + 10°$ for half class $= 159°\,20'$

$R = \sqrt{(A^2 + B^2)} = \sqrt{(543\cdot3 + 162\cdot5)}$

$= \sqrt{705\cdot8} = 26\cdot57$

$L\% = R/n \times 100, n = 76$. Thus $L\% = 2,657/76$ $= 35$ significant at 10^{-4}.

Till fabric measured at Carlingill.

Frequencies recorded at other sites

θ	f1	f2	f3	f4
0–20	6	1	9	6
20–40	4	0	15	8
40–60	6	1	6	9
60–80	6	15	6	3
80–100	5	25	5	1
100–120	13	6	0	2
120–140	6	2	3	2
140–160	4	0	2	8
160–180	0	0	4	11

f1 Tebay, near Lune Bridge   f3 Hell Gill drumlin
f2 Gaisgill, north of Howgills f4 Goodwife stones

Exercise 2.17. Slope creep measurement

Place painted markers, which can be stones from the area, on the slope, record their positions relative to a fixed point and re-measure their positions later. These observations can provide evidence of the movement of stones on screes and the sorting action of thaw–freeze activity in periglacial areas, such as the high summits of the mountains of the Lake District, Snowdonia and the Scottish Highlands. This technique records surface movement only, but it is also necessary to establish movement in depth.

Pissart used this method to record downslope movement on scree in the French Alps. On a 200 m line perpendicular to the slope 65 marked blocks were set out. Of those found afterwards 12 had remained stationary, and 49 had moved between 5 and 600 cm, the mean value being 64 cm. Size affected the distance moved. Experiments were also made by means of marked stones in sorted polygons at a height of 2,800 m in the French Alps. Six pebbles were placed 15 to 25 cm from the centre of a polygon 160 cm wide. These pebbles moved 7, 7, 10, 15, 16 and 21 cm in 17 years towards the margin of small polygons within the larger one.

Two important processes operating on most slopes are soil creep and rain wash. Some geomorphologists associate different slope profiles with these two processes, suggesting that convex or waxing slopes are typical of creep, while concave or waning slopes are characteristic of rain wash. There is support for this argument in that concave slopes are more common near the base of the slope, where the rain is more likely to have become concentrated into rills. The processes causing creep are associated with wetting and drying of the soil and thawing and freezing. These processes can operate effectively near the top of the slope. Both changes of state cause swelling and contraction of the material, first lifting the particles upwards and then allowing them to settle lower down the slope. They are thrust upwards at right angles to the slope and settle back vertically, thus moving downslope. The lack of permanent moisture on the upper slopes aids these processes. Permeable sandstones and chalk, which are usually dry, have typically rounded convex slopes.

One method devised to study downslope movement of soil is the Young pit, so called after Young who first devised this rather laborious method of recording the action of creep on a slope. A pit is dug into the hillside and markers are fixed to one side of the pit in the undisturbed soil and their positions noted relative to a fixed surface point. The pit is filled in and left for at least six months or a year before being re-excavated. The new position of the markers is then ascertained relative to the fixed point. One difficulty of this method is the disturbance of the soil in digging the pit.

This problem can be overcome to a certain extent by using a Rudberg pillar, so called after the Swedish geomorphologist who developed the method. A vertical hole is augured into the ground and filled with a series of short concrete cylinders one on top of the other. The site is then left undisturbed for as long as possible. The cylinders are then revealed by digging and their movement gives an indication of soil movement at depth, as illustrated in Figure 2.17a.

Experiments of this type have revealed much valuable data concerning slope processes, showing that movement is generally greatest at the surface, falling off quickly with depth. The movement due to creep penetrates further into the ground than that associated with rill or rain wash, which only influences the surface particles of soil, and which is not effective on well vegetated slopes. Creep can occur on slopes covered by grass or forest as well as on bare slopes.

It is rather easier to record rain wash because it is a surface phenomenon. It can be measured by fixing a soil tray into the ground so that material moving down the slope by rain wash is left in the tray, which must be protected from the surface by a cover as shown in Figure 2.17b. The amount of material accumulating in the tray can be weighed to provide a measure of movement over the width of the tray in the time between observations.

Another method that can be used to establish the change of ground level on a slope is to use erosion pins. They are thin metal rods driven firmly into the ground with a washer fixed to mark the ground level when the pins were inserted. After some time the position of the washer relative to its initial position can be measured. Alternatively, the height of the pin above ground level can be measured at the beginning and end of the experiment. The results give a direct measure of the amount of erosion or deposition at that point on the slope. A series of markers provides values along a slope profile. Such measurements will only be satisfactory on actively changing slopes, such as unvegetated slopes of reasonably incoherent material. Valuable information has come from this method in small scale experiments, for example on bare waste tips (Schumm, 1956, 1967).

The plan of a stream in the horizontal plane is worthy of study. Some streams are straight, others follow a sinuous course, while yet others meander or are braided. A compass traverse is the best method of recording the detail of the stream course in the field, if maps are not adequate. The cause of meandering is still not fully understood, although it has been suggested that it is associated with a type of spiralling flow called *helicoidal* flow. The water flows downwards at the outside of a bend and upwards at the inside, thus enhancing erosion on the outside and deposition on the inside, as shown in Figure 2.18a. The form of the meander fits a sine curve, which is the curve that allows the most efficient use of energy, thus fulfilling the law of conservation of energy. Meandering seems a natural form of water flow in a stream, and is not necessarily related to the load being carried, as melt water streams with no load and ocean currents meander at times. (See Plate 13.)

Exercise 2.18. Stream course analysis

Analyse the stream course shown in plan in Figure 2.18b, which shows a portion of a meandering stream. The meander length is l, and the amplitude is a, the stream width is w and the mean radius of curvature is r. Calculate the sinuosity by dividing the stream length by the valley length and calculate the ratios l/a, which for the example is 2·2, and l/w, which is

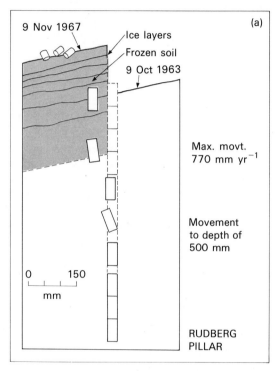

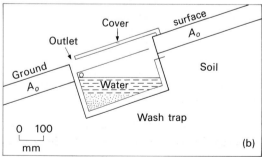

Figure 2.17 (a) Diagram to show a Rudberg pillar after 4 years movement, and the effect of frost heave. (b) Diagram to illustrate trap method of estimating slope wash.

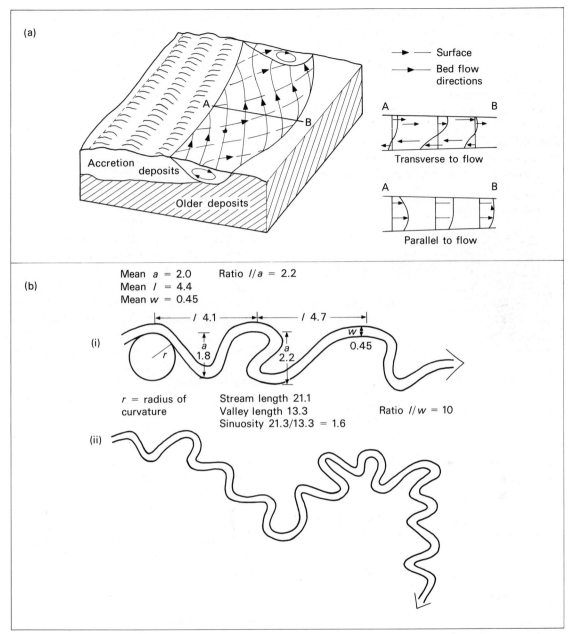

Figure 2.18 (a) Diagram of helicoidal flow in a meandering river. (After J. R. L. Allen, 1970) (b) Diagram of a meandering stream course to illustrate morphological relationships, (i). Diagram (ii) refers to exercise 2.18.

about 10. The sinuosity of the example is 21·3/13·3 = 1·6. These ratios are of interest in working out past discharges.

The type of channel material is relevant because meandering only develops fully in alluvial streams with beds and banks that are coherent but can be easily eroded. The process of braiding occurs where the stream banks are coarse and gravelly. This bed type is found in heavily laden glacial melt water streams, which have rapidly changing discharges.

67

Plate 13 Meanders on the River Cuckmere in Sussex, where the river cuts through the gently rising ground of the chalk of the South Downs before entering the sea. It has formed a wide flood plain with well-developed meanders.

Deposition takes place at times of low flow, when shoals form in the channel, causing the stream to diverge around them, thus widening the channel and producing distributaries.

The bed material can be related to the morphology in explaining the processes of fluvial action. When a short section of the stream long profile is surveyed, the cross profiles should also be recorded at intervals and notes made of bed and bank material. Samples can be analysed to assess size, angularity and rock type. The position and spacing of riffles, which are elongated shoals of coarse material in the centre of a gravelly river, should be noted. It has been found that the riffles are spaced between 5 and 7 times the river's width apart. The reason for this spacing is not fully understood, but it is thought to be related to the general tendency of rivers to meander. Once the riffles have formed with

intervening deeper, quieter sections variations in the flow develop. Greater erosion on the bank near the riffle may initiate meandering, and the process tends to be self-generating until the ideal sine curve is achieved and equilibrium established. The meanders then move downstream.

Each river has an optimum cross profile to accommodate the water it transmits. The form depends on the material. Coarse material tends to produce wider, shallower streams than fine material. It has also been shown that the sinuosity of meanders can be partially related to the amount of fine sediment in the banks and bed. Bed and bank erosion can be measured in relatively small streams. Erosion pins can be fixed into the banks at suitable positions above the water level, at and below the water level, and the length of pin protruding can be measured at intervals. The measure-

Plate 14 A braided river bed in northern Iceland. The braided nature of the bed results from the variable and coarse load carried by typical meltwater streams. The Eyja Fjordur river frequently changes its channels and bridging these glacial rivers is difficult. A bridge is seen in the middle distance.

ments should be related to the flow of the river and to weather conditions. In some instances erosion is much more rapid for a given flow when the banks had previously been saturated by heavy rain or weakened by strong freeze–thaw action in winter, thus reducing the strength of the material. The bed of a river is scoured during the rising phase of discharge and fills again as the flood subsides. The scouring can be recorded when the river dries out at times of low flow. At this time a chain can be placed in the river bed in a vertical hole drilled into the stream bed. The top or bottom end of the chain is firmly anchored to a fixed point. After the flood has subsided the chain can be re-excavated and the length that has been disturbed can be measured, giving a value of the amount of scour during the flood.

Another important aspect of field-work in a river is to measure the load it is carrying. The load is carried in several ways: the bed load is carried along the bottom, and consists of the larger particles; the solution load is carried in chemical solution; and finally the suspension load is carried in the body of the water. It consists of the finer particles that only settle slowly to the bottom and are prevented from doing so by turbulence.

The solution load can be measured by chemical analysis of samples of river water. Calcareous rocks are susceptible to solution in weak carbonic acid, formed by the addition of carbon dioxide to water. The amount of dissolved calcium carbonate gives a measure of the rate of lowering of a limestone surface. A value of 0·44 mm/year has been recorded in the limestone area of northwest Yorkshire. The variation of calcium carbonate in the water from area to area and from time to time in any one area provides useful data in karst geomorphology. The Palintest method provides a simple and reasonably accurate method of determining water hardness. The method can be used in the field and uses tablets with EDTA (di-sodium salt ethylenediamene tetra-acetic acid). The tablets are added to a measured volume of water. In 100 ml of water one total hardness tablet is equivalent to 20 ppm total hardness. Tablets are added until the solution changes colour from reddish-purple to pure blue. Other tablets can be used to record calcium hardness, with a colour change from pink to violet (Douglas, 1968). Solution is important in most drainage basins, and more material is often removed in solution than by other methods.

The measurement of suspension load is not too difficult. Samples of water can be collected from various depths in the channel. The usual procedure is to use an integrated sample, which is obtained by lowering the sampling bottle slowly throughout the whole depth of the stream, because the sediment concentration is likely to vary throughout the depth. The sampler must be fixed to a rigid pole and be enclosed in a device that keeps it facing upstream, as shown in Figure 2.19.

The problem of measuring bed load is very difficult and expensive. The proportion of material carried as bed load varies greatly with the type of river. A value of 10 per cent is used in America, but other measurements suggest that it may often be much less; on the other hand in glacial rivers it may amount to 50 per cent. It is generally reckoned to be a fairly small proportion of the total load carried by the river, especially if the solution load is included in the total. The load carried by a stream is closely associated with the amount of water flowing in it, its discharge, which has already been considered under hydrology.

Coastal geomorphology

The coast provides an environment in which the operation of processes can be observed over a short time interval, and in which there is a very close association between the morphology, the material and the processes operating. Although the three elements of the geomorphological whole are closely associated they will be considered separately for convenience.

Morphology. Morphology includes the form of both the solid foundation of the coast and of the superficial layer of material that forms the beach. In most areas of hard rock the foundation of the coast is in the form of cliffs and a shore platform. Drift cliffs occur widely where glacial deposits or other superficial material reaches the coast. Coastal geomorphologists have done more work on the beaches than on the solid rock coasts, but it is useful to examine the morphology of both cliffs and coastal platforms. (See Plate 15.)

The nature of the cliff profile depends both on the material forming it and on the balance between the subaerial and marine processes that shape it. Chalk, for example, forms vertical cliffs at Beachy Head and other exposed sites, but at Reculver, for instance, the profile is less steep because of the dominance of subaerial processes here. Subaerial processes are usually dominant on drift cliffs, with the sea playing the role of transporting agent. Waves remove material carried to the cliff foot by mass movement, including rotational slumping, mud flows and landsliding. (See Plates 16, 17.)

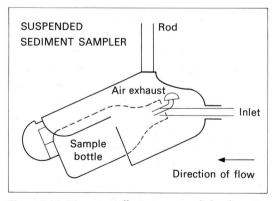

Figure 2.19 Diagram to illustrate suspended sediment sampler.

70

Plate 15 Till cliffs at Filey, East Yorkshire. Thick till overlies the Jurassic rocks at the cliff foot, and it has been eroded into badland relief by subaerial processes.

Steep cliff profiles are not easy to survey in the field, but the cliff foot and shore platform can be surveyed more easily. The gradient of the shore platform and its degree of smoothing can be recorded by levelling. The nature of the bedrock should be examined and mapped along the profile, with details of the dip of the rocks being recorded. The presence of seaweed often provides information concerning the degree of erosion currently in progress. One of the most important processes causing smoothing and lowering of the rock platform is the scouring action of beach material, particularly large particles, as they are carried to and fro across the surface by waves. The nature and thickness of the beach material should be noted, because if the layer of beach material is thick and continuous it protects the platform from erosion. Thus platforms are eroded where much beach material cannot accumulate. The position and widths of the

platforms and their degree of smoothing should be related to the direction of longshore transport of material and its availability. They should be developed best where more beach material is removed than reaches the site.

On eroding coasts where long-term observations are possible useful measurements of erosion rates can be made fairly simply. The distance between the cliff top and some fixed point inland can be measured at intervals. Such observations have, for instance, been carried out along the rapidly eroding cliffs of Holderness, where the average erosion rate exceeds 1 m/year. Erosion on these drift cliffs is usually spasmodic, with major cliff retreat taking place by sudden slumping or rotational shear sliding of large blocks of material. These processes are often initiated by excessive rainfall or snow melt, causing increasing pore water pressure that weakens the material, especially where there is much clay. Wave

71

Plate 16 Durdle Door, Dorset. An arch has formed in the vertical and steeply dipping Portland and Purbeck beds of the Jurassic. A beach has accumulated in the shelter of the headland. Note the steepness of the beach of coarse material.

action may also be effective if the cliff foot is rendered unstable by the removal of material from the base by destructive wave action. The effects of surges that raise the general water level are particularly dangerous in these areas. Under extreme conditions levels several metres above the normal can occur.

Along much of the coast there is a beach consisting of mobile sediment. Beach changes can be measured and related both to the type of material and the processes causing the change. The coastal zone can be divided into a number of distinct units, each of which forms part of the whole coastal system. Some of these units are illustrated in Figure 2.20. They are arranged both along the coast and at right angles to the shore. Those normal to the shore include the backshore zone (b), which is the part normally above high water level, but which is still influenced by the sea, such as the cliffs or backshore dunes. The foreshore (a) covers the area between tide marks. It is exposed at low tide but under the sea at high tide. This zone reacts most to the changing

wave conditions and is the easiest to measure. Below water is the nearshore zone (h), which can be subdivided into the swash–backwash zone, the surf zone, which ends seawards at the breakpoint of the waves, and the zone seaward of the breakers in which material is still moved by the waves, the offshore zone. Along the shore these units may be more varied. On a steep coast they often consist of alternating bays with beaches and rocky headlands. There may be spits across river mouths (e), barriers, tombolos tying islands to each other or the mainland, and other depositional features.

On a low coast there is often a barrier island, backed by salt marsh in a lagoon and carrying dunes on its crest, which alternates with tidal inlets with various types of tidal bars, deltas and channels through it and on either side of it. Some of these are indicated on Figure 2.20.

When studying coastal processes it is important that the whole system is considered; this can be defined as the stretch of coast along which material can move freely

Plate 17 Looking towards Durdle Door and Bats Head, Dorset. Cliffs are formed in the Jurassic and lower Cretaceous rocks. In the foreground and middle distance the cliffs are of chalk, being steeper where the exposure is greater in the middle distance. The rocky reef in the foreground has acted as a shelter behind which a tombolo is forming. Note the pattern of wave refraction and the crossing wave crests in the foreground. Caves are forming at the base of the chalk cliffs.

alongshore. On a low straightish coast it may be a very long stretch, as material can move along and across tidal inlets, but on a steep rocky coast each small bay may be an independent system in which beach material is trapped. The latter is probably the best system for field-work over a short time interval. It can usually be identified by the fact that adjacent bays have different types of beach material, indicating that material does not move along-shore from bay to bay. Examples occur on many crenellate coasts, such as Donegal, or Cornwall. (See Plate 18.)

Beach profiles can be measured by levelling, using a dumpy level on a tripod and a levelling staff. The distances can be recorded tacheo-metrically by reading both the upper and lower readings given by the upper and lower cross wires of the level eye piece, while the central cross wire gives the level when the bubble is centred. The difference between the upper and lower readings, when multiplied by

the constant of the instrument used, which is usually 100, gives the distances directly between the staff and the level. This method avoids difficulties of measuring with a tape over wet sand or mud, and also has the advantage of providing a horizontal measure-ment.

Plate 18 Barrier island. The low foreshore and dune-covered backshore of the barrier island of Fire Island along the exposed Atlantic side of Long Island near New York. The relatively narrow foreshore is the result of erosion which is also evident in the steep frontal dune slope and bare sand exposures.

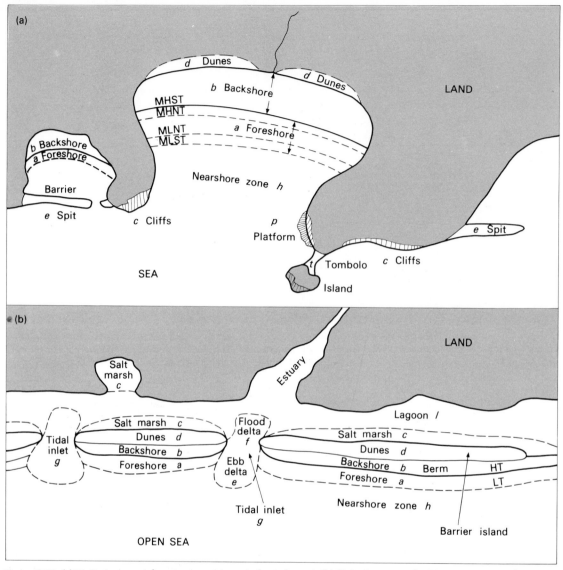

(a)

d Dunes

b Backshore

d Dunes

LAND

MHST
MHNT

MLNT a Foreshore
MLST

b Backshore
a Foreshore

Barrier

Nearshore zone h

e Spit

c Cliffs

p
Platform

t Tombolo c Cliffs

e Spit

SEA

Island

(b)

LAND

Salt
marsh
c

Estuary

Lagoon l

Salt marsh c

Dunes d

Tidal
inlet
g

Salt marsh c

Dunes d

(Flood)
delta
f

Backshore b

Backshore b Berm HT

Foreshore a

Ebb
delta
e

Foreshore a LT

Nearshore zone h

Tidal inlet
g

Barrier island

OPEN SEA

Figure 2.20 (a) Units in a coastal system on a steep, indented coast. (b) Units in a coastal system on an open, low coast with barrier islands.

If a level and staff are not available there is another simple way of measuring the beach profile, illustrated in Figure 2.21. The only equipment needed is two poles about 2 m long, and divided into suitable divisions, only the upper one need be subdivided into small divisions, usually of centimetres. The two poles are held 2 m apart, the distance being measured with one of the poles. The difference

in level between the two poles is recorded by sighting from the back pole along the top of the front one to the sea horizon. The process is repeated every 2 m along the width of the beach perpendicular to the shoreline. A marker should be fixed at the back of the beach where the profile starts so that it can be repeated along the same line at each low tide. The marker also provides a means of keeping

74

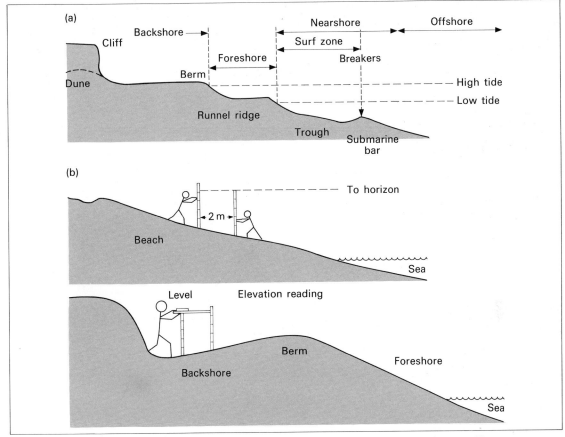

Figure 2.21 (a) Diagram to define beach terminology. (b) A simple method of measuring beach profiles.

the profile straight as the poles are kept in line with the marker as each is moved down the beach. This method needs two people, but it can be used by one person if longer poles are used and these are inserted into the beach to a fixed depth so that they will stand upright without being held. When the beach has ridges or a backshore berm that slopes landwards, the seaward pole may be higher than the landward one. If this occurs then another pole can be extended horizontally from the top of the back pole to the seaward pole, and the elevation read directly on the seaward pole where the horizontal one touches it. If there are abrupt changes of slope these can be recorded by placing the seaward pole at the position of change.

Exercise 2.19. Beach profile surveying and plotting

The data given in Table 2.13 provide the results of a beach profile levelled using tacheometry. Calculate the reduced levels by adding the backsight values to the reduced level to give the height of instrument and subtracting the foresight and intermediate sight values from the height of instrument to give the new reduced level for each staff station. The letter B refers to a back intermediate sight and F to a fore intermediate sight. Calculate the return levelling to provide a check for the survey, which returns to the bench mark from which it started. Calculate the distances for stations 1 to 9, which were

Table 2.13 Beach profile levelling data

Backsight	Intermediate	Foresight	Height of instrument	Reduced level	Distance	Remarks
0·72				24·45		bench mark
7·74		3·09				
10·31		2·38				
4·28		9·69				
3·12		12·77				
3·47						
3·95		9·37			1	
4·43						
	2·86					
	B 3·28				2	
	3·70					
	6·90					
	F 7·17				3	
	7·44					
2·06		7·47				
2·81		8·03			4	
3·56		8·59				
	3·99					
	B 4·27				5	
	4·56					
	4·60					
	F 4·90				6	
	5·20					
	4·94					
	F 5·48				7	
	6·02					
2·41		4·70				
3·16		5·42			8	
3·92		6·14				
	6·11					
	F 6·47				9	
	6·83					
5·05		3·43				return
13·89		1·64				levelling
11·92		2·38				
1·95		7·11				
		3·70				bench mark

surveyed across the foreshore. Plot the profile on graph paper using a suitable vertical exaggeration.

Profiles provide valuable information on the changing morphology of the beach. Loss and gain of beach material can be assessed; changes in gradient and profile shape can be found. A beach gaining material often has a generally convex up profile, while an eroding one often has a concave profile. The nature of the berm and features such as ridges and runnels can be identified. Some of the relationships between beach profiles and processes are analysed later.

The profile across the beach only gives a two-dimensional picture of beach morphology. The three-dimensional form should also be considered. The types of features that can occur include ridges and runnels, which are aligned nearly parallel to the shore, rip current channels, beach cusps and other rhythmic beach forms, which are aligned nearly perpendicular to the shore.

The exact alignment of ridges and runnels is important. On some beaches, such as Blackpool where they are parallel to the shore they maintain their positions, often related to the tide levels, over considerable periods of time. Whereas on other beaches, such as those of south Lincolnshire, they lie at a slight angle to the coast, and on any one profile they appear to move landward as they are transferred down coast in the direction of longshore movement.

Recently the pattern of rhythmic features that occur along some beaches has been examined. These features include small beach cusps, often only about 10 to 40 m from horn to horn. Larger features also occur on some sandy beaches. These may be 500 to 700 m from horn to horn. The horns point seaward and separate bays in which finer sediment usually collects. These features sometimes move along the shore and their movement can be related to temporary erosion and accretion on the backshore as the protruberances pass. The larger protruberances are often associated with rip currents and their channels. These are zones where fast-flowing currents pass seaward through the breakers, forming part of a nearshore cell circulation. (See Plate 19.)

Material. The morphology of the beach is intimately related to the material, and both should be recorded together. Waves behave differently on different materials, and shingle beaches have different characteristic profiles from sand beaches. Both size and shape of material are significant, the shape being more important for shingle beaches. Few beaches have much fine sediment, but sometimes there is enough shelter for the finer sediment to collect. This occurs, for example, in the runnels of a ridge and runnel beach, or where protection from wave action is provided close offshore by offshore tidal banks that emerge at low tide, as on the south Lincolnshire coast near Gibraltar Point. Offshore sea ice provides similar protection in polar climates, and beaches in this zone tend to have mixed material.

Most beaches, however, consist of sand. Shingle is commoner in higher latitudes where glaciation has provided much coarse material. Flint forms resistant beach material and forms the bulk of the large shingle structures of Orfordness, Dungeness and Chesil beach in southeast England. The type of material is an important criterion in assessing the size of a coastal system. Where the geology is complex as along parts of the coast of Cornwall and Brittany, it is possible to relate the beach material to the rocks that outcrop in and around the bay. In some bays on the west coast of Ireland beaches consist of granite boulders, with adjacent ones being sandy, while some, far away from the rivers, have organic sand in the form of calcareous foraminiferal or siliceous remains. This material must have been derived from offshore and occurs where sediment cannot reach from inland sources.

The size of the sediment can best be obtained by sieving a sample from the beach. A cumulative frequency curve can be plotted from the results of the sieving. The median size is the point where 50 per cent of the sample is coarser and 50 per cent finer. The degree of sorting can be estimated by the spread of the sample values. Beach sands tend to be fairly well sorted, and the larger grains accumulate in zones where the energy is highest, in the swash–backwash zone and near the breakpoint of the waves. The simplest sorting measure is that suggested by Trask, which is the square root of the 75th and 25th percentiles on the cumulative frequency graph. There are more elaborate sorting measures that take in more of the tails of the distribution and these are often diagnostic of the environment of sedimentation. The one suggested by Folk

Plate 19 Point Reyes Beach, north of San Francisco, California. The long swell of the open Pacific Ocean provides a wide surf zone, through which fairly regular seaward-directed rip currents are flowing, breaking the beach zone into cellular circulation patterns, indicated by the pattern of swash on the foreshore. The constructive waves have brought sand to the foreshore from where it has blown inland to form the dunes seen behind the beach.

and Ward is often used, and it is given by

$$\frac{\phi_{84} - \phi_{16}}{4} + \frac{\phi_{95} - \phi_5}{6 \cdot 6}$$

where ϕ is $-\log_2$ mm. (See Figure 2.22.)

When samples are collected for sieving it is important to note the exact position on the profile and to take only the surface layers, because size often varies downwards. A section through the beach can be dug and this will reveal the nature of this variation, and also the character of the bedding and dip of the layers.

The disturbance depth of the sand by the waves can be measured by inserting a column of artificially dyed sand and marking the point by inserting thin pegs a measured distance from the column. Alternatively a washer can be fixed on the peg to mark the beach level before the tide covers the point, and then it can be located when the tide has again fallen and the depth of disturbance recorded. The depth of disturbance can be recorded in different sized material and under differing wave conditions, thereby establishing some material–process relationships.

The shape of the beach pebbles can be recorded by means of Cailleux's roundness index which provides a measure of the amount of attrition the pebble has undergone (see page 61). The source of the sediment can sometimes be determined by this means, if, for example, the stones get progressively rounder from a river mouth.

Shape measurement is most easily carried out on shingle sized particles. Waves rapidly round most types of rock so that beach pebbles on the whole tend to be much rounder than those from other environments. An example of the analysis of stone roundness is provided by observations made on raised beaches and deltas in Baffin Island. Samples of 50 stones from a raised beach and a delta were measured. The significance of the difference between these two values can be tested statistically using either a parametric or non-parametric test. A very quick and easy test, called the

Kolmogorov–Smirnov test, can be applied to this type of data. It is a non-parametric test so that the normality of the distribution does not have to be assumed.

Exercise 2.20. Beach pebble roundness analysis

Plot the data given in Table 2.14 in groups of 50 roundness grades as a cumulative frequency diagram and superimpose the two graphs. The maximum difference between any group is noted and if it exceeds a value given in the appropriate table (Siegel, 1956) then the roundness of the pebbles can be considered to be significantly different at the appropriate level of confidence. Where $n_1 = n_2 = 50$, $N = n_1 n_2 / (n_2 + n_2) = 25$. For $N = 25$ for the 99 per cent confidence level the maximum difference must equal or exceed 32 per cent and for the 95 per cent confidence level the value is 26·5 per cent. The result of this example shows that the waves have signi-

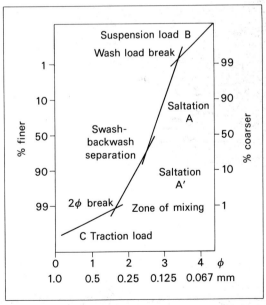

Figure 2.22 Cumulative frequency curve of sediment size distribution to illustrate relationships between size distribution and method of transport. The graph is drawn on probability paper, on which a normal distribution plots as a straight line. (After G. S. Visher, 1969)

79

ficantly rounded the beach pebbles relative to those of the delta, even though the time available was short, because the uplift of the land soon carried the beaches out of reach of the waves. Thus past marine activity could be deduced to have been present at a level well above the present sea level in this area of active glacial isostatic recovery.

The sizes of pebbles can also prove worthy of analysis in some areas. Perhaps the most impressive example of size sorting along the British coast is at Chesil beach in Dorset, where pebbles range from potato-sized ones at the southeast end near Portland to pea-sized ones at the Bridport end in the northwest. The reason for this sorting is not fully understood, but one process that is related to material size is wave energy. The larger material accumulates in zones where the energy is highest. The wave energy depends in part on exposure and in part on the depth of water offshore. This relationship can be substantiated in the Chesil beach area by correlating the distance offshore to the 10 fathom line and the size of the beach pebbles. The results for 8 points give $Y = 5 \cdot 61 - 1 \cdot 845X$, where Y is the size of the pebbles and X is the distance offshore, the correlation coefficient is $-0 \cdot 9526$, which is highly significant, indicating that the larger pebbles are where the wave energy is greatest at the most exposed southeast end of the beach. (See Plate 20.)

Near perfect sorting can occur on this beach because it is a relict feature to which little if any new shingle comes. The material that is there can become adjusted to the waves that form the beach. There is little longshore movement under these conditions, allowing sorting to take place.

Most beaches are of sand, and the size of the sand must be obtained by sieving samples from the beach. The mean size provides a useful measure, as already suggested, for relating process to morphology. One interesting relationship is that between material size and beach slope. Table 2.15 gives figures of this relationship.

Table 2.14 Stone roundness classes

Roundness class	Raised beach	Delta	Frequencies
0–99	0	1	
100–199	2	9	
200–299	6	10	
300–399	7	13	
400–499	12	9	
500–599	10	3	
600–699	7	2	
700–799	3	2	
800–899	3	1	
900–999	0	0	

Exercise 2.21. Beach material and beach gradient

The data provided in Table 2.15 refer to beach material mean diameters and beach gradients. These data can be used to consider the effect of transformation on the degree of correlation. The values for the gradient are given in cotangent values. They can be used directly, alternatively they can be converted into logarithmic values or into degrees and decimals of a degree. Use all three values to calculate the correlation coefficients and regression lines, using sand size as the independent variable X and the gradient as the dependent variable Y. Plot the regression lines.

The data show that as the material gets coarser so the beach slope gets steeper. In fact in coarse shingle the beach gradient can stand as steep as 1 in 2 or 1 in 3, as recorded on Chesil beach at Abbotsbury, where the shingle mean diameter is $1 \cdot 25$ cm.

The reason for this relationship is that as the material gets coarser so the percolation rate increases, and more of the advancing swash is absorbed into the beach reducing the backwash. The beach slope must be steeper to compensate for the reduced backwash in proportion to the swash. When the sand is very fine only a small part of the swash can percolate into the beach so the volumes of swash and backwash are more nearly equal,

Table 2.15 Beach gradient and sand median diameters

Gradient	Degrees	Median diameter (mm)
1:90	0° 38′	0·17
1:82	0° 42′	0·19
1:70	0° 49′	0·22
1:65	0° 53′	0·235
1:50	1° 09′	0·235
1:38	1° 30′	0·30
1:13	4° 24′	0·35
1: 7	7° 18′	0·42
1: 5	11° 18′	0·85

and the swash slope gradient has a flatter equilibrium value.

Although the size of the material is the most important variable determining the slope of the beach in the swash–backwash zone, the gradient is also affected by the wave size. For any given sand size observations of beach gradient along one profile will show that the beach slope varies under different wave conditions. When the waves are long the beach becomes flatter, and when the waves are steeper the slope is also less. Variations in the wave dimensions operate in a similar way to changes in material size, in that with steeper waves of the same length the volume of backwash will be larger relative to the swash, and the same applies as the waves become longer.

Processes. Waves, tide and wind are the most important processes affecting changes on the coast. The waves are the most variable in their influence; on coasts exposed to the open ocean they are often generated a long way from the beach where they eventually break. In sheltered, small seas, such as the Irish Sea and North Sea the waves are more often generated locally and are much smaller. On exposed coasts, therefore, the wind and waves may be unrelated, whereas in enclosed waters they are more likely to be associated with one another.

The waves arriving at a beach are usually a mixture of many waves of different lengths

and heights. The combination of different sized waves is called a wave spectrum. Measurement of the whole spectrum is difficult and requires a wave gauge to record the continuous fluctuations of water level, which may be done by recording pressure changes. The fluctuations can be converted into a wave spectrum by Fourier or spectral analysis, which indicates the amount of energy present in waves of different periods. It is usually adequate, however, to measure the height of the waves as they pass a pole or breakwater, remembering that there are bound to be variations in height. When the waves have been recorded for a short time the individual heights can be noted and the mean of the highest $\frac{1}{3}$ of the waves is taken to give the significant wave height.

The position at which the waves break should also be noted and this depth depends on the water depth in relation to the wave height. A wave normally breaks when the depth is $\frac{4}{3}$ the height of the breaker at the breakpoint. The breaking of waves is brought

Plate 20 Shingle at Abbotsbury on Chesil Beach, about in the centre of the structure. The larger pebbles were brought from the southeast end of the beach at Portland. They illustrate the change in size along the beach, with the largest stones at the southeast end.

Plate 21 Wave patterns including long swell waves superimposed on shorter, locally generated seas in New Zealand.

about by changes that take place as the wave moves into shallow water. The wave velocity and wavelength decrease rapidly when the depth is less than half the wavelength. At the same time the height, after an initial decrease, starts to increase, while the period remains constant. The orbits of the water particles within the wave change from open circles to open ellipses, and increase in dimensions. Thus as the wave moves into progressively shallower water the waveform moves more slowly, while the water within the wave is moving faster and faster. At a critical point the two velocities become the same, the water then overtakes the waveform and the wave breaks.

Different types of breaker are recognized. Waves can plunge to break, when their form is

lost in the process of breaking, or they can spill. This occurs when the wave advances with a foaming crest across the beach as surf. This type of breaker is used by surf riders as they are carried forward on the crest of the breaking wave. A third type is the surging breaker, which occurs on a very steep beach as the wave does not break before it becomes swash, moving up the beach as a coherent sheet of water.

The wavelength is related to the wave velocity and period. The period, which is the time taken for the waveform to move one wavelength, can most easily be recorded in the field. The passage of ten waves past a fixed point can be timed and the result, divided by ten, gives the wave period in seconds. This value can be related to the deep water wavelength by means of the equation $L = 1.56T^2$, with L the wavelength in metres and T the wave period in seconds.

As waves move into shallow water from offshore their deep water properties change with the exception of the period. The wavelength and velocity decrease as the depth decreases from a value about half of the wavelength in depth. In deep water wave velocity depends on wavelength and period, according to $L = CT$, where C is the velocity; but in shallow water the depth affects wave speed and wavelength.

An important result of decreasing depth is that the wave crests bend and the waves refract, turning their crests to become more nearly parallel to the bottom contours. This results in variations in wave energy received at different points along the coast. Energy tends to be concentrated on headlands and opposite submarine ridges, while it is dissipated in bays and opposite submarine valleys. Some patterns of wave refraction are shown in Figure 2.23. (See also Plates 20, 65.)

Exercise 2.22. Wave refraction

Figure 2.24 shows two inlets, both of which have been partially closed by a sand spit. (a) shows a drift aligned spit and the inlet is drift deflected. The waves approach from the southwest and are short, and hence little refracted. They reach the coast at a considerable angle along the length of the spit, thus causing material to move alongshore. In (b) the spit is swash aligned and the inlet is swash deflected. Waves also approach from the southwest, but they are longer and more refracted, reaching the beach along the spit with their crests parallel to the spit alignment. Draw orthogonals for both situations and explain the alignment of the spits and the factors that controlled their development and morphology.

The distribution of wave energy along the coast can be observed from the shore by

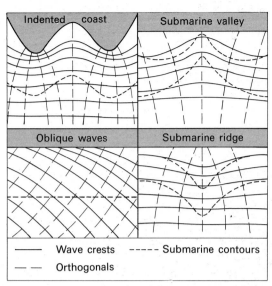

Figure 2.23 Examples of wave refraction, showing wave crests and orthogonals relative to submarine contours.

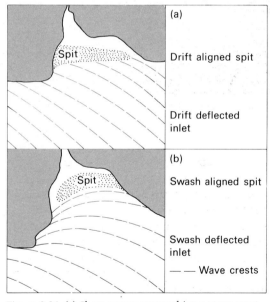

Figure 2.24 (a) Short waves approaching a steep coast obliquely with little refraction. (b) Long waves approaching a flat coast with more refraction.

noting the variation in wave height, the waves being higher in zones where energy is concentrated. Field observations can be confirmed by drawing refraction diagrams, which illustrate the pattern of orthogonals—lines drawn at right angles to the wave crests. Where orthogonals converge on the coast waves will have more energy and be higher, because the energy between orthogonals remains nearly constant if they are equally spaced in deep water. In order to draw refraction diagrams it is necessary to have a chart showing the offshore relief to a depth equal to half the wavelength. The direction of wave approach also influences the amount of refraction, and diagrams are drawn for different lengths and approach directions.

Wave steepness is another significant property from the point of view of coastal geomorphology. It is the ratio of wave height to wavelength. A steep wave is usually destructive on the foreshore especially on a sandy beach, while a long, low swell is constructive. Swell waves are generated far from the coast and have lost energy in travelling across long stretches of sea, resulting in a reduction of wave height. These long crested swells have a typical period of 10 to 15 seconds and occur on exposed coasts. The shorter, steeper waves, known as *sea*, occur in the generating area of strong winds. When this area is near the coast the waves that reach the coast are short crested and steep. They are frequently accompanied by strong onshore winds, which enhance their destructive effect on the beach. The direction of wave approach is important and can be observed by sighting along the crest of the wave with a compass as it approaches the beach. (See Plates 21, 65.)

Beaches that face the open ocean with long swells breaking on them can have flatter profiles than the beaches in enclosed seas, such as the Irish Sea or North Sea, where waves are generally shorter. Small lakes will have even steeper beaches because the waves are very short for the same size of material. Only in the open ocean can long waves be formed; here the generating winds can blow over a long stretch of open water creating waves of 300 m or more in length. In the more restricted waters of the Irish and North Seas the waves will usually only be about 45 m long, and the beaches will tend to be steeper.

Repeated surveys of the beach profile combined with observations of the nature of the waves and wind allow the effect of these variables on the beach profile to be assessed. The nature of the waves influences their effect on the movement of beach material. Some waves are destructive, carrying material seaward, while others tend to build up the foreshore. These movements represent transfer of material perpendicular to the beach and must be differentiated from movement of material alongshore under the action of longshore currents associated with waves approaching at an angle to the shore.

Longshore movement is the most important in the long term because it normally acts dominantly in one direction over a long period of time. It is thus responsible for the major changes involving coast erosion and accretion. Erosion takes place where more material leaves an area than enters it, and the reverse applies to accretion. Thus the direction and strength of longshore currents should be measured wherever possible. This can be done in the field fairly simply by timing the passage of a floating object between two fixed points a known distance apart. There are complications, however, in that the current at the surface is not necessarily the same as that near the bottom, and the bottom currents influence the sediment movement. Bottom currents can be measured with current meters at one point over a short time interval. Longer-term tidal currents can be recorded by releasing Woodhead sea bed drifters that move close to the bottom. They consist of orange plastic saucers, suitably weighted, to which are attached waterproof cards for returning when they are found stranded on the beach or caught by fishermen. These drifters provide information of the residual movement of tidal currents, but the exact route traversed and the time taken cannot be established in this way, although they do give useful information of sediment movement.

Beaches normally build up during the summer when waves on the whole tend to have a lower steepness value and are more constructive in their action. During the winter, which is usually stormier, with higher, steeper waves, the beaches are combed down and lose sand by destructive wave action. Beach profiles are, therefore, often divided into summer profiles, which are generally convex up with a high berm and fairly steep swash–backwash slope, and a winter or storm profile with a narrower berm or no berm and a generally flatter swash–backwash slope as shown in Figure 2.25.

The wind also exerts an influence on the movement of sand through its effect on the water. When the wind is blowing strongly onshore, as it often does during a gale that generates steep, destructive waves, it will enhance the destructive effect of the storm waves. Water is blown shorewards on the surface and this creates a return seaward flow near the bottom. An offshore wind tends to have the reverse effect, and when it accompanies long swells it will make these more constructive by lowering their steepness as it absorbs energy from the advancing waves. Both waves and winds must be considered and they are often related.

Useful analyses can be made to test the relationships between beach changes and waves and wind. A series of beach profiles was surveyed in a small bay on the coast of County Durham once a week and during the intervals the waves and winds were recorded. The beach changes were recorded by super-imposing two successive profiles so the cut and fill on the beach could be measured and the gradient obtained, as shown in Figure 2.25.

Exercise 2.23. Beach changes in relation to wind

Table 2.16 gives the results for onshore winds and offshore winds and for the upper and lower foreshore with respect to cut and fill on the profile. Use the chi square test to ascertain whether there is a significant relationship between cut and fill on the beach and the wind direction, and between the changes on the upper and lower beach.

The to and fro movement of sand up and down the beach takes place only to a limited depth. Even on an open exposed coast, such as that of

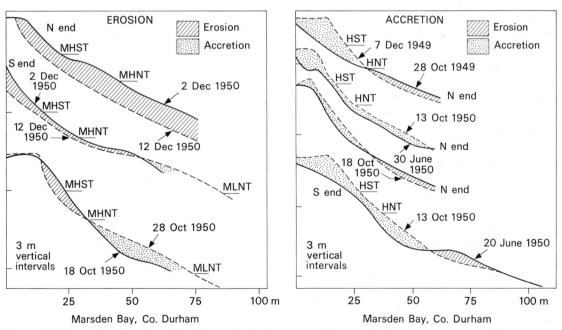

Figure 2.25 Examples of beach profiles characteristic of storm conditions and calmer, constructive conditions.

Table 2.16 *Beach changes: Marsden Bay, County Durham*

	Onshore wind	Offshore wind	Observed frequencies
Fill	4	13	
Cut	13	3	

	Upper beach	Lower beach	Observed frequencies
Fill	4	11	
Cut	13	2	

California, the movement goes down to only about 10 m, while in the more sheltered areas of eastern England, as shown in the second part of Exercise 2.23, the sand often moves only from the top to the bottom of the foreshore.

The effect of longshore movement of material on the coast is responsible for most of the major problems of coastal erosion, and it also explains many features of deposition. The coast of Holderness provides an example where the effect of longshore movement can be readily appreciated. The larger waves approach this coast from a northerly direction and the general movement of material is southwards along the coast. To the north of the drift cliffs of Holderness the headland of Flamborough stands out as a major pro-truberance that prevents beach material from the north reaching the beaches of Holderness. The waves, in attempting to build their equilibrium gradient, find themselves short of sand and take it from the cliffs, which consist of easily eroded glacial deposits. Some of this material travels southwards to build the sandy spit of Spurn Head, but much is carried offshore and further south into the Wash and onto the beaches of Lincolnshire.

The tide affects the foreshore and nearshore zone and is very important locally. Tides are much more regular and predictable than waves, and tide tables are available that give the times of high and low water and the range. Around the British coasts and much of the Atlantic Ocean the tide is semi-diurnal in type, having two high waters and two low waters every lunar day. In other areas, particularly around the Pacific Ocean, the tide is mixed or diurnal, while in a few places the tide follows the sun and not the moon, as at Tahiti for instance.

The tidal range helps to determine the width of the foreshore. The widest foreshore occurs where the tidal range is large and the material fine. At Blackpool, for example, where the sand is fine and the tidal range large, the beach is over 1 km wide at low spring tide. The British tides are on the whole large compared with many other coasts. Some areas, such as the Mediterranean, Baltic and Gulf of Mexico are virtually tideless.

Abnormally high tides occur in places when peculiar meteorological conditions create surges. These usually accompany violent storms, such as hurricanes or very deep depressions, that move at a critical speed. The North Sea is liable to surges, sometimes with disastrous consequences. In early 1953 the tide rose up to 4 m above predicted high tide levels, and flooded much low-lying land in eastern England and Holland. The effects of such surges on coastal morphology are well worthy of study, because more change can occur in one day of surge conditions than years of normal conditions. For example, one low cliff 2 m high in East Anglia was driven back more than 30 m overnight during the 1953 surge.

2.5 Biogeography

Introduction

Plants and soil provide the basis for local scale biogeographical studies. The study of pedology in the local area is concerned with the recording of the soil profile and changes of soil in relation to relief and parent material. On the local scale climate is a less important variable. Vegetation is closely related to soil, but it is also strongly influenced by the climate and weather, even on a local scale in some areas where, for example, exposure produces strong contrasts of the microclimate. Human interference has also modified vegetation practically everywhere to some extent. In countries such as Britain the vegetation is almost entirely man-controlled. Nevertheless the physical geographical restraints still apply.

Conditions in Britain range from the relatively dry and fertile southeast, where there is often a water deficit and irrigation can be necessary, to the wetter, bleaker west and northwest, where water and rock conditions allow upland peat to form in places. Although over much of the country the natural vegetation would still be forest, Britain has a relatively small area of natural forest at present. There is, however, sufficient variety of both soil and vegetation to make local studies in biogeography rewarding. Some of the relationships between the different aspects of physical geography have already been mentioned, such as the effect of climate on valley asymmetry through vegetation and soil. Therefore, it is necessary to consider similar interrelationships when observing the biogeographical elements of the environment, as they link very closely, on one hand, to geology and geomorphology, and on the other hand, to climatology. In both instances there are important feedback relationships in the whole complex ecosystem. Thus it is only for convenience that the different elements of the ecosystem are studied separately. They must always be considered as a whole system if they are to be fully understood.

Soils

Soils form the medium in which vegetation grows and are themselves derived from rocks, through the influence of climate, biogeography and time. Soils must be studied in the field to appreciate their essential characteristics, and the essential part they play in the distribution and character of vegetation. Field-work on soils involves a study covering the soil character both in depth and spatially.

Soil profiles may be studied by digging soil pits or by using a soil auger to obtain a vertical sample of the soil. The variations in soil character in different positions on slopes and in different aspects, on valley floors and upland plateaux, must be considered. The term *soil catena* is used to describe changes of soil type in relation to slope. The changes depend mainly on changes in gradient, hydrological conditions and vegetation.

The equipment necessary for a simple soil study includes a notebook, a measuring tape, a map and/or aerial photograph to locate the positions studied, a spade, a hand lens, a knife, an auger, if possible a colour film camera, a pH recording apparatus, a bottle of dilute hydrochloric acid, a Munsell colour chart and sheets of polythene. Turves from the pit site can be placed on the polythene for eventual replacement to avoid unnecessary disturbance of the ground. The pit should extend down to the parent rock if possible and the profile should be cleaned with the knife when the pit has been dug. The profile is recorded in the notebook by reference to the different soil horizons, which can be measured with the tape.

The description of the profile should include its site, described in terms of slope, elevation, aspect and vegetation. Drainage qualities should be noted, parent material described and the climate commented upon. The nature of the horizon boundaries should be recorded, for example as sharp, indistinct or diffuse, the latter covering more than 7·5 cm, and their shape should also be recorded as smooth, wavy, irregular or discontinuous. The recording of the soil colour can usually give useful information, and should include mottling where this exists. The Munsell colour chart provides a standard system for comparison.

Soil texture can be described in terms of the triangular diagram on which the percentages of sand, silt and clay are entered, as shown in Figure 2.26. This information can either be obtained by sieving or an assessment can be

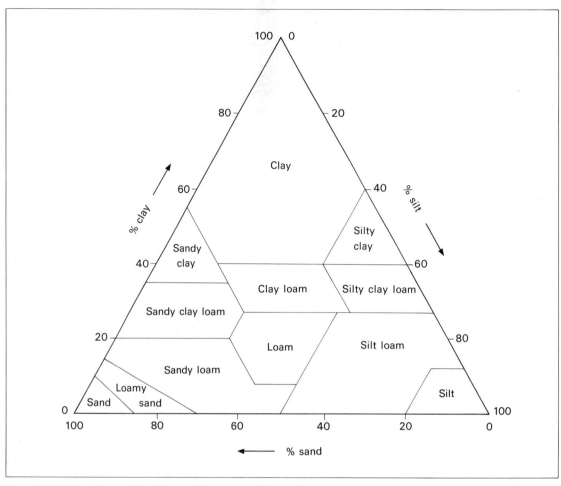

Figure 2.26 Triangular graph to show soil types according to the proportion of clay, silt and sand.

made in the field. Sand can be seen as individual grains and is not sticky or plastic. Silt feels slippery or soapy, but not sticky, and can be moulded when moist. Clay is often plastic, sticky and cohesive, but hard and lumpy when dry. Soil with much humus feels greasy.

The structure of the soil is very important from the point of view of fertility. Soil structures include columnar, prismatic, platy, blocky and crumby, the latter occurs under grass mostly, but not under forest.

The measurement of the pH of the soil gives an indication of its acidity, and can be done fairly simply with a test indicator. The carbonate content can be measured by using dilute hydrochloric acid, the greater the reaction the higher the amount of calcium carbonate present. Soil organisms and root amounts should also be recorded. The organisms include both plants and animals, the latter varying from large, such as rabbits and moles, through earthworms to bacteria.

The soil profiles are usually divided into horizons, the upper one is the A horizon, consisting sometimes of humus, litter and the eluvial zone, from which matter is washed out downwards. The B horizon is the illuvial zone, into which material is washed from above. The C horizon is weathered bedrock and passes into the solid rock below. Figure 2.27 illustrates the terminology of the soil profile.

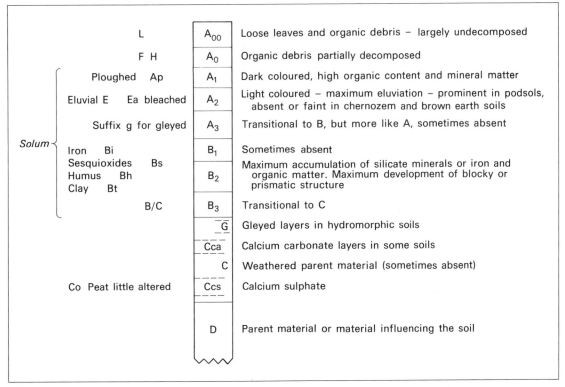

L	A_{00}	Loose leaves and organic debris – largely undecomposed
F H	A_0	Organic debris partially decomposed
Ploughed Ap	A_1	Dark coloured, high organic content and mineral matter
Eluvial E Ea bleached	A_2	Light coloured – maximum eluviation – prominent in podsols, absent or faint in chernozem and brown earth soils
Suffix g for gleyed	A_3	Transitional to B, but more like A, sometimes absent
Iron Bi	B_1	Sometimes absent
Sesquioxides Bs Humus Bh Clay Bt	B_2	Maximum accumulation of silicate minerals or iron and organic matter. Maximum development of blocky or prismatic structure
B/C	B_3	Transitional to C
	G	Gleyed layers in hydromorphic soils
	Cca	Calcium carbonate layers in some soils
	C	Weathered parent material (sometimes absent)
Co Peat little altered	Ccs	Calcium sulphate
	D	Parent material or material influencing the soil

Solum brackets A_1 through B_3.

Figure 2.27 Soil profile terminology. (After L. F. Curtis, F. M. Courtney and S. Trudgill, 1976)

In wet areas minerals are washed downwards to form buried layers, while in dry areas they are carried upwards by capillary processes to form eluvial layers, thus enriching the surface of the soil. Similar changes can occur on slopes in response to changes in water supply on the slope, steeper gradients being better drained than either high level plateaux or low level valley bottoms. These variations give rise to the typical soil catena, which can range from a podsol on the upper flatter slopes, brown forest soil on the steeper gradient mid-slope, to gley and peat on the impeded drainage of low slopes and valley bottoms, as illustrated in Figure 2.28. Thus the soil catena relates soil type to slope.

A podsol forms mainly in cool temperate areas under coniferous forest and has a highly leached whitish-grey A2 horizon and an illuvial B horizon. The brown forest soil forms under subhumid temperate conditions. The A horizon is dark and the B horizon is granular and there is little evidence of clay illuviation. They are usually fertile soils. Gleys develop where drainage is poor, they are often grey and mottled, due to reduction of iron. A fertile soil is the black earth soil, or chernozem. It has a very dark A horizon, hence the name, and forms under subhumid temperate grassland, occurring at the boundary between leached and unleached soils. These soils are usually rich in calcium carbonate, which is one of the first minerals to be precipitated by evaporation. The calcium carbonate is higher in the profile as the climate becomes drier. The black horizon of the chernozem may reach 2 m thickness, for example in the Ukraine, but it is usually $\frac{1}{2}$ m. The A horizon is increasingly granular towards the B horizon, which is not sharply demarcated, but becomes increasingly aggregated and browner, often with calcium carbonate nodules. The black colour is due to grass vegetation, and the fertile granular structure to calcium carbonate. The high root

89

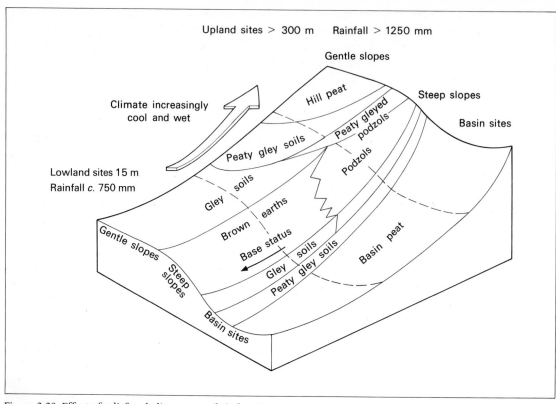

Figure 2.28 Effect of relief and climate on soil. (After E. Crompton, in L. F. Curtis et al., 1976)

content contributes to the humus and the black colour. These soils are about neutral in pH value, giving way to more acid prairie soils as the climate becomes wetter and chestnut soils as it becomes drier. The latter are more basic, especially in the B horizon, where calcium carbonate and gypsum can accumulate. These soils in turn grade into the soils of the arid climates. There is a continuum of soil types, as reflected in the complex modern soil classification schemes. Some soils are associated with particular rock types. Calcareous soils are basic and include rendzina soils, forming on the steeper slopes of limestone areas. Other soils are associated with drainage conditions, such as gleys, bog and fen soils of impeded drainage areas.

Relief is only one of the many variables that affect the characteristics of the soil. Vegetation and climate play an important role in determining the nature of the soil. The humus content of a soil is dependent in part on the vegetation it supports and on the climate and drainage conditions. It exerts an important influence on the soil fertility and structure. Humus is organic matter in the soil and is usually found in the uppermost horizon, called the A or sometimes O horizon. This horizon may be subdivided into two subhorizons, one of which is called the litter layer or L horizon, as it consists of fresh leaves, twigs and other plant debris in a fresh state. The lower subhorizon consists of more decayed, yet still recognizable, plant remains. This layer of decomposed litter is called the F layer, while beneath is the H layer of completely decomposed and structureless organic matter. The letters stand for O = organic, L = litter, F = fermentation and H = humified.

The humus layer is more distinct under forest vegetation and can be distinguished by the type of forest. Mor is an acid material

90

under coniferous forest and heath, or where soil is badly drained. The layer of humus is often distinct in the absence of earthworms to stir the soil. Mull, on the other hand, is a more basic material forming under mixed or deciduous forest. The organic material is more mixed with the mineral soil due to earthworm activity and other soil fauna. The stirring process goes on more actively under grass, so that there is usually no distinct humus horizon under grassland. The humus in these soils is derived from root decay and so is already mixed with the mineral soil, and is the basic type.

The soil forming variables can be summarized by the equation $s = f(cl, o, r, p, t, \ldots)$ where s is the soil, f means a function of, cl is local climate, o is organisms, r is relief, p is parent material, and t is time. It is useful to assess the relative importance of the different variables from site to site. The variable 'time' refers to the period over which the soil has been forming. It may be short in some circumstances, for example on newly erupted volcanic material, coastal flats or areas exposed by retreating glaciers. A study of progressively younger material provides a valuable insight into the effect of time.

The parent material provides the initial mineral elements of the soil, although weathering can alter these with time, especially the feldspars, which weather chemically into clay minerals. The parent material affects the permeability and hence the drainage which exerts considerable influence on the soil. Relief also affects drainage conditions. Climate plays a large part through precipitation effects on soil moisture. Water is a basic variable, determining whether there is a moisture deficit or surplus. In the former instance chemicals move up in the soil and in the latter they move downwards. Moisture surplus soils are usually leached of their nutrients in the upper layers, while moisture deficient soils may develop hard pans, such as laterite, through excess evaporation leading to deposition of mineral matter in the upper layers. Climate influences the degree of chemical reaction through temperature. The nature of the soil, however, particularly its colour, affects the heat holding capacity of the soil.

Vegetation

Vegetation is so closely related to soil and other environmental variables that it is necessary to consider its role in the physical geography of the small drainage basin. The environmental variables can be divided into physical variables, including edaphic (soil), geological (rock and relief), and climatic (water, temperature, light, atmosphere, wind), and biological variables, including plants, animals, man and fire. All these variables are closely interrelated with complex feedback loops. Another problem is to draw the line between biogeography and botany. Biogeographers are more concerned with the plant as a whole organism, and with its influence on its environment as well as the environmental elements that control its growth. Botanists, on the other hand, are concerned more with the way the plant functions as a growing organism. In a small drainage basin the gross climate is likely to be similar throughout, but local variations of aspect and exposure can often be related to vegetation, and relief often shows a close relationship with the plants. Vegetation mapping provides a useful preliminary step in studying these relationships.

Basic vegetation categories can be distinguished fairly readily by non-specialists, as indicated in Table 2.17, which gives groups of woody plants, basically trees, and herbaceous plants, which fall into three main categories: grasses, herbs, and lichens and mosses. Characteristics of their leaves can be observed according to the five categories given. The stratification of vegetation shows that plants live in communities and not as individuals. These are the ecosystems that coexist in sympathy with their environment. The coverage can also be recorded, and symbols used to map the major plant associations.

Table 2.17 Lifeform classification of plants (Raunkiaer, in Selby, 1971, p. 255)

Phanerophytes (Ph) exposed plants—buds high above ground. Includes most trees and shrubs

Chamaephytes (Ch) ground plants—herbaceous or low woody plants with buds close to ground. Common in dry and cold climates

Hemicryptophytes (H) half-hidden plants—grasses, herbs, rushes and sedges die back in winter. Common in cold and moist climates

*Geophytes (G) earth plants—die back below soil level because of cold or drought. Includes bulbs, corms, tubers, rhizomes and rootstocks, bud entirely protected

Therophytes (Th) summer plants or annuals, produce seeds and remain inactive during unfavourable seasons. Common in deserts.

Epiphytes (E) exist above ground growing on trunks or branches of woody plants

*Hydrophytes (HH) water plants, bud immersed in water

*Geophytes and hydrophytes can be called cryptophytes—hidden plants.

Relative abundance areal percentage

	Ph	Ch	H	G	Th
World	46	9	26	6	13
Temperate	15	2	49	22	12
Arctic tundra	1	22	61	15	1
Humid tropics	61	6	12	5	16
Arid tropics	9	14	19	8	50

Vegetation categories for mapping (*ibid.*, p. 303)

Lifeform categories		Structural categories	
Basic lifeforms	Special lifeforms	Height stratification Class height:	Coverage
Woody plants			
Broadleaf evergreen	B Climbers (lianas)	C 8 = >35 m	c = continuous (>76%)
Broadleaf deciduous	D Palms	P 7 = 20–35 m	
Needle-leaf evergreen	E Stem succulents	K 6 = 10–20 m	i = interrupted (51–75%)
Needle-leaf deciduous	N Tuft plants	T 5 = 5–10 m	
Aphyllous	O Bamboos	V 4 = 2–5 m	p = parklike, in patches (26–50%)
Semideciduous (B+D)	S Epiphytes	X 3 = 0·5–2 m	
Mixed (D+E)	M	2 = 0·1–0·5 m	r = rare (6–25%)
	Leaf characteristics	1 = <0·1 m	b = barely present, sporadic (1–5%)
Herbaceous plants	Hard (sclerophyll) h		
Graminoids	G Soft w		a = almost absent, extremely scarce (<1%)
Forbs	H Succulent k		
Lichens, mosses	L large (>400 cm²) l		
	small (<4 cm²) s		

Aphyllous—leaves nearly absent—typical of semiarid areas, e.g. Casuarina Australia
Graminoids—grasses, reeds, sedges
Forbs—broadleaf herbs, includes most flowering plants and ferns

Time rarely permits a complete coverage so that sampling is necessary. This can be carried out as a transect, usually along a slope profile, where other observations have been made. Quadrat sampling involves a square sample area, the size of the square being determined in relation to the number and size of species. Sample quadrats of different sizes can be used to assess the optimum size, which will be where the number of species recorded does not increase rapidly with the size of the quadrat. Dimensions of $\frac{1}{4}$ to 4 m are usually adequate for herbaceous vegetation. The quadrats should be placed randomly to ensure a statistically valid sample.

Plants are adapted to grow in a very wide range of environments, from the extreme cold of the polar areas, where tundra vegetation makes the most of the short summer growing season, to the hot, wet equatorial forests, where continuous growth of luxuriant vegetation is possible. At other extremes are the plants adapted to withstand the extreme aridity of the deserts to the underwater environment in which seaweeds flourish. The adaptation of plants to the wide range of environments provides a useful link between the different aspects of physical geography, including the climate, the soil, and the rock type and relief. In any one small area the climate will not exert such a strong control as the rock type and relief, although aspect can sometimes be shown to affect the vegetation considerably. The vegetation of limestone, for example, provides a strong contrast to that characteristic of sandstone and grit. Where a range of rock types outcrop within a small area it is possible to sample the different rock types to study the vegetational pattern in detail. The sample design should be arranged to provide data that can be analysed statistically, often by analysis of variance, which allows several variables to be considered simultaneously, and their effect evaluated on the vegetation pattern.

Interactions between biogeography and the environment

As has already been stressed physical geography consists of a number of separate branches that interact in a very complex way to make the natural environmental ecosystem, which consists of four major elements. These are: (a) the relief or morphology; (b) the vegetation and life, flora and fauna; (c) water in all its forms; and (d) the climate and weather. The interactions between these four components can be shown by means of overlapping circles as in Figure 2.29. The interaction between the elements includes heat transfer between water and climate, both elements being responsible for the distribution of heat throughout the earth. Climate and vegetation also interact closely through the different forms of plant and animal life adapted to the different climates and habitats. The interaction between relief and climate is through weathering, which also involves soil formation, relief, and climate; each affects the others by means of feedback connections.

Further breakdown of the major elements is suggested. Water can be broken down into that in storage, as ground water or glaciers, for instance, and that moving in the hydrological cycle, in rivers and rainfall. Vegetation consists of communities and individuals, while relief can be subdivided into solid rock formation and superficial drift features, including soils, while drainage basins provide natural units of relief. Interactions are suggested in both directions. For example, vegetation depends on water and in turn affects soils through fertilization, but also derives nutrition from them, while at the same time modifying water storage through transpiration. Similarly water and relief interact through erosion and regulation in each direction.

(a) Soil erosion. One of the major problems of land use for agriculture is soil erosion. It tends to occur when farming methods developed in one environment are used in another in which they are unsuitable on account of differences in the controlling variables. These variables include soil character and its erodibility, vegetation, climate and relief. Soil erodibility depends both on the nature of the soil and the processes to which it is subject, such as raindrop impact and surface runoff. The degree of permeability is, therefore, important and a measure of the size of the soil particles may help to assess erodibility. Climate is also important because raindrop impact is an important aspect of soil erosion, while rainfall intensity affects direct runoff. Vegetation is probably the single most important element,

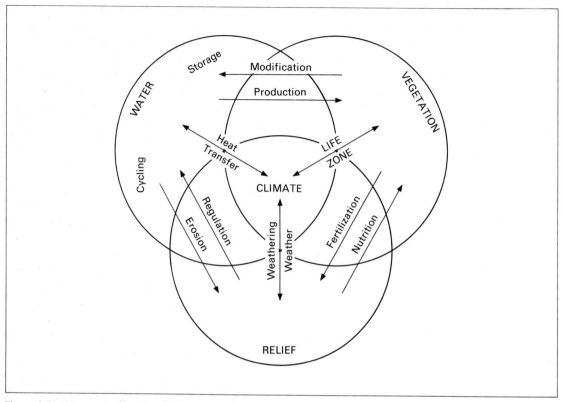

Figure 2.29 Diagram to illustrate the interaction between relief, vegetation, climate and water.

because it is the modification of the natural vegetation and its replacement by farm crops or grazing for animals that usually initiates soil erosion. A complete cover of grass or plants, such as heather or bracken, is next best to forest for preventing erosion, while over-grazed pastures and bare ploughed fields are the most liable to erosion. Slope is also important, the steeper slopes naturally being the most vulnerable. (See Plate 22.)

Soil erosion can take the form of sheet wash, which removes the whole top layer, or gullying, which cuts deeper into the ground along specific channels. Remedial methods include the control of water supply to the slope and also control of water within the gully to lessen its erosive capacity, by building dams or making the channel of suitable dimensions to carry the water. (See Plate 23.)

An example of gully control structure in the Colorado Front Range illustrates the measures that can be used to rehabilitate a watershed catchment. The first step is to assess the likely storm flow volume, and in the particular example a 5 cm precipitation event was used. Infiltration curves were established for different vegetation–soil complexes in the area to obtain the peak flow. The total area of the drainage basin was measured from air photographs and the area of each soil–vegetation type was also measured so that the contribution of each to the peak flow could be evaluated.

Before the study was carried out check-dams had been constructed 20 to 25 years earlier, and the sediment accumulated behind the dams gave an indication of their success. The channel gradient and cross sections were recorded before and after treatment. The gully area investigated is situated in the Colorado Front Range where the precipitation is 35 cm/year, but it comes in heavy summer thunderstorms, which cause the erosion. Forest covers 80 per cent of the basin, which has an area of 145·5 ha falling from 2,440 m to 2,280 m in 2·4 km. The check-dams were so spaced that the top of one dam formed a horizontal plane with the toe of the next one upstream. Most of the check-dams installed were successful, but a few failed. The failure was partly due to destruction of the structure and partly bypassing through valley widening. Gullies develop either continuously or discontinuously; in the latter instance a series of head cuts work upstream. These are critical points and the dams were built on these sites. The dams must be able to accommodate the highest peak flow and must be well keyed into the bed and banks.

Lack of success of the dams was due to one or more of the following reasons: the critical location was not realized, hydraulic requirements were disregarded, inadequate construction materials, adverse effect of willow plantings neglected, and no maintenance applied. The failures were largely due to lack of understanding of the physical systems involved. The hydraulic requirements of the channel must be respected and adequate materials used. The adverse effect of willow plantings in the channel was due to some of them being washed out. They then grew in a dense stand where they accumulated in a zone of deposition. The barrier formed in this way caused the stream to cut a new channel around them.

Most soil erosion abatement schemes, such as the gully control just discussed, involve the slowing down of runoff, and the provision of conditions in which soil can be deposited rather than eroded. Contour ploughing, terracing, and crop management all help to reduce overland flow and hence soil removal. The planting of suitable vegetation, either grass or forest, is important. Its effectiveness has been clearly demonstrated in the Tennessee Valley, which is the classic example of coordinated erosion control. In one small catchment with an area of 6·9 km² in east Tennessee, which had suffered severely by

burning and overgrazing, due to forest removal and soil deterioration, remedial measures included reforestation and protection from fire. The same volume of runoff from a given sized storm occurred, but instead of reaching a peak flow with 1·5 hours lag time, the time increased to 8 hours and the relative discharge was reduced to 15 per cent of the value before the rehabilitation had been carried out. The reduction for a large flood would, however, be considerably less.

(b) Floods and their control. Soil erosion and flooding are closely associated, because the main cause of soil erosion is sheet wash and gullying, which occur under conditions of heavy precipitation when there is excessive runoff as a result of inadequate infiltration usually due to vegetation deterioration. Excessive runoff also leads to flooding. The example just quoted illustrates the relationship. There is also the problem that damage is transferred downstream as well as taking place in the fields and upper valley directly. The stream sources, the watershed areas, are vital to a well considered river system. It is in these upland basins that precipitation is likely to be heaviest because of the effect of relief on rainfall. The normally heavier precipitation of the upland areas, at least in low and middle latitudes where temperatures are not too low, leads to forest growth. The effect of forest vegetation on various elements of the hydrological cycle is therefore relevant to flood protection schemes.

A well managed upland forest reserve area provides essential water supply and flood protection at the same time. This is now recognized in most advanced countries, but in developing countries the need to conserve forest watersheds is less well understood. As long ago as 400 BC Plato realized the importance of forest in protecting the uplands and in providing a steady and good water supply for use in the lowlands. Plato stated

There are mountains in Attica which can now keep nothing more than bees, but which were clothed not so very long ago

95

Plate 22 Death Valley, California. Badland erosion is evident on the barren hill slopes, with incipient gullying on the steep slopes, giving an intricate fretted appearance to the landscape.

with fine trees, producing timber suitable for roofing the largest buildings; the roofs hewn from this timber are still in existence. There were also many lofty cultivated trees, while the country produced bountiful pastures for cattle. The annual supply of rainfall was not then lost, as it is at present, through being allowed to flow over a denuded surface to the sea. It was received by the country in all its abundance, stored in impervious potter's earth, and so was able to discharge the drainage of the hills into the hollows in the form of springs or rivers with an abundant volume and a wide distribution. The shrines that survive to the present day on the sites of extinct water supplies are evidence of the correctness of my present hypothesis. (H. C. Pereira pp. 25–6 (1973))

It has taken over 2,000 years to begin to repair

the damage of which Plato speaks, by some reforestation in the hills of Greece and Turkey. The lesson is learnt painfully and very slowly, but at last it is becoming appreciated more widely and put into practice. The problems of large areas in the lower reaches of rivers are intensified or ameliorated according to the degree of control and management put into practice in the important small upland watershed of the type considered in most of this chapter.

Pereira (1973, p. 27) quotes the instance of the headwaters of the Bermejo River in the high Andes where soil erosion in an area of only 4 per cent of the watershed of the Parana River causes serious problems in Buenos Aires. The main branch of the Parana, the Paraguay River, is a clear stream until it is joined by the muddy Bermejo, draining from overgrazed mountain pastures in the north of Argentina. The river carries 80 million metric tonnes of

Plate 23 Soil erosion in Tennessee. Gullies are extending headwards into the gently sloping land as the surface vegetation deteriorates.

clay of very fine size that does not settle until it is flocculated in contact with sea water 1,200 km away near Buenos Aires. The dredging of this fine material from the long approach to Buenos Aires up the River Platte can cost up to US$10,000,000/year. The Bermejo River supplies 80 per cent of the sediment, which comes from high, steep country from 300 to 2,000 m in height. One basin of 25,000 km² alone supplies 65 million tonnes, a loss equivalent of 2,500 tonnes km^{-2} yr^{-1}. The cause of the problem was the importation in 1556 of 8 Andalusian cattle, which increased greatly in numbers. Their excessive grazing over a long period in an area of low seasonal rainfall has caused very severe soil erosion. The soil has been deposited in the delta area at a rate of 3 km² yr^{-1} new land. Between 1873 and 1879 the delta advanced at a rate of 46 m yr^{-1} and from 1900 to 1964 at 84 m yr^{-1}. The erosion is caused by flood flows in torrents cutting into their banks. The fine clay in the water provides a problem in the water supply. Water treatment costs another US$2,000,000/year. Major restoration works are needed to solve this problem, beginning with watershed control.

Flood control and conservation start in the watershed areas, and the influence of forests in these areas provides a useful example of the value of biogeography and its interaction with other aspects of physical geography. Forests have been observed to increase the precipitation by causing dripping from trees in misty conditions. Near San Francisco on the Berkeley Hills 25 cm of precipitation was recorded in each of 4 rainless summers due to drips from pine trees. Forests in the northeast of European USSR have been shown to increase stream flow through their influence on snowfall, the snow being trapped by the trees. Where the proportion of forest cover rose from 10 per cent to over 80 per cent the water yield increased from 3·0 to 5·8 m³ s⁻¹ km⁻² in one group and from 4·5 to 7·5 m³ s⁻¹km⁻² in another. There is no evidence that forests modify the amount of rain.

The effect of fire can be severe. For example an uncontrolled fire burnt out two catchment areas in the Snowy Mountains of Australia, with areas of 41 and 224 km². Gauging of the streams had been undertaken previously so a comparison between the conditions before and after the fire could be made. Sharp flood peaks occurred after the fire. Rain storms that before produced a peak flow of 60 to 80 m³ s⁻¹ produced peaks of 370 m³ s⁻¹. The increase in sediment load was 100 fold. At 95 m³ s⁻¹ a concentration of 14·4 per cent by weight produced a yield of 115,000 tons/day. One creek produced a sediment yield of 1,000 times the value before the fire, taking into account both the higher peak flow and the great sediment concentration.

The Tennessee Valley provides another example. One small catchment of 40 ha had been felled for timber and cultivation, as well as being overgrazed, burned and misused. Soil erosion became active. A stream gauge was installed before remedial measures were undertaken in 1946. Surface runoff was reorganized by using gently sloping stream courses, contour furrows were ploughed, pines were planted throughout the basin, with Black Locust in the gullies, totalling 100,000 trees. The result was the elimination of floods and soil erosion, peak flows were reduced by 90 per cent and the sediment load by 96 per cent. Clean, controlled flow replaced muddy floods, and the trees probably used about half the water in this instance.

Sound management must be based on sound knowledge, and research is being carried out into the relationships between forests and the hydrological variables. Watershed experiments are of two types, long-term comparisons between stream flows under different vegetation and land use, and detailed intensive studies of the energy balance in relation to the annual water budget. By 1966 39 experiments

had been reported concerning the effects of forest cover on stream flow from western USA, Japan, East Africa and South Africa.

More detailed studies are necessary to provide an assessment of the effects of forest in terms of various components of the hydrological cycle. In a forest catchment most of the difference between precipitation and runoff is evapotranspiration. Penman's work showed that over Britain the average evapotranspiration loss is 35 cm, ranging from 33 cm in Scotland to 61 cm in southeast England. Detailed studies of the microclimate of forests involves the building of recording towers up to 32 m high, such as those used by the Institute of Hydrology in the pine forests of the Breckland, Norfolk, where wind speed, water vapour, air temperature, radiation from the sky and the forest are recorded. Similar observations are being made in the forested valleys of the Severn, where preliminary results suggest that for 1969 and 1970 sheep pastures received more rainfall and yielded 20 per cent more runoff than a forested catchment. Thus if water demand is the main aim forests may not prove beneficial. If, on the other hand, flood control is the main aim under consideration, then forests should be preserved or planted, as they provide the optimum conditions for flow control.

Urbanization, on the other hand, provides almost the exact opposite of the forest situation, in that absorbant vegetation is replaced by concrete, pavements, roadways and roofs, all of which produce immediate runoff and also absorb much of the incoming radiation. Artificial drainage systems are included in the design of cities and these must be able to deal with very rapid runoff. The peak flow of the city drains must then be accommodated in the natural drainage system, and this may cause problems.

At the early stage of urbanization there are added problems of exposed construction sites that are very liable to rapid sediment removal. This problem also applies more widely in road building programmes, such as motorway construction. In the Thames valley above

Teddington weir, a catchment of 1 million ha, about 6,000 ha have been covered by impermeable roads and buildings between 1939 and 1962, but this is less than 0·5 per cent of the agricultural area, so that it has had relatively little impact on river flow.

Detailed studies made in small catchments in Devon have shown that during the construction phase the amount of sediment provided as a result of building operations is very great compared with conditions before construction starts. Once the houses have been completed, roads built, gardens planted and lawns seeded, then sediment supply is lower than it would be under agricultural activity. Nevertheless the large areas of impermeable surface do markedly affect the shape of the flood hydrograph in urbanized catchments.

2.6 Conclusions

This chapter has aimed to give some suggestions concerning field-work on the small scale in small upland catchments and coastal locations, and some methods of analysing the type of data that could be collected have been suggested. Some of the methods of analysis have used data collected over a number of years and in a wide range of environments. The principles illustrated by these examples are as important as the methods of analysis and probably more so. The methods are a means to an end, which is the understanding of the complex interactions that constantly take place between the different aspects of the ecosystem, which involves all the fields of physical geography.

The hydrological cycle provides a useful thread that ties together many of these interrelationships, as water plays a vital role in so many fields. The last section illustrated by means of some examples the practical importance of understanding the relationships in the field of applied physical geography or environmental conservation.

The complexity of the natural system will have become apparent from the examples of interaction between the different aspects of physical geography and within any one aspect. The principles of feedback, equilibrium and stability are all important. Equilibrium will only be maintained when the feedback relationships are negative, i.e. the changes lead to stability. Under conditions of positive feedback changes are self-generating, leading to greater and greater modification and instability, such as the vicious spirals of the deleterious effects of soil erosion. Such changes can cause other problems in areas very distant from the initial source of the trouble.

Even the worst effects, however, cannot go on for ever getting worse. There comes a point when new controls are imposed. This brings in the very important principle of thresholds. A threshold must be exceeded for a new set of conditions to apply and a major change to occur. For example a slope will remain stable until a threshold is exceeded and then it may suddenly collapse by slumping or landsliding, when a new stability will be achieved, until another threshold is crossed. An extreme event such as a major flood may be the immediate cause of crossing a threshold that is being approached so that particular flood will have spectacular consequences compared with one that occurred before the threshold was reached. The concepts of feedback, thresholds, and complexity are important in physical geography, and they apply to all the branches of the subject. They must be borne in mind when the natural system is being manipulated for human benefit. It is essential to work with nature and not against it, thus the first necessity is to understand the way in which the complex natural system operates.

References and further reading

Allen, J. R. L. (1970) *Physical processes of sedimentation.* Allen and Unwin, London.
Benedict, J. B. (1970) Downslope soil movement in a Colorado Alpine region: rates, processes and climatic significance. *Arctic and Alpine Research* **2**, 165–226.

*Chandler, T. J. (1965) *The climate of London*. Hutchinson, London.

*Chorley, R. J. (ed.) (1969) *Water, earth and man*. Methuen, London.

*Curtis, L. F., Courtney, F. M. and Trudgill, S. (1976) *Soils in the British Isles*. Longman, London.

Douglas, I. (1967) Man, vegetation and sediment yield of rivers. *Nature* 215, 925–8.

Douglas, I. (1968) Natural and man-made erosion in the humid tropics of Australia, Malaysia and Singapore. *Symposium of River morphology. General Assembly Bern*, 17–29.

Faulkner, R. and Perry, A. H. (1974) A synoptic precipitation climate of South Wales. *Cumbria* 1, 127–38.

French, H. M. (1976) *The periglacial environment*. Longman, London.

Gregory, K. J. and Brown, E. H. (1966) Data processing and the study of land form. *Geografiska Annaler* N.F. 10, 237–63.

Gregory, K. J. and Walling, D. E. (1968) Instrumented catchments in South Devon. *Transactions of the Devon Association for Advances in Science Literature and Art* 100, 247–62.

*Hanwell, J. D. and Newson, M. D. (1973) *Techniques in physical geography*. Macmillan, London.

King, C. A. M. and Buckley, J. T. (1968) The analysis of stone size and shape in arctic environments. *Journal of Sedimentary Petrology* 38, 200–14.

King, W. B. R. (1957) Water supply and geology. *Science Progress* 180, 609–18.

Langbein, W. B. and Schumm, S. A. (1958) Yield of sediment in relation to mean annual precipitation. *Transactions of the American Geophysical Union* 39, 1076–84.

Leopold, L. B., Wolman, M. G. and Miller, J. P. (1964) *Fluvial processes in geomorphology*. Freeman, San Francisco.

Manley, G. (1944) Topographical features and the climate of Britain. *Geographical Journal* 103, 241–63.

Manley, G. (1945) The effective rate of altitude change in temperate Atlantic climates. *Geographical Review* 35, 408–17.

*Miller, A. A. (1953) *The skin of the earth*. Methuen, London.

*Morisawa, M. (1968) *Streams: their dynamics and morphology*. McGraw-Hill, New York.

Ollier, C. D. and Thomasson, A. J. (1957) Asymmetrical valleys of the Chiltern Hills. *Geographical Journal* 123, 71–80.

*Pereira, H. C. (1973) *Land use and water resources*. Cambridge University Press, Cambridge.

*Rosenburg, N. J. (1974) *Microclimate: the biological environment*. Wiley, New York.

Savigear, R. A. G. (1965) A technique of morphological mapping. *Annals of the Association of American Geographers* 55, 514–38.

Schumm, S. A. (1956) The role of creep and rainwash on the retreat of Badland slopes. *Columbia University Department of Geology Technical Report* 12, 693–706.

Schumm, S. A. (1967) Rates of surficial rock creep on hillslopes in Western Colorado. *Science* 155, 560.

*Selby, M. J. (1971) *Surface of the earth. Vol. 2 Climate, soils and vegetation*. Cassell, London.

Visher, G. S. (1969) Grain size distribution and depositional processes. *Journal of Sedimentary Petrology* 39, 1074–106.

*Young, A. (1972) *Slopes*. Oliver and Boyd, Edinburgh.

CHAPTER 3

Regional and continental scale studies

3.1 Introduction

Variety increases in all aspects of physical geography in the larger area that constitutes the regional or continental scale environment. The British Isles provides an example of the regional scale of study, and in Britain the variety is exceptionally great compared to its size, because of its marginal position from many points of view. Britain is situated at the western extremity of the large land mass of Europe and Asia, marginal to the ocean and the continent, a position that has a very marked effect on the weather and climate. It is also marginal in respect to latitudinal zones, lying between the cold climates of the north and the low latitude deserts and tropics. This situation affects the biogeography of Britain through the climate and hydrology.

Geomorphologically also Britain is marginal, lying near the junction of the continental crust and oceanic crust. Its position near the edge of a trailing plate is very significant in terms of the development of the landscape through time, and its geological

Plate 24 Tropical forest in Uganda. Note the tall trees, one of which has a buttress root system, and the various other layers of vegetation, including many lianas and smaller trees and ground plants.

character. The geological variety is exceptional for so small an area. In Britain there are examples of rocks of nearly all geological ages, and indeed many of these are named after British localities or peoples, such as the Cambrian, Silurian and Devonian.

The landscape of Britain shows evidence of this marginal character in the relics of different climates that have influenced the shaping of the country in the past. There is some evidence of earlier tropical conditions, but clearer evidence of the more recent glacial and periglacial events in the landforms of much of the country. Different geomorphological processes have acted on an extremely wide range of rock types, so that the range of British scenery is very wide, adding greatly to the interest of the country from the point of view of physical geography.

To appreciate the physical geography of Britain fully, however, it is necessary to look further afield to other areas of comparable size, set in very different environments. The contrasts affect all aspects of geography. In the high latitudes there is the contrast between the seasons, between the long, dark and cold winters, and the short, light and cool summers. In low latitudes the range of temperature and daylight through the seasons is small, but contrasts are often provided by wet and dry seasons. The basic climatic influences, in their turn, control other aspects of physical geography, including hydrology, vegetation and landform processes.

Although the interaction of the different elements of the physical environment is essential to an understanding of its physical geography on the regional and continental scale also, each section is devoted for clarity to a different aspect of physical geography. Meteorology on the regional scale is concerned with the characteristics of air masses, and the conflicts between them, which give rise to frontal phenomena in the zones where different air masses come into contact with one another. The variations of weather associated with air masses and frontal systems give rise to climates that characterize different regions on this scale. The climate also influences the regional scale hydrology, which is considered in the following section which is devoted to meteorology. Hydrology on the regional scale involves longer time intervals than the local scale and the annual river regimes provide appropriate data on the regional scale. There is a close link between the climate and hydrology at this scale. In some areas the annual regime is based on fluctuations of ice rather than liquid water. The study of glacier budgets exemplifies this aspect of hydrology, which again is closely related to climate, and also to relief.

In the field of biogeography on the regional scale, variety is large. Ecosystems of different climatic zones provide examples of this, ranging from the life of the cold tundra zones to that of the hot, wet tropical forests. This aspect also illustrates clearly the influence that human interference can have on the natural ecosystem. New Zealand provides an excellent example of problems associated with the introduction of alien species into a delicate environment.

Regional physical geography can also be applied to the oceans, both in the small area of the North Sea and the larger area of the Southern Ocean, ranging from the regional to the continental scale, and this aspect is considered in the next section. The biogeo-graphical aspects are related to the physical ones in the oceans, which are too often neglected by physical geographers. The chapter ends with a consideration of geomorphology on the regional scale, with particular reference to Britain. The final section deals with some aspects of the time element in physical geography on the regional scale.

3.2 Meteorology

In this chapter some of the dynamic aspects of meteorology, basic to an understanding of weather and climate in different global situations, are considered. A basic characteristic of the air, on which its behaviour depends, is its state of stability. This in turn depends on its moisture content, temperature and the history of its formation and movement. The different types of air are called *air masses*. Because of its marginal position Britain experiences a wide range of different air masses. It lies in the zone of conflict between different air masses so that it also provides a setting in which to examine the nature of frontal systems and the weather associated with them in the temperate zone. Some areas are dominated by specific pressure types, such as the equatorial low pressure zone, and the cells of high pressure characteristic of some low latitude areas that give rise to arid conditions. In other areas alternating pressure systems give rise to marked seasonal variations, such as those that occur in the monsoon areas.

Stability and instability

In order to appreciate the qualities of stability and instability in the atmosphere, it is necessary to examine changes of temperature and moisture content with height. The relationship between pressure, height and temperature can be examined on the (T/ϕ)gram, which is short for the relationship between temperature and pressure. These variables are plotted against each other on a

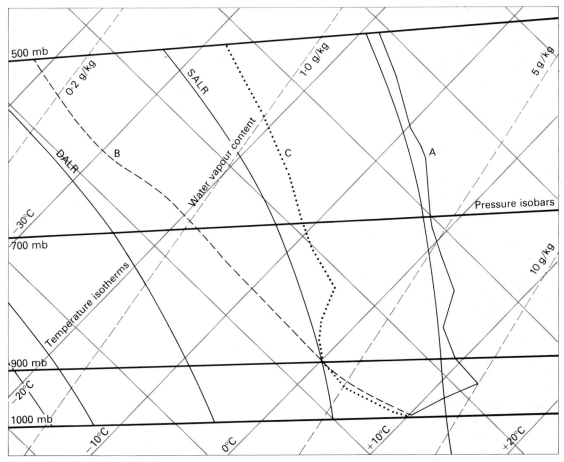

Figure 3.1 T/ϕ gram for plotting upper air data. Sample path curves are shown.

graph of the type shown in Figure 3.1. The lines running nearly horizontally across the graph refer to the pressure in millibars. They are equivalent to lines of elevation as pressure decreases with height. The lines that run from bottom left to top right are isotherms, lines of equal temperature. The straight lines at right angles to these, running from bottom right to top left, are the dry adiabatic lapse rate (DALR) lines. The set of curving lines that run at a steeper angle across the graph in roughly the same direction are the saturated adiabatic lapse rate (SALR) lines. The dashed lines nearly parallel to the isotherms are the water vapour content lines in g/kg.

The term *adiabatic* describes changes in temperature that result from air moving up or down without loss or gain of heat. When the air moves up it moves to zones of lower pressure, this allows it to expand, and the work done in expanding lowers the temperature of the air, and is called an adiabatic change. The dry adiabatic rate of change, called the DALR for short, is 9·8 °C/km increase in height, and is constant throughout the atmosphere. If the rising air is saturated, however, condensation will take place as it rises and its temperature falls, because colder air can contain less moisture than warm air. Latent heat is given off, due to condensation, and this heat is added to the air, thus the rate of fall of temperature with height is reduced. Hence the SALR curves are steeper than the DALR. They vary according to temperature,

105

being steeper as the temperature increases, because warm air can contain more moisture than cold air. The rate varies from 4 °C/km at high temperatures to 9 °C/km at a low temperature of −40 °C (or °F).

The static lapse is the rate at which temperature falls off with height at any one place and time and it is called the *environmental lapse rate* curve. It must be distinguished from the dynamic lapse rate, which is the adiabatic lapse rate. The dynamic lapse rate is that of a parcel of air raised from the surface under specific surface moisture and temperature conditions. These curves are called *path curves* on the (T/ϕ)gram. The relationship between the environment curve and the path curve determines the stability of the air.

If the path curve lies to the left of the environment curve the air is stable, because at any given height the rising air is colder than the surrounding air. It is, therefore, denser and will tend to subside, making the air stable. If, on the other hand, the path curve lies to the right of the environment curve the rising air on the path curve will be warmer than the surrounding air and less dense. Because it is lighter the air will continue to rise and it will become unstable.

The rising air, if it contains moisture, will rise at the DALR until it reaches condensation level after which it will rise at the SALR. Eventually the path curve and the environment curve may cross, and this point indicates the upper limit of instability in the case of initially unstable air. This will set an upper altitudinal limit to cloud development, while the cloud base will be at the condensation level. Thus (T/ϕ)grams are useful for indicating the type and vertical thickness of clouds in conditions of instability when cloud formation is the result of rising air currents. Such clouds, e.g. cumulus and cumulo-nimbus, are convectional clouds. When the unstable layer is thin, strato-cumulus clouds may form. (See Plate 25.)

In the examples shown in Figure 3.1 curve A represents air that is saturated. The path curve will follow the SALR from the +10 °C isotherm. At all heights the rising air is cooler than the surrounding air and would tend to subside, giving stable conditions. Curve B represents air that would become saturated at 950 mb. Its path curve would run along the DALR to the 950 mb level and then it would follow the SALR. At all heights this curve lies to the right of the environment curve indicating extreme instability. The air would go on rising and intense convectional activity would be expected, leading to the build up of massive cumulo-nimbus clouds and thunderstorms. The cloud base would be at 950 mb, and the tops would reach to great heights, up to perhaps 10,000 m. Curve C represents air that would become saturated at 900 mb.

Exercise 3.1. (T/ϕ)gram analysis

Draw the path curve for environment curve C with ground temperature at 9 °C and estimate the level of the cloud base and its thickness. Suggest the type of cloud. The situation given by curve A could represent an early summer morning as the environment curve indicates that there is an inversion, that is temperature increases with height between 1,000 and 950 mb before decreasing as is normal in the atmosphere. The conditions are, therefore, likely to be misty if it is calm, as the air is saturated. During the day, however, the air will warm up at ground level and the lower layers will no longer be saturated. The path curve will now lie for part of its length along the DALR as the surface temperature rises. When the surface temperature reaches 13 °C the inversion will disappear, and a further rise is likely in summer. Consider what would happen, and draw the appropriate path curves, if the temperature rose to 16 °C and then to 21 °C. Describe the type of cloud and weather that might occur in the afternoon as the temperature rises.

Regular observation of upper air conditions forming the basis of the (T/ϕ)gram was a great step forward in the forecasting of weather

106

Plate 25 Towering cumulus over Texas, showing active convection leading to condensation and great vertical cloud development.

conditions. The state of stability of the air can be readily ascertained by means of (T/ϕ)grams and from this information the occurrence of showers, thunderstorms and inversions could be accurately forecast. Under some conditions, such as those of conditional instability, the weather depends on a trigger to initiate vertical movement of the air. This trigger could be a mountain range that forces the air to ascend into an unstable zone. Surface heating may also trigger off instability under suitable conditions. The (T/ϕ)gram can also be used sometimes to indicate the nature and position of frontal surfaces between different air types, which are often called *air masses*.

Air masses

When air remains stationary over a large area of uniform character for a number of days, it acquires the characteristics of its source area over a wide stretch. These conditions apply in stable high pressure areas. The air tends to move out of the source area and is known as an *air mass*, which undergoes modification as it moves according to the nature of the area over which it is moving relative to that of its source area. Air masses are defined by the nature of their source areas and their subsequent modifications. The source areas are defined by temperature and nature of the surface, providing six main types: the Arctic, Polar and Tropical, each of which can occur as a maritime or continental type, giving the main types of the higher and middle latitudes. These are mA (maritime Arctic), cA (continental Arctic), mP (maritime Polar), cP (continental Polar), mT (maritime Tropical) and cT (continental Tropical). All six types can reach Britain under specific synoptic situations.

Air masses are often associated with characteristic wind directions. Thus mA air reaches Britain usually as a northerly wind in winter. Its source area is the Arctic basin which is very cold and ice covered in winter. In its source area, therefore, it is a very cold, fairly stable air mass. It reaches Britain by crossing the sea between the Arctic and Europe, moving south usually by a direct route. On reaching Britain it is a cold, northerly wind, and as it moves over the sea, which is warmer than its source area, it is warmed in the lower layers. This makes it unstable by the time it arrives. It is associated with cold weather, and snow showers are common in areas exposed to this very cold, northerly wind, the east coast of Scotland and England being particularly vulnerable. Because of its instability visibility is usually good, as particles do not become trapped in the lower layers as occurs in stable conditions.

The cA and cP air masses are rather similar. They develop in winter over the northern part of Europe and Asia where there is a stable, very cold anticyclone at this time of year. The cA air is the coldest that can reach Britain, coming from the extremely cold conditions in Siberia and northern USSR at this time of the year. Mean winter temperatures in the interior of Siberia are about $-50\,^{\circ}$C with extremes as low as $-68\,^{\circ}$C. The cP air mass forms a little further south, but is still very cold in winter, owing to the extreme continentality of interior Eurasia. These air masses can reach Britain in winter from the northeast, east or possibly southeast. They have moved mainly across the cold, snow-covered continent, and are thus fairly stable. Nevertheless even the short

crossing of the North Sea can modify them quite a lot; they arrive in Britain with much more cloud than over the continent, owing to the moisture picked up in the short sea crossing. Their temperature is also raised substantially, thus indicating the importance of the distinction between the continental and maritime influences. Arctic air reaches Britain when pressure is high over Scandinavia and air moves clockwise round it to reach Britain from the east. Temperatures are usually below freezing, but not as low as $-18\,^{\circ}$C that often occurs on the continent. Slight snow showers may occur along the east coast when sufficient moisture has been picked up in the sea crossing. The strong east winds and the high humidity often make the temperature feel colder. Continental Polar air is very similar although not quite so cold, especially in the upper levels, as the Arctic air. It also reaches Britain as an east or southeast wind when there is a high pressure area over northern Europe.

The cT air, by contrast, brings the warmest conditions to Britain. It is more likely to reach Britain in summer, but occasionally brings abnormal warmth at other seasons. The air is warm and dry, originating from the high pressure area of northern Africa over the Sahara desert. When pressure is fairly high over the continent, in France or Spain, and low over the western approaches the air arrives from the southeast or south. The cT air produces heat wave conditions in summer, and it may produce relatively very high temperatures of $21\,^{\circ}$C to $24\,^{\circ}$C in autumn and spring, although it is very rare in winter. Temperatures may rise to $35\,^{\circ}$C in summer, and the air is usually very dry, although again the short sea crossing adds some moisture.

The warm, settled spells of south or southeast weather usually break up in summer with thunderstorm activity. This occurs when cooler, moister air is drawn in from the west, moving around a low pressure centre to the southwest of Britain. The colder air pushes under the warm and forces it to rise, which, with the added moisture, causes a very unstable situation and active thunderstorms break out. The thundery fronts are usually slow moving, and the advancing cold air is slowed near the ground by friction, while it advances aloft, giving a very unstable situation and causing rapid convection currents. Once the air is cleared by the thunder activity, the cooler air behind the front takes over.

The cooler air represents another important air mass as far as Britain is concerned. It is the mT air mass, formed in the high pressure area around the Azores over the Atlantic Ocean. It is a warm, moist air mass, which reaches Britain from the southwest. The air is generally moving towards cooler latitudes as it travels northwards so that its lower layers are cooled and it is rendered more stable. The air is, therefore, associated with layer type cloud, such as stratus, or even with fog at some times of the year. High pressure to the south or southeast and low pressure to the north or northwest of Britain is common, so the strong southwest winds associated with this pattern occur frequently. The mT air brings mild, moist conditions in winter, while in summer it is drier, especially well inland, with warm conditions and little precipitation. In winter orographic precipitation is a feature of mT air on the higher parts in the west of the country, as demonstrated by the local variations discussed for the two stations in south Wales. The eastern part of the country is in the rain shadow for southwest winds and so is usually much drier than the western hills.

The most common air mass to reach Britain is the mP air. It is variable as it can approach Britain at all seasons and from a wide range of directions from north–northwest to south–southwest. In winter it forms over the cold northern Atlantic Ocean and reaches Britain after passing over progressively warmer water. Its lapse rate, therefore, increases as it is cold above and warms up below. As a result it is unstable, and is associated with cumulus clouds and often showers, even thunder may occur in the western parts of the country when the air is forced to rise suddenly over the western mountains of Scotland and Wales. Its degree of instability varies with the route it

follows. Sometimes it travels more or less directly from the northwest to reach the country as a cool or cold, showery northwest wind in the rear of a depression moving eastwards to the north of Scotland, when pressure is low to the north and high to the south. If the low pressure lies to the west of Ireland the air may travel a considerable distance to the south before reaching Britain. It then arrives as returning mP air, and is almost indistinguishable from mT air near ground level. It will be cooler above, and when the lower stable layer is eliminated in summer, by surface heating, convection may occur leading to showers. The mP air normally reaches Britain in the rear of the frontal zone of a depression, forming the cold air in contact with the warm mT mass ahead of the cold front. Therefore, under different synoptic situations, Britain can receive six very different air masses. This explains the great range of the British climate, despite its moderate mean values of temperature and precipitation.

Exercise 3.2. Air mass recognition

Figure 3.2 illustrates a number of synoptic situations representative of the different air masses that can influence Britain. The nature of the weather depicted on the maps and the pressure patterns allows the identification of the air mass involved on the occasions illustrated, all of which are drawn from the exceptional year of 1976. Name the air mass concerned and outline the weather characteristic of it.

The six air masses that can reach Britain represent all the major northern hemisphere air masses, with the exception of the equatorial air mass that occupies a narrow belt just north of the equator. The thermal equator lies north of the geographical equator so this air mass is usually mainly in the northern hemisphere even in the northern winter when it extends furthest south. The greater amount of land in the northern hemisphere and the

Figure 3.2 Examples of air masses and associated weather from the Daily Weather maps of 1976. Isobars, temperature, wind and weather are shown. (a) 24 January, 0600. (b) 31 January, 0600. (c) 28 November, 1800. (d) 8 March, 1800. (e) 25 November, 1800. (f) 28 June, 1800.

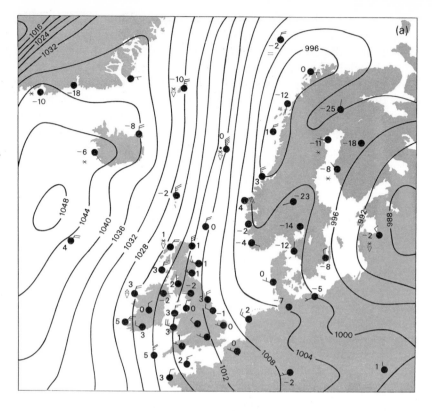

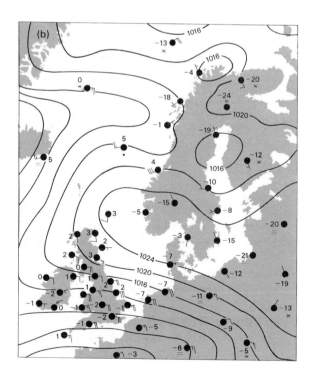

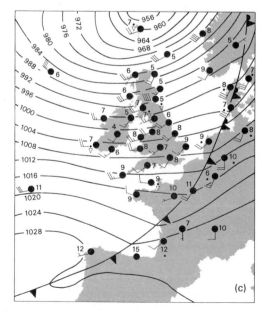

110

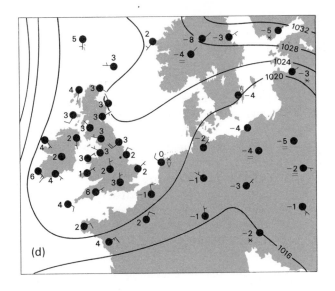

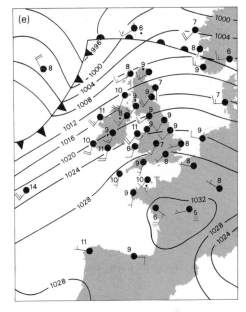

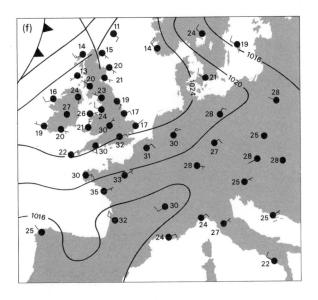

very cold southern continent of Antarctica explain this position of the thermal equator. The air masses of the southern hemisphere are similar to those of the north, although the continental ones are restricted in area.

Another important influence of the characteristics of the air masses is the seasonal one, an influence that increases in importance away from the equator. It is important in Britain because the wide range of air masses is affected by the seasonal swing of the pressure systems.

Britain has four distinct seasons, whereas further north and towards the continental interior the durations of spring and autumn become very short, and the main seasons are more extreme. In the highest latitudes this is the result of continuous darkness in winter and perpetual light in summer, while in the continental interior there is no maritime influence to temper the continental extremes of temperature. The mean range of temperature between summer and winter in England is about 11 °C, from 5 °C to 15 °C, compared with a range of nearly 42 °C at Schefferville in Quebec, from −24 °C to +17 °C, and even greater ranges in Siberia. The latitude of Schefferville is similar to that of northern Britain, but the climate is very different, one being an east coast continental climate and the other a west coast maritime one.

Winter weather in Britain is particularly variable. Mild winters occur when westerly winds prevail, and these are usually wet, as a constant procession of depressions passes eastwards across the country, bringing an alternation of damp mT and mP air masses. On the other hand a blocking anticyclone can become established over Scandinavia, bringing prolonged cP or cA air from the east over Britain, with prevailing very cold conditions. This situation is difficult to dislodge once it is firmly established. It often continues throughout spring, giving the easterly winds and cool conditions with low precipitation that are often typical of spring in England. Sometimes, however, the anticyclone shifts to draw in air from further south, bringing markedly warm weather with the cT air mass.

Summers are often dominated by the westerly air masses, with cool and changeable conditions, although the depressions tend to be less deep and precipitation less heavy, except for the occasional thunderstorm. The conditions under which summer thunderstorms can occur as mT air replaces cT air have already been mentioned. Occasionally, however, an anticyclone becomes established for a long period and drought can occur in Britain. The period from the winter of 1975 to the autumn of 1976 was the driest on record since the eighteenth century in some areas in the south of England, with only 72 cm of rain being recorded at some stations. Once the air has become thoroughly dry and the ground moisture used up there is little water available to moisten the air, so precipitation is much lighter than when the ground and air contain more moisture. One reason for the drought of 1976 was the movement of the jet stream north to Iceland which allowed a ridge of high pressure to develop over Britain. The pattern was maintained by feedback as the Atlantic temperatures increased. The final breakdown of the drought was associated with a cold outbreak over Canada that caused the ridge to retreat and the jet stream to move south to its normal position. This allowed a low pressure system to become established. The autumn rains were particularly heavy owing to higher temperatures over the Atlantic allowing the air to hold more moisture. There is no evidence of climatic change, but feedback relationships are important.

August is usually a wet month, partly because of the liability to heavy thunderstorms and partly because at this time of the year the sea is becoming warmer and so evaporation from the water surface allows more moisture to enter the atmosphere, with heavier rain as a result. In this month the amount of heat arriving on the earth from the sun is decreasing and cloudier conditions than those of July are common. The autumn is variable, some seasons being characterized by quiet, foggy conditions, especially in November, while others are stormy, with much rain as depressions follow one another across the country.

The weather in Britain is variable from season to season and at any one season it varies greatly from year to year. There is also considerable variation in different parts of Britain. Differences are greatest between north and south in summer, and between east and west in winter. The isotherms tend to run north–south in winter and east–west in summer, owing to the influence of continental conditions in the east in winter especially.

Frontal weather

When two air masses with different characteristics come into contact, a plane of separation divides them from each other; this plane is known as a front. At the surface of the earth it can be represented by a line on the weather map, but it also continues upwards as a plane, sloping in such a way that the colder, denser air underlies the warmer, lighter air. The frontal plane normally develops a waveform, and under some conditions the waves increase in amplitude and a circulation around a developing low pressure centre is initiated. These are the depressions that bring much of the precipitation to areas in middle and high latitudes. In Britain the front commonly separates mP air on its cold side from mT air on its warm side. The depressions of middle latitudes are mesoscale features, having a diameter when fully developed of 1,500 to 2,000 km and a life span of 4 to 7 days. The depressions move generally eastward as they develop; a warm front advances ahead of a cold front, as first warm air penetrates northwards in the growing wave to be replaced later by the cold air at the cold front. The cold front moves rather faster than the warm and so overtakes it and lifts the warm air off the ground to form an occlusion. The frontal surface is rarely very sharp as the warm front passes, often spreading over a distance of 100 to 200 km in which there is a large temperature gradient. The cold front is sometimes more sharply defined and as it passes cold air can replace the warm over a short period, and clearing skies follow the usually dull, humid and cloudy weather of the warm sector.

The occlusion can be either warm or cold,

Plate 26 Cirrus cloud. The high level wispy cloud usually indicates an advancing depression, as it precedes the arrival of a warm front, as warm air rises up over the colder air ahead of it.

depending on the relative temperature of the air ahead of the warm front and that behind the cold front. From ground level depressions can often be recognized by the type of cloud and weather conditions associated with the frontal types. Ahead of the warm front the frontal surface slopes up ahead of its ground position, and the warm air rises over the cold. The rising warm air causes condensation and cloud forms, firstly high cirrus cloud, then cirro-stratus and stratus, from which gentle rain will fall except in winter when snow may fall. The frontal surface inhibits instability so sometimes the first sign of an approaching depression is the flattening of cumulus clouds that had formed in the unstable cold air ahead of the warm front. As the warm front approaches rain may intensify, but it eases off or stops as the front passes, giving mild, damp conditions in the warm sector in winter and perhaps clearing skies in summer. The cold front is usually steeper than the warm front, as the cold, dense air pushes under the warm air, forcing the warm air to rise sharply. This situation is unstable and results in heavy, showery precipitation. It is often short-lived and is followed by clearing skies and showery weather, often with a brisk northwest wind and good visibility in the unstable mP air.

Depressions go through a cycle of development and decay, often starting as gentle waves on a trailing front, intensifying and developing well-defined warm and cold fronts, and then becoming occluded as the warm air is lifted off the ground (see Figure 3.3). Eventually the fronts die away as the air becomes thoroughly mixed and all that remains of the depression is a weakly circulating area of low pressure. Depressions often form families, a new one developing on the trailing occluded front of the mature one. Under these conditions each member of the family tends to be centred south of its predecessor. Each depression is an individual and differs slightly from the others, some are more vigorous and their intensity is associated with the degree of contrast between adjacent air masses. They tend to be more vigorous in winter when the contrasts tend to be greater and the general atmospheric circulation in the middle latitudes is stronger. (See Plate 26.)

Exercise 3.3. Weather map analysis

Depressions and their associated fronts must be thought of as three-dimensional phenomena, since the fronts are planes extending up into the air. The weather map on Figure 3.4 gives the pressure, temperature and barometric tendency, present weather and wind direction and strength.

Use the pressure readings to draw isobars; isallobars, which are lines of equal barometric tendency, can be drawn by using the barometric tendency values. A vigorous frontal system is influencing the weather. Insert the fronts by reference to the temperature, pressure and other data provided. The analysis of the weather situation can be carried into the third dimension by plotting (T/ϕ)grams for the four stations, Lerwick in the Shetland Islands, Stornoway in the Hebrides, Valentia in southwest Ireland and Liverpool. The upper air data in Table 3.1 provide evidence of the characteristics of the air masses involved in this weather situation. They also enable the slope of the warm front to be measured, as it appears very clearly on the Stornoway (T/ϕ)gram.

Depressions form along the Polar front in specific zones, of which that part of the Atlantic off eastern North America and the ocean off eastern Asia are particularly important. These zones vary, tending to move south about 10 degrees in the northern winter. In the northern summer other frontal zones develop over Eurasia and middle North America, owing to the low pressure over the land masses in summer. There is also an Arctic frontal zone lying roughly along the north coasts of North America and Eurasia. The Canadian Arctic front tends to move south in the northern winter. There is also a southern Polar front at around 40 to 50 °S in the southern winter and slightly further north in the southern summer.

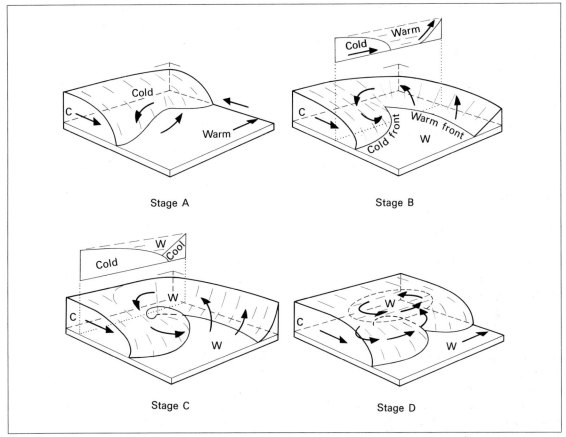

Figure 3.3 Block diagrams to illustrate stages in the development of a depression in the northern hemisphere. (a) early development, (b) mature stage, (c) occluded stage, (d) decaying stage. (After R. G. Barry and R. J. Chorley, 1976)

The depressions associated with the southern fronts tend to move in a more or less constant sequence from west to east, as this frontal belt occurs round the continuous, stormy southern ocean. The development of frontal depressions is closely associated with the circulation of the upper air, which is a global phenomenon and will, therefore, be considered in the next chapter.

Fronts are responsible for supplying much of the precipitation of the middle and high latitudes, apart from the convectional precipitation associated with the unstable air masses. The orographic effect on rainfall must be triggered either convectionally or frontally in most instances. Frontal rainfall, especially that associated with the warm front, supplies the continuous, steady downpour that usually provides more of the total precipitation than the shorter-lived, though often heavier, precipitation associated with thunderstorms in temperate oceanic regions, such as the British Isles.

3.3 Hydrology

Introduction

The importance of water in all fields of physical geography has already been stressed. An international hydrological decade was set up, from 1965 to 1975, for the purpose of filling some of the gaps in our knowledge

Figure 3.4 Weather map for 4 January, 1949, 0600.

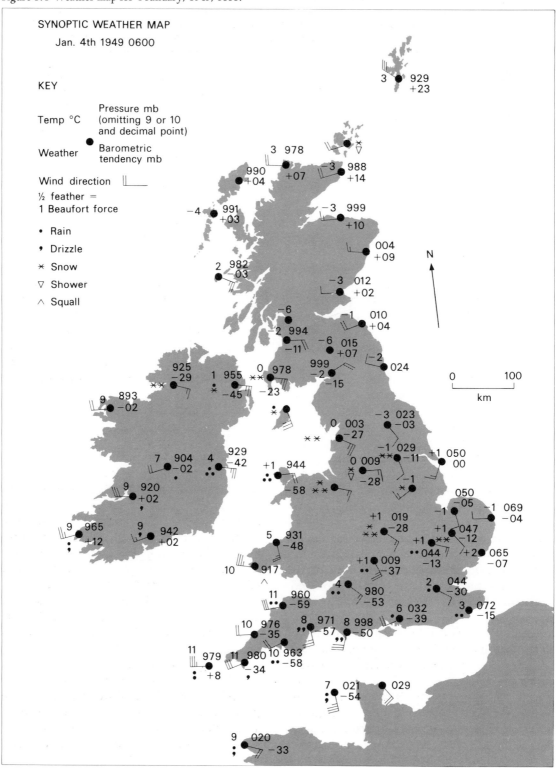

Table 3.1 Upper air data for 4 January 1949

Height	Lerwick 0300 Temperature (°C)	Stornoway 0900 Temperature (°C)	Liverpool 0900 Temperature (°C)	Valentia 0300 Temperature (°C)
1,000	+ 2	− 2	+ 1	+10
950	+ 1	− 1	0	+ 7
900	− 3	− 4	− 1	+ 4
850	− 7	− 5	− 2	+ 2
800	−11	− 7	− 4	0
750	−13	− 9	− 6	− 2
700	−17	−15	− 8	− 6
650	−21	−19	− 9	− 8
600	−26	−24	−14	−10
550	−30	−27	−18	−16
500	−36	−25	−23	−21
450	−43	−31	−30	−27
400	—	−36	−37	−34

about water. The length of time allocated to this world-wide collection of data concerning water illustrates the importance of long-term records in many aspects of hydrology, including rainfall, river flow and the behaviour of glaciers, which have great variability.

The annual cycle of variability can be analysed for different areas, providing a regional approach to hydrology in terms of annual river regimes. The units are months on this scale, and the areas involved include quite large drainage basins as well as smaller ones. The larger ones usually show a less extreme pattern when they include within their area a number of different types of regime. Diversity of volume in different parts of the same regime zone is well illustrated within the British Isles, which offers a greater diversity within a relatively small area than most other regions of comparable size. New Zealand provides another area with considerable contrasts.

There is an additional variable in New Zealand that brings in another aspect of hydrology that will be considered in this chapter. This is the annual pattern of variation associated with water in its solid state, the mass balance of glaciers. Ice masses in different parts of the world display striking differences in their annual rhythms, and any one glacier can vary greatly from year to year.

The mass balance of glaciers is relevant beyond hydrology; it is significant in geomorphology through its impact on glacial processes. Water supply problems are also involved, as glaciers in many parts of the world provide an important source of water during the warm season of the year when they melt most.

Annual river regimes

Some relationships concerning relief and rainfall can be studied on the medium scale by comparing hydrological data provided by the Surface Water Yearbook of Great Britain. The examples in Table 3.2a (p. 120) illustrate three very different conditions. The first two drainage basins have nearly the same area of 855 km² yet their hydrological characteristics are very different. The first drainage basin is part of the River Stour in East Anglia. The highest point of the basin is 122 m and its outlet is at GR 62/042340. The average discharge of the stream is 1·67 m³ s⁻¹. The values given for mean and gross mean illustrate the amount of water that is abstracted for use, the gross being the estimate of natural runoff. This value is given converted into cm/unit area. The year given, 1955–56 water year,

117

provided almost the mean total precipitation of 60·2 cm, compared with the long-term average of 61·5 cm.

Exercise 3.4a. Annual rainfall and runoff

Plot the runoff and rainfall month by month to illustrate the hydrological pattern of the water year. The figures illustrate the very low proportion of the rainfall that runs off in the dry, eastern part of England. The annual rainfall of 61 cm only produces 10·2 cm of runoff. The extreme values show that the monthly means hide a considerable variation, from 21·3 m³ s⁻¹ to 0·16 m³ s⁻¹. The extreme variability of river flow makes control and full use of river water difficult, especially in the southeastern part of the country, where precipitation is much lower, and runoff is a much smaller proportion of it.

The River Beauly in the north of Scotland illustrates conditions at the opposite extreme of British conditions. The basin is of almost identical size as the Stour basin, with an area of 850 km². Its highest point is 1,182 m and the gauging station is situated at 44 m at GR 28/426405 in the county of Inverness. The relief is great and the average discharge is 44 m³ s⁻¹, the highest recorded on 17 February 1950 was 476 m³ s⁻¹ with a minimum of 3·5 m³ s⁻¹ in July 1952. The year illustrated was rather wetter than the long-term average, with annual amounts of 229 cm precipitation in 1955–56 and an average of 215 cm. By plotting the rainfall and runoff month by month the relationship between the two can be appreciated. The close response of the two is apparent; even in the warmer summer months runoff is maintained at a high level in the cooler, wetter northwest.

The River Dove in Derbyshire provides an intermediate situation. Plot the data in a similar way and examine and explain the relationships. All three graphs show that there is a strong relationship between the rainfall and runoff, but each station has a different response between the variables.

Exercise 3.4b. Regression analysis of rainfall and runoff

Use each set of 12 values for separate regression analyses between runoff and rainfall, using the rainfall as the independent variable, X, and runoff as the dependent one, Y. Calculate the correlation coefficients for each set of data. A detailed analysis of this type illustrates a number of significant facts concerning the relationships between rainfall and runoff, with respect to different environmental conditions. The variety within such a small area as Britain is notable, while the variety from year to year is apparent, as well as the seasonal effects. August was particularly wet in the year under consideration, but apart from this abnormality the usual pattern of river flow in Britain is shown. Flow in Britain has a winter maximum and a summer minimum, because although precipitation is normally fairly evenly spread throughout the year, evaporation is always much greater in summer. This compensates for the normally drier conditions in spring, when evaporation is low. 'February fill-dyke' is an expression that depends much more on lack of evaporation than on heavy precipitation, and August is often one of the wetter months.

All the rivers of Britain have certain elements in common. They usually have a period of high flow in the winter when heavy precipitation is not counteracted by high evaporation. This is referred to as a simple pluvial oceanic regime. Simple river regimes have only one peak of flow during the year and one season of low flow. This pattern can result from a variety of climatic conditions. The simple regimes apply mainly to fairly small drainage basins that are influenced predominantly by one type of climate.

In large river basins tributaries may come from a variety of different climatic zones, providing large discharges at different seasons so that the flow is generally more even over the year. A regime is called complex in the first degree if the river has two peak flows and two low flows during the year. In larger and more complex basins there may be three or more peaks during the year, providing a regime complex in the second degree. An analysis of the pattern of river flow can thus give useful information concerning climatic controls.

Exercise 3.5. Annual river regimes

Table 3.2b illustrates some of the possible river regimes, including all three categories mentioned. Plot the figures which give the monthly coefficients to provide a visual impression of the annual cycle of river flow. The coefficient is obtained by dividing the mean monthly flow by the mean annual flow, allowing direct comparison of different sized basins. Name the type of regime and look up the location of the station in an atlas.

The simple regimes include climates that provide one major peak of flow. One is the glacial regime, where maximum flow occurs in the warmest season when glacier ablation reaches its maximum, while during the winter the precipitation falls as snow and does not run off. This regime produces a high summer maximum. It is the opposite of the oceanic pluvial regime, which has already been exemplified in the British rivers, where flow is low in summer and high in winter, despite fairly uniform precipitation. Another simple regime is the tropical rainfall type, which has one rainy season, usually in summer in the northern hemisphere. The Mediterranean area produces a winter maximum and very low flow in the dry summer season. The Blue Nile at Atbara illustrates the extreme tropical regime, with no runoff at all for seven months of the year and only three months with appreciable flow. The various snow regimes respond to the main snow melt season, being earlier in lower latitudes and lower elevations. The mountain snow regime has a maximum in early summer, June in the northern hemisphere, while the plain snow regime has an earlier maximum, usually in April or May.

The first degree complex regimes occur when there is an early maximum through winter rain, where a later maximum is produced by melting snow from high level tributaries. Some Mediterranean regimes produce double maxima related to heavy autumnal rains, a second peak being produced by late winter rains, or in the mountain areas melting snow may provide a spring maximum. The central continental type applies in central Europe and the Appalachian region of America. Here there is an early spring maximum, partly due to melting snow, while the dry autumn, following a very hot summer, leads to low flows at this season. Some equatorial rivers produce two peaks where the rains occur twice a year, because the thermal equator passes across the region twice with the seasonal oscillations. The monsoon climate in some areas can produce similar double peaks, for example in Japan, where the monsoon brings rain at two seasons.

The regimes that are complex in the second degree include large rivers that have tributaries coming from a variety of different regimes. The Rhine provides an example. It has glacial connections in some of its upper tributaries, the main stream draining from the Rhine glacier has a typical glacial regime. Other tributaries produce mountain snow regimes, while in its lower course tributaries bring in the pluvial type, such as the Moselle.

The Mississippi and Missouri also illustrate the more complex regime, differing from the European type in having a high summer flow due to rain at this season, despite the heat. The Rocky Mountains provide a source of glacial and snow melt water in spring and early summer, giving peaks in April and June at Kansas City, the June snow melt adding to the summer rain. There is low flow during the winter season. The Mississippi has several pluvial sources as well as the snow melt and glacial ones. The Ohio provides a pluvial and pluvial–snow–oceanic regime, and the Arkansas and Red River provide a maximum in May. At Baton Rouge, near the mouth, the time taken for the flow to pass down the river

Table 3.2a Monthly hydrological data for British rivers

Month	Discharge (m³ s⁻¹)	Discharge (km² m⁻³ s⁻¹)	Runoff (cm)	General rainfall (cm)		Extreme discharges (m³ s⁻¹)	
				1955–6	1881–1915	Max.	Min.

River Stour: drainage area 855·36 km²							
Oct.	1·819	0·00213	0·838	10·41	6·60	14·043	0·210
Nov.	0·195	0·00023	0·432	1·78	5·84	0·437	0·159
Dec.	0·277	0·00032	0·635	4·32	6·10	1·424	0·159
Jan.	5·694	0·00666	2·489	8·38	4·32	19·746	0·159
Feb.	3·762	0·0044	1·803	1·78	3·81	21·306	0·190
Mar.	0·900	0·00105	0·965	1·78	4·57	4·596	0·159
Apr.	0·326	0·00038	0·483	2·29	3·56	1·424	0·159
May	0·035	0·00004	0·305	1·78	4·57	0·357	0·210
Jun.	0·220	0·00026	0·254	5·33	5·59	0·270	0·210
Jul.	0·226	0·0012	0·254	6·60	5·84	0·644	0·210
Aug.	0·369	0·00043	0·406	11·68	5·84	1·711	0·210
Sep.	0·387	0·00045	0·381	4·06	4·83	1·331	0·210
Year	1·196	0·0014	9·220	60·20	61·47		

River Beauly: drainage area 850·18 km²							
Oct.	58·971	0·0690	19·761	27·43	19·81	93·337	19·065
Nov.	25·873	0·0300	9·296	9·14	22·61	48·796	10·242
Dec.	71·038	0·0840	28·651	41·15	26·16	278·026	22·554
Jan.	68·626	0·0810	21·336	27·43	24·64	211·640	29·789
Feb.	54·584	0·0640	15·113	15·75	19·30	163·128	14·525
Mar.	44·030	0·0520	12·598	10·67	18·29	246·535	12·057
Apr.	26·356	0·0310	7·341	11·94	13·46	74·613	10·809
May	34·753	0·0410	11·633	17·78	11·94	72·627	11·348
Jun.	23·547	0·0280	5·512	8·89	10·67	44·257	13·476
Jul.	25·902	0·0300	9·246	18·80	14·73	143·836	11·178
Aug.	44·910	0·0530	15·367	23·11	16·76	143·269	20·909
Sep.	32·739	0·0385	9·881	18·03	16·76	77·734	17·022
Year	41·761	0·049	165·684	229·108	215·138		

River Dove: drainage area 399·2 km²							
Oct.	1·488	0·0037	0·991	7·11	10·41	5·560	0·851
Nov.	4·086	0·0102	2·642	7·11	9·40	16·483	1·038
Dec.	6·088	0·0152	4·089	10·16	10·92	18·270	0·982
Jan.	13·677	0·0343	9·169	15·75	8·38	44·825	4·511
Feb.	7·456	0·0187	5·588	3·56	7·37	17·504	5·305
Mar.	9·453	0·0237	6·350	4·32	8·13	28·937	4·511
Apr.	4·814	0·0121	3·124	4·57	6·10	6·809	3·972
May	3·092	0·0077	2·083	2·79	6·86	3·915	2·230
Jun.	2·743	0·0069	1·778	7·87	7·11	3·546	2·230
Jul.	4·814	0·0121	3·226	13·46	8·64	20·370	2·014
Aug.	17·263	0·0432	11·582	20·83	9·65	56·173	7·064
Sep.	10·800	0·0271	7·010	8·89	6·86	30·384	5·447
Year	7·280	0·0042	57·582	106·43	99·82		

To convert cusecs (cubic feet per second) to m³ s⁻¹ multiply by 0·0283726.
To convert square miles to square kilometres multiply by 2·592.

120

Table 3.2b River regimes

Month	1	2	3	4	5	6	7	8
Jan.	0·31	0·10	0·17	1·58	1·40	0·00	0·21	0·21
Feb.	0·28	0·09	0·13	1·95	1·59	0·00	0·14	0·18
Mar.	0·28	0·09	0·22	1·50	1·89	0·00	0·10	0·19
Apr.	0·45	0·14	0·48	1·01	1·28	0·00	0·07	0·43
May	0·89	0·65	1·27	0·85	0·98	0·00	0·09	1·62
Jun.	1·78	1·90	2·18	1·57	0·66	0·00	0·25	2·70
Jul.	3·14	3·14	2·24	0·42	0·58	1·94	1·15	2·48
Aug.	3·78	3·16	2·13	0·34	0·52	6·00	3·42	1·69
Sep.	1·55	1·76	1·40	0·38	0·53	3·31	3·47	1·08
Oct.	0·79	0·51	0·76	0·76	0·57	0·75	1·95	0·71
Nov.	0·43	0·22	0·45	1·14	0·79	0·19	0·76	0·42
Dec.	0·37	0·13	0·25	1·56	1·22	0·00	0·37	0·27

Month	9	10	11	12	13	14	15	16
Jan.	0·33	0·73	0·96	0·84	0·68	2·54	2·83	2·79
Feb.	0·29	0·71	1·26	1·23	0·69	2·52	2·89	2·79
Mar.	0·46	0·95	1·55	1·43	0·93	2·69	2·95	2·71
Apr.	3·81	1·35	1·29	1·61	1·15	3·03	2·80	2·39
May	2·91	1·60	1·03	1·38	1·63	3·43	2·78	2·28
Jun.	0·82	1·52	0·33	1·09	1·77	3·85	2·99	2·48
Jul.	0·66	0·69	0·21	0·91	1·03	3·79	2·88	2·37
Aug.	0·55	0·40	0·18	0·81	0·64	3·60	2·67	2·08
Sep.	0·52	0·38	0·77	0·65	0·56	3·32	2·47	1·85
Oct.	0·63	0·92	1·32	0·62	0·88	3·00	2·38	1·76
Nov.	0·47	1·45	1·17	0·70	1·10	2·76	2·51	2·00
Dec.	0·41	1·28	1·92	0·63	0·92	2·68	2·85	2·77

1 = Arve at Chamonix
2 = Matter Visp at Randa
3 = Veneon at Bourg-d'arud
4 = Saone at Auxonne
5 = Weser at Hameln
6 = Nile at Arbara
7 = Blue Nile at Khartoum
8 = Reusz at Andermatt
9 = Volga at Wiazowaia
10 = Verdon at Quison
11 = Trebbia at Rivergaro
12 = Oder at Pollenzig
13 = Durance at Mirabeau
14 = Rhine at Kehl
15 = Rhine at Coblenz + Moselle
16 = Rhine at Emmerich

Values for rivers 1 to 13 inclusive apply to monthly coefficients.
Values for River Rhine, 14, 15 and 16 give mean water level in metres.

is seen in the delayed maximum, which now occurs in April and May. The minimum occurs in September and October. On the whole the regime shows greater contrasts than European rivers, although they are both predominantly pluvial–snow types in their main characteristics.

Very long rivers, such as the Nile, especially those flowing mainly north–south, cross several climatic zones so that their tributaries bring in water at several seasons. The Nile in fact receives water from both sides of the equator—it crosses the Sahara desert and enters the Mediterranean zone. The large lakes in its upper course moderate the flow. In Egypt there is a minimum in May and a maximum in September owing to the domination of the Blue Nile coming from Ethiopia. The river provides the annual autumn flood.

Glacial systems

Water in its solid state forms a significant part of the hydrological cycle on the medium scale. The annual regime of the valley glacier

Plate 27 The Jostedalsbreen in central Norway is a plateau ice cap. It is the last remnant of the large Scandinavian ice sheet and drains through a series of outlet glaciers to the fjords below. The upper part of the ice cap is in the accumulation area of the system and snow still covers the surface in late summer.

exemplifies the glacial hydrological system on this scale. The glacial hydrological system forms a link between the climatic system and the glacial system, as well as directly affecting other aspects of the hydrological cycle, such as the river regimes in glacial basins. Because the climate varies from year to year and place to place, glaciers are never constant in size. Their masses oscillate in response to climatic variations, although with a time lag that depends on the size of the glacial system involved. The time lag varies from 10 to 30 years for a normal valley glacier to several thousand years for the major ice sheets, such as the Antarctic ice sheet.

The response of a glacier to climatic variations depends both on the nature of the climatic fluctuation and on the nature of the glacier itself. Those aspects of the glacier that are important from this point of view include its thermal characteristics and its morphology. Ice can be classified as cold or temperate. A cold glacier is one that has a temperature that is everywhere below the pressure melting point, while a temperate one is at its pressure

melting point throughout. The situation is complicated by the fact that many glaciers have both properties in different parts. So it is often better to refer to the ice rather than to the glacier as a whole. The important distinction is that cold glaciers are frozen to their beds, while temperate ones have melt water at their bases and can, therefore, slip over their beds on watery films, thus moving more easily.

From the morphological point of view the relative area of the glacier surface at different elevations is of particular importance in its response to climatic variations. A glacier with a relatively large area at higher elevations is more likely to advance than a glacier that has a large area at a low level. This is because snow accumulation generally increases upwards, while ablation increases towards lower levels.

The balance between accumulation and melting (or ablation) is critical in explaining the way in which a glacier behaves. The balance is usually called the mass balance or budget of the glacier system; the health of the glacier and whether its snout will advance or retreat depends on it. A healthy glacier with a

Plate 28 Austerdalsbreen drains southwards from Jostedalsbreen through an ice-fall, the lower part of which is seen in the foreground. The glacier below shows well developed ogives and a conspicuous medial moraine. A trim line is seen on the far side of the glacier, which has cut deeply into the gneiss that forms the walls of the glacial valley.

Plate 29 Briksdalsbreen drains northwards from Jostedalsbreen as a short, steep outlet glacier. It ends in a moraine dammed lake. It responds rapidly to changes in mass balance and has left a conspicuous trim line, exposing its lateral deposit of till. The glacier has also cut deeply into the gneiss.

positive mass balance will advance, while one with a negative mass balance will retreat or decay. The mass balance is the difference between accumulation and ablation measured over one hydrological year, which starts at the end of the ablation season in autumn. The glacier has its minimum mass at the beginning of the budget year and during the early part of the budget year, during the winter, mass is added in the form of snow, while in the later part of the budget year melting causes loss of mass. (See Plates 27, 28 and 29.)

From the geomorphological point of view in particular there is an important distinction between the net glacier budget and the gross glacier budget. The net budget is the difference between the accumulation and ablation of the glacier system. This determines whether the glacier will advance, stagnate or retreat.

The gross budget is the sum of accumulation and ablation. Upon it depends the relative activity of the glacier. At one end of the scale are the very cold glaciers of the polar regions, where very low temperatures limit the amount of moisture the air can hold, and hence snowfall is light, and accumulation small. The low temperatures also limit ablation, so that losses are small and largely achieved by calving where glaciers reach the sea, as in Antarctica, the Canadian Arctic and Greenland. Glaciers of this type have small gross budgets and they are relatively inactive and move only slowly as a result. For instance one such glacier in East Baffin Island moved at only about 3·5 m/year. Their geomorphological activity is further reduced by their thermal character, in that they are often cold-based glaciers, except where the ice is very thick. This does not apply to valley glaciers of the type under discussion.

At the other end of the scale are the well nourished glaciers of the more temperate areas, including southern Iceland, central Norway, western USA and western New Zealand. These areas all lie in the westerly wind belt and are fed by plentiful precipitation, much of which falls as snow at the higher elevation in winter. Height of the glacier system is, therefore, very important in those glaciers that lie near the regional snow line. They receive heavy snowfall in winter. For example, in Iceland a pit dug on the ice cap of Vatnajokull 6 m deep only penetrated $2\frac{1}{2}$ years of snowfall, which had already become quite dense. On the glacier snout dirt cones nearly 8 m high indicated that this was about the amount of annual ablation taking place near the glacier snout. Ice melted at the rate of several centimetres each day in midsummer. In order to maintain the ablation zone very rapid flow is necessary in these glaciers to balance the high gains and losses. These glaciers are geomorphologically active, and being temperate they usually have melt water at their base which further enhances their capacity to flow. Glaciers of this type move at about 1 m/day, often flowing faster in summer and rainy weather when the amount of water at their bases is increased.

Amongst the fastest flowing glaciers are those of the west coast of New Zealand, such as the Franz Josef glacier that moves at up to 5 m/day. It has the advantage of a very wet climate, relatively high temperatures and a very steep gradient. It flows from the highest area of the Southern Alps at about 3,000 m to within the rain forest near the west coast. This glacier contrasts strongly with those on the east side of the Alps lying in the rain shadow so that they receive much less precipitation. The eastern glaciers move sluggishly, their snouts are static and covered by a thick layer of superficial moraine. The western glaciers are clean on the surface and the positions of their snouts oscillate rapidly, particularly with variations in precipitation. There is a five year lag in the case of the Franz Josef glacier.

The importance of the gross budget of glaciers has been recognized in the development of indices of glacier activity. These indices are based on the variation of ablation with height, and one is called the ablation gradient.

The vertical gradient of the net balance curve at the equilibrium line altitude is a measure of glacier activity. b_n, the net glacier balance, is zero at the equilibrium line altitude and the activity index is d_m/d_e in mm/m. This is the ratio of the elevation change to mass change. For a unit of elevation, if the mass change is large, the gradient is low and the glacier is temperate in type. Where the mass change is small the gradient is steep and the glacier polar. In the former case there is a large change from high positive balance above the equilibrium line altitude, to a large negative change below it, while a small change occurs in the latter case. The values range from 1 mm/m in high polar glaciers to 10 mm/m for those in temperate maritime areas.

Polar glaciers tend to have a steeper gradient than temperate ones, because there is less variation with height on a polar glacier

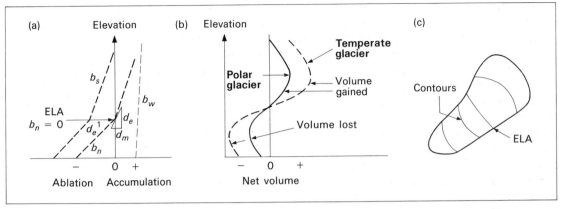

Figure 3.5 Diagram to illustrate glacier activity index. b_s = summer balance, b_w = winter balance, and b_n = net balance. ELA = equilibrium line altitude. Net balance line gradient at ELA = d_m/d_e in mm/m^{-1}. (After J. T. Andrews, 1975)

owing to the overall lower temperatures and precipitation. The contrast is illustrated in Figure 3.5. The ablation gradient is closely associated with mass loss at the equilibrium line, which is the line separating the accumulation zone above it from the ablation zone below it. Maps have been produced to indicate the distribution of glacier activity as established by means of these criteria. They help to explain geomorphological change due to glacier activity. (See Plate 30.)

The relationship between glacier activity and mass budget is yet another example of the connections between different aspects of physical geography. In this instance the hydrological aspect of water in its solid state relates closely to geomorphological aspects of glaciation. Features of active glacier erosion, such as deep U-shaped valleys, are associated with glaciers having a high activity index. Sluggish glaciers with a low activity index can protect the surface on which they rest. Examples of the former occur in western Scotland and Norway, and of the latter in eastern and central Baffin Island and north Greenland. (See Plate 31.)

Changes in glacier activity vary over a fairly long time scale. From year to year each individual glacier changes relatively slowly, and there is often less year to year variation on any one glacier than differences between glaciers in the same area. This fact indicates the importance of the area–height pattern of the individual glaciers in their response to climate in their net budgets. This point is illustrated by reference to two examples of small glaciers in northern Norway that are situated close together. Table 3.3 gives figures for the different elevation zones of the two glaciers in steps of 50 m. One glacier, Trollbergdalsbreen, covers a range from 900 m to 1,300 m. The areas within each height range are given as well as the winter and summer net balances in volume of water equivalent, thickness of water equivalent and the amount per km^2 for each height range.

Exercise 3.6. Glacier budgets

The values given in Table 3.3 are plotted in Figure 3.6. The total net mass balance for the glacier in volume water equivalent is given as -0.735 million m^3. The reason for the negative balance is apparent when the area–height curve for the glacier is examined. The bulk of the glacier area lies between 1,000 m and 1,100 m, which is within the zone where the summer balance of ice loss is large and the winter gain small. Plot the values for Engabreen in a similar form to those for Trollbergdalsbreen from the values given in Table

Plate 30 Glen Docherty, Wester Ross, Scotland. The glaciated valley shows the typical U-shape, with Loch Maree in the distance. In the foreground a stream has dissected recent morainic deposits to reveal the characteristic unsorted and unstratified nature of the morainic till. The hummocky nature of the moraine surface is also seen in the middle distance.

Plate 31 Kirkstone Pass, English Lake District. The hummocky ground in the centre of the picture is moraine of the late-glacial ice advances. Ice passed through the col at the head of the Pass eroding the valley and depositing the moraine. Note the large erratic boulders on the moraine surface.

Figure 3.6 Winter, summer and net balance in relation to elevation on Trollbergdalsbreen in 1974. The areal distribution and areal net balance for every 50 m elevation interval are shown. (After G. Østrem, 1975)

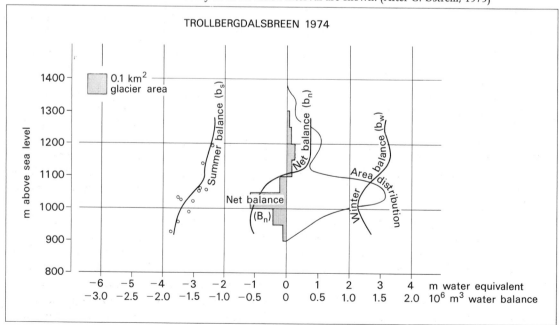

Table 3.3 Glacier balance data

Trollbergdalsbreen 1974

Height (m)	Area (km²)	Winter balance		Summer balance		Net balance	
		B^w ($\times 10^6$m³)	b^w (m)	B^s ($\times 10^6$m³)	b^s (m)	B^n ($\times 10^6$m³)	b^n (m)
1,250–1,300	0·005	0·016	3·13	0·012	2·38	0·004	0·75
1,200–1,250	0·080	0·250	3·14	0·190	2·38	0·060	0·76
1,150–1,200	0·173	0·559	3·23	0·427	2·47	0·132	0·76
1,100–1,150	0·128	0·409	3·19	0·336	2·63	0·073	0·56
1,050–1,100	0·556	1·400	2·52	1·521	2·73	−0·121	−0·21
1,000–1,050	0·615	1·397	2·27	1·991	3·24	−0·594	−0·97
950–1,000	0·199	0·472	2·37	0·698	3·51	−0·226	−1·14
900– 950	0·063	0·165	2·62	0·228	3·63	−0·063	−1·01
900–1,300	1·819	4·668	2·57	5·403	2·97	−0·735	−0·40

Engabreen 1974

Height (m)	Area (km²)	B^w ($\times 10^6$m³)	b^w (m)	B^s ($\times 10^6$m³)	b^s (m)	B^n ($\times 10^6$m³)	b^n (m)
1,500–1,594	0·124	0·465	3·75	0·230	1·88	0·235	1·87
1,400–1,500	2·508	10·308	4·11	4·038	1·61	6·270	2·50
1,300–1,400	9·348	38·140	4·08	16·172	1·73	21·968	2·35
1,200–1,300	8·548	30·345	3·55	19·489	2·28	10·856	1·27
1,100–1,200	7·600	25·384	3·34	19·608	2·58	5·776	0·76
1,000–1,100	4·662	14·685	3·15	13·893	2·98	0·792	0·17
900–1,000	2·460	6·642	2·70	8·364	3·40	− 1·722	− 0·70
800– 900	0·940	1·701	1·81	4·080	4·34	− 2·379	− 2·53
700– 800	0·500	0·625	1·25	2·412	4·73	− 1·787	− 3·48
600– 700	0·370	0·363	0·98	2·153	5·82	− 1·790	− 4·84
500– 600	0·270	0·203	0·75	1·663	6·65	− 1·460	− 5·90
400– 500	0·210	0·054	0·25	1·715	7·94	− 1·661	− 7·69
300– 400	0·165	0·041	0·25	1·600	9·41	− 1·559	− 9·16
200– 300	0·220	0·055	0·25	2·310	10·50	− 2·255	−10·25
80– 200	0·095	0·019	0·20	1·190	12·55	− 1·171	−12·35
80–1,594	38·020		3·39		2·59		0·80

3.3. Engabreen is a larger glacier extending over a height range of over 1,500 m, from 80 m to 1,594 m. The largest area lies within the range 1,100 m to 1,400 m, well up in the zone of high winter balance and low summer balance. The result is that the total net mass is very different. Calculate the total winter balance, summer balance and net balance for the year, and explain the result by reference to the graph. Under the same climatic conditions very different mass balances occur in two neighbouring glaciers with very different relief.

Small glaciers respond to variations in climate much more quickly than larger ones, but some glaciers have an occasional unusual form of movement called a *surge*. Only a few of a large number of valley glaciers surge, and the conditions that lead to these sudden, very rapid advances are not fully understood. Some glaciers surge periodically, while others never surge and a few may surge more or less continuously. The Franz Josef glacier in New Zealand, with its very rapid flow, has been suggested as an example of a glacier that surges most of the time. Glaciers that surge have a reservoir in their upper reaches in which ice can accumulate. At a certain stage the upper ice mass becomes unstable and the rapid forward surge is initiated in the accumulation basin. This zone of rapid flow moves down the glacier, for example in the surge that affected the Tikke glacier in British Columbia. Movement started in the upper reaches in 1963, extended throughout most of the 10 km glacier the following year. The surge reached the lower half of the glacier in 1965 and the snout was affected in 1966. It advanced several kilometres down the valley quickly at this time. Surge velocities may reach between 10 and 100 times the normal ice flow, and speeds of 350 m/day have been recorded. These very rapid movements cause spectacular frontal advances, distortions of moraines on the glacier surface also occur. Intense fracturing of the ice causes a chaotic mass of broken ice, which advances with considerable noise down the valley. As the surge passes down the glacier in a few months to a few years the upper part comes to rest at a much lower level, while the snout builds up in height. After the surge the advanced snout tends to remain *in situ*, gradually melting away as a mass of stagnant ice.

One trigger that may set off some surges is the conversion of the basal layer of ice from cold to temperate ice, so that melt water suddenly lubricates the base of the glacier allowing more rapid flow. This generates heat,

Plate 32 Strangford Lough, Northern Ireland. Drowned drumlins are shown, their parallel alignment and rounded form is apparent, with the steeper stoss slope.

and thus by positive feedback, enhances the flow further until the surge is initiated. Surges occur at roughly 15 to 100 year intervals in different glaciers, being roughly periodic in any one glacier. Other triggers that have been suggested are climatic changes, or earthquakes, which dislodge unstable snow onto the glacier surface by starting avalanches. It is likely that different glaciers react differently and that surges are not all initiated by the same process. Again they illustrate the great variability amongst glaciers in their response to hydrological and other controls.

There is a relationship between the characteristics of the glacier and its hydrology, in terms both of the glacier form and its influence on the landscape. An advancing glacier tends to have a high, steep snout and a strongly convex profile, sometimes with clean ice on the surface. The extent of its activity will depend mainly on its gross budget. Advancing glaciers often plough up material in their advance, some of which is overridden, but some may be moulded into terminal moraines. There is good evidence from the valleys of Norway that terminal moraines formed during short periods of advance. Two advances occurred, in the early 1900s and 1920–1928, and moraines were formed, during a period of general glacier retreat from the historical maximum attained about 1750, during the Little Ice Age.

Valley glaciers can retreat in two distinct ways. Some continue to advance as the position of their snouts retreat upvalley, and their thickness is reduced by excessive ablation. Others stagnate and decay *in situ*. The first situation is more likely to occur in the type of glacier that is fed from a high cirque. In this situation its supply continues, even if at a reduced rate, when the snow line gradually rises, owing to its height–area relationship, which gives it a wide height range and a considerable area at higher elevations. The second situation may occur where a glacier is fed from a small plateau ice cap, such as the Jostedalsbreen in Norway and Vatnajökull in Iceland. The accumulation area is at about the same level in this situation and when the snowline rises above the plateau level the supply of excess accumulation is suddenly cut off and the glacier outlets starve and the ice decays. (See Plates 30, 31.)

Landforms associated with dead ice have been increasingly recognized. They include many fluvioglacial forms. Long winding eskers, representing the deposits left in subglacial streams, have much more chance of survival in stagnant ice. Kame terraces can build up along the margins of the dead ice and deltas can form at the snout if this ends in a water body. A steep ice-contact slope marks the position of the ice front. Such features are conspicuous in parts of Ireland, Scotland, Scandinavia and Baffin Island. (See Plates 34, 35.)

An actively retreating glacier snout is usually associated with a trim line or lateral moraine, marking its former position on the valley side, and an uneven morainic frontal area that sometimes shows evidence of the forward movement of the ice in the form of a fluted surface. (See Plates 28, 29.)

The hydrology of glaciers is influenced both by their climatic environment and their physical setting. Geological conditions also influence them to a certain extent, by controlling in some measure the amount of material they have on their surfaces, and as englacial and subglacial load. This in turn affects their albedo and their response to climatic variables. The relief influences the flow as well as the mass balance. (See Plate 32.)

3.4 Biogeography

Introduction

On the regional scale biogeography is concerned with the environment of living organisms in areas about the size of the British Isles. Within the area the basic climatic controls are roughly similar, although many minor variations occur, such as those considered in the local environment. The ecosystem includes the soil and the plants that live in it, as well as the animals that feed on the plants and each other. In this chapter a selection of different environments will be considered to illustrate the type of relationships and controls involved in the ecosystems typical of each environment.

The polar environment exerts particularly extreme conditions on its ecosystem, and the adaptations to the particular environmental factors have provided an unique living world. There is also a great contrast between north and south polar biogeography. The differences are in part related to the two environ-ments. The Arctic Ocean with the surrounding tundra lands contrasts with the glacial Antarctic fringed by ice shelves terminating in deep sea around much of its periphery so that land-based life and shallow sea organisms are restricted. Polar ecosystems are dominated by the seasonal variation between continuous darkness and extreme cold in winter, and the continuous daylight of the short, cool summer into which nearly all life activities must be concentrated.

The second example is taken from New Zealand. The isolation of this temperate land

Plate 33 (left) The sandur or outwash plain of Skeiðararjökull in southeast Iceland. The river Skeiðará is seen issuing from the glacier snout in its normal summer flow and it can be compared with Plate 57, p. 172. Moraine dammed lakes lie in front of the glacier, where retreat has been rapid in the last 50 years. The paler outwash in the valley train of Morsarjökull in the right foreground has not built up to such a high level as the darker outwash of Skeiðararjökull, hence Morsá flows along the junction of the two sandar.

Plate 34 (lower left) Kame terraces in west Baffin Island. The large boulders in the foreground are part of an ice contact deposit, and several levels of kame terraces can be seen on the opposite side of the valley.

Plate 35 (lower right) Kettle holes in the outwash deposits of west Baffin Island. Kame terraces are well developed in the valley, in which lumps of dead ice have melted to form the kettle holes. The rounded nature of the stone on the kame terrace indicate effective erosion in rapidly flowing streams.

until comparatively recently allowed peculiar flora and fauna to develop, an ecosystem in balance with its environment and within itself. The part played by the recent invasion of the islands by species first imported by the Maoris and later by European immigrants illustrates very well the dangers of upsetting a natural ecosystem. The effects of these introductions have been widespread throughout the field of physical geography, affecting the hydrology and the geomorphology as well as the biogeography. This example illustrates the need to understand the ecosystem in all its aspects before it is upset by the introduction of new species of plants and animals to which the environment is not adapted.

The third example is of another extreme environment that requires of its living organisms special adaptations to rigorous conditions. The deserts of the world illustrate this environment. They include the very cold deserts of the Antarctic, the temperate coastal deserts of eastern continental seaboards, such as north Chile and Peru, as well as the hot deserts of the tropical latitudes. The American southwest, the Sahara and central Australia illustrate these latter conditions.

They form a complete contrast to the fourth example, which is the equatorial rain forest. This is an environment in which life can flourish at its most exuberant without any controls of moisture and temperature, such as constrain the first and third examples. The tropical zone is the most prolific in creating life, but it also has its problems when exploited by man, and lacks the seasonal aspect so important elsewhere.

Arctic biogeography

The dominant life forms of the south and north polar zones differ markedly. In the south the dominant life is based on the open ocean because on the Antarctic continent itself there is practically no land life. However, some of the marine life makes use of the land for breeding purposes, although it cannot obtain sustenance from the land. The southern ocean will be considered in the next section as an example of an oceanic region of regional scale dimensions.

In this section the life of the northern polar zone will be considered. In the north large expanses of arctic tundra exist in North America and Eurasia almost completely encircling the globe in high latitudes, and extending to within a fairly short distance of the north pole.

The soils of the Arctic provide a medium in which plants must survive and respond within the short time span of the arctic summer. The typical tundra soil is characterized by poor drainage, owing to the presence of permafrost, which renders the ground impermeable due to the presence of ground ice. Only the top layer, up to 2–3 m in thickness and often much less, melts out in summer to form the active layer in which periglacial processes are active. These processes form patterned ground in frost-susceptible soils, in which the silt content is high. Silt facilitates the formation of ground ice, because water can move up into the soil with relative ease by capillary action from below. Coarser material lacks adequate capillary action and clay is too impervious.

Not all slopes in the Arctic are badly drained, and where rocks outcrop a thin lithosol or skeletal soil will form, as a result of active frost weathering. On steep, well-drained slopes, where the active layer is thick and little ground ice forms in winter, an arctic brown soil will form. On completely waterlogged areas a bog soil occurs. The arctic brown soil deserves its name as its most characteristic feature is its colour. Slight podsolization occurs in the B horizon, while the A horizon shows crumb structure at the surface. The soil is very acid on the surface with pH of about 3·5, due to leaching by melt water. The tundra and bog soils are variable, usually acid and mottled through gleying. The soil is grey-blue owing to a low rate of accumulation of organic matter and slow humification owing to the lack of organisms and anaerobic conditions. Soil horizons are generally thin.

The characteristic vegetation of these arctic areas is tundra, which occupies the zone north of the taiga and boreal forests. The tundra extends southwards to the zone where the temperature in the warmest month is 10 °C and the mean temperature of the coldest month is well below 0 °C. The taiga is a transitional zone, with some stunted trees and bushes, separating the bare tundra, with only very small dwarf trees, from the continuous coniferous forests of the boreal forest zone to the south. Tundra is a Finnish word meaning barren wasteland or hostile ground. It applies to vegetation that grows beyond the tree line either at high latitude or high altitude. In the high latitude areas permafrost is one of the factors that is related to the distribution of tundra vegetation, while at high altitude cold winds and exposure are often sufficient to prevent tree growth, and permafrost may also occur. In the high altitude Colorado Rockies the border zone forest is referred to as *krumholtz*, which literally means *crooked wood*. It refers to the distorted shape of the trees, which often grow nearly horizontally owing to the force of the wind. Wind also limits tree growth in the arctic tundra, the last remnants of the forest surviving in the sheltered hollows.

The major characteristic of tundra vegetation is its low growth and its miniature character. The trees are less than 30 cm high, including dwarf birch, *Betula nana*, and dwarf willow, *Salix arctica*. There are many beautiful flowers amongst the tundra vegetation. Carpets of white *Dryas octopetala* cover the ground in July, and even close to the glaciers the delicate arctic poppy flourishes with its yellow flowers. Pink saxifrages are common and a miniature willow herb, only 5 cm high, flourishes in some areas. Mosses and lichens are particularly common and extensive. The reindeer moss which forms the staple diet of the caribou and reindeer is of particular importance. (See Plates 36, 37.)

Lichens are the first colonizers when rocks are first exposed by ice retreat, for example. Their growth rate is very slow in the harsh

Plate 36 *Dryas octapetala* is a characteristic tundra flower. It has given its name to the first and third late-glacial periods, when it grew in now temperate areas.

environment and for this reason they provide a useful chronological time scale. Their size and cover indicate the extent of time that the ground has been ice free, for example. The study of lichenometry has developed as a means of dating in the tundra zone. The dimensions of the 10 largest thalli are measured in a sample area and when a growth rate curve has been established the age of the substrate on which they were growing can be established fairly accurately. A study of the lichen cover from aerial photographs on the uplands of Baffin Island has revealed that during the Little Ice Age the snow cover was much more extensive than at present. These conditions lasted nearly long enough to initiate a new major ice advance.

The short growing season, rarely more than two months, results in the predominance of perennial plants, and the lack of nitrogen fixing plants limits the soil fertility. Growth is very slow and it probably takes up to 50 years for a heath vegetation to develop. It is partly

Plate 37 Caribou on the tundra of northern Canada. The animals live on the reindeer moss, which is nourishing because of its high sugar content.

for this reason that disturbance of the tundra through economic developments, such as pipeline construction and other building operations, are viewed with disquiet by environmentalists. Productivity is very low, often only about 1 per cent of that in a temperate area. There are five main types of tundra plants: lichens, mosses, grasses, cushion plants and low shrubs. They all have the ability to remain dormant for long periods and possess underground runners and rhizomes. The shrubs have small leaves to withstand drought, as precipitation is low in cold climates, mostly falling as snow in winter.

The vegetation has a high sugar content, which makes it nutritious for animals, and it is surprising how many species of large animals can survive in this harsh environment. In the tundra there are very large herds of caribou in North America, which are very similar to the reindeer of northern Europe and Asia. These animals migrate seasonally and feed on the reindeer moss as they move. They live in harmony with the arctic wolves when left in

their natural state, in that the wolves cull the weak and sickly caribou, leaving the fit animals to maintain the vigour of the herd. The wolves also feed on lemmings, the small rodents of the arctic. Lemmings have a well known cycle of increasing numbers followed by a very sudden reduction of numbers. The increase in numbers puts severe pressure on the land to provide food and as a result many die. They do not, as is sometimes suggested, rush in masses to the sea to drown. Musk oxen and polar bears are other very large mammals of this region. The musk oxen are now restricted in numbers, but still survive in places on the tundra vegetation. Polar bears are also protected by conservation laws. They obtain much of their sustenance from the sea and are as at home on the polar ice floes as on land, feeding on the rich life of the arctic ocean. The animals are all well adapted to their environment. (See Plate 37, 38.)

The sea animals have an insulating layer of blubber or fat, while the land animals have extremely well-insulating coats, provided by

133

Plate 38 Tundra vegetation on the high level fjell on the Archaean gneiss of the Baltic Shield in Dovrefjeld, Norway.

short hairs protected by longer guard hairs. The caribou has tapering hairs, thicker at the end, so that they trap much air, on the same principle as a wet suit. Another mechanism to withstand the winter cold is hibernation.

Migration is typical of many arctic birds. Some of these, such as the arctic terns, travel from north polar to southern high latitudes to get the maximum light and warmth.

Temperate latitudes—New Zealand

New Zealand is an isolated island group far from the nearest large land mass. Its biogeography illustrates well the effect of prolonged isolation on the development of life forms, and also the effect of recent introductions from other temperate areas, particularly Britain during European colonization in the past 150 years.

New Zealand is nearly antipodal to Britain, although it lies in slightly lower latitudes so that the North Island can support tropical species, such as mangroves. There are some connections between the flora and fauna of the three southern hemisphere peninsulas and islands. The pattern of dispersal could have been achieved in various ways: (a) dispersal across land connections in the far south; (b) dispersal across wide ocean gaps; or (c) dispersal from northern sources through the tropics followed by die-back in the north. Dispersal by land connections does not seem possible for the late Mesozoic and Tertiary. There is no fossil evidence for interchange of terrestrial vertebrates between South America, Australia and New Zealand since the late Cretaceous. Dispersal across the water gaps is more plausible and the west wind drift links the three southern oceans. Dispersal could be by water, wind or birds. Island stepping stones would help such a dispersal route, and these include Tierra del Fuego, Tristan da Cunha, Gough Island, the Falkland Islands, the Macquarie Islands and the Crozet Islands. The third possibility could have occurred, for example, in the spread of trees, such as the *Nothofagus*, southern beech. This tree, which occurs both in South Island, New Zealand, and southern Chile, is related to the northern beech, *Fagus*, and is known in the form of fossil pollen from Kazakhstan and west Siberia in the late Cretaceous and early Tertiary, and cool tropical mountain forms are also known from New Guinea and New Caledonia.

New Zealand's only endemic land mammals are two species of bat. The fauna include large flightless birds, the extinct Moa, *Diornis*, and kiwis, *Apteryx*. The Moas were exterminated by the early Polynesian invaders. Most of the fauna probably reached New Zealand during the Mesozoic from Eurasia. The flora could have come via the ocean routes as seeds and pollen spores, using the islands as staging posts. The Antarctic could have been part of the route in earlier times before it drifted into high latitudes and its ice sheet developed during the early and middle Tertiary period. Mammals seem to have dispersed by the land routes from the north and later died back in the areas of their origin, becoming adapted to their new habitats in southern lands. Thus when Europeans first reached New Zealand they came to an area with an unique fauna and flora adapted into an efficient ecosystem in harmony with the environment.

The effects of the introduction of flora and fauna from Europe into one small area of New Zealand illustrate some of the problems associated with the disturbance of the natural ecosystem. The area lies on the east side of the Southern Alps in the South Island. It consists of the Lake Coleridge catchment and the Avoca and Harper valleys, an area of steep slopes mainly of greywacke. The area is influenced by active faulting and has been heavily glaciated, creating oversteepened slopes of easily weathered rock. The pre-

cipitation averages 82·6 cm near the mouth of the catchment, increasing to 150 cm at the head of the lake, and is probably considerably more at higher levels, although records are lacking. Dry nor-westers cause high evaporation in the lower levels, while heavy rain may be falling on the watershed. Snow and glacier ice melt help to supply summer flow in the rivers, which are important for hydroelectricity generation and water supply.

The soils are all immature as a result of recent glaciation, being skeletal in type with no profile development. They are less than 30 cm thick over wide areas, and are stony, with steep slopes predominating. The natural vegetation that existed before European occupation was very soon disturbed by firing. It can be reconstructed from isolated remnants, and consisted of low tussock grassland on the lower slopes, with high tussock grassland in the shadier and damper patches. The tussock grassland gives a close sward of tussocks and aniseed grows in their shelter, providing effective protection to the soil. The primitive grassland provided a palatable and nourishing pasture, apart from the impenetrable thickets of matagouri, a prickly shrub. The highest zones were covered by subalpine vegetation above 1,220 m, dominated by snowgrass, *Danthonia flavescens*, which again provides excellent protection to the soil. Between the two types of grassland there lay a belt of dark green mountain beech forest, almost entirely composed of *Nothofagus cliffortioides*. Under natural conditions this vegetation provided a complete cover over the whole area. It was not affected by grazing animals, and this allowed the soil to develop. The only eroding agents were avalanches and earthquake-induced slipping, events that were rare and relatively ineffective.

When European farmers settled the High Country they fired the natural vegetation, partly to dispose of the matagouri and other spiny shrubs. However, the introduction of grazing animals, particularly sheep and deer, had even worse results. The sheep lived mainly on the lower ground, while the deer

Plate 39 A gully and alluvial fan formed in a storm in 1951 in the foothills of the southern Alps of Canterbury, New Zealand. The natural tussock grassland has given place to scrub through over-grazing by sheep, deer and rabbits.

Plate 40 Shingle slides and landslides in the Avoca Valley, Canterbury, New Zealand. The southern beech, *Nothofagus*, clothes the lower hillside leading to the sub-alpine vegetation and bare rocky summits above. Deer have caused the degeneration of the beech forest leading to the increase of erosion.

spread into the forest zone. The early fires recorded by Samuel Butler in 1860 caused serious deterioration of the tussock grassland, especially the red tussock and inter-tussock palatable plants. Weeds could gain a foothold, and these included gorse, bracken and sweet briar. Overstocking and the introduction of rabbits and hares further depleted the vegetation. (See Plates 39, 40.)

The result of the invasion of the grazing animals on ground already weakened by fire was to break the former complete vegetation cover. The soil became exposed and soil erosion ensued. The most spectacular forms of erosion were gullying on the steep slopes, and shingle deposition below the gullies in the valley bottoms. Sheet wash, frost heave and wind erosion also took their toll, although in a less conspicuous way. These processes operate most effectively in the upper part of the catchment where frost is common, occurring up to 200 nights each year. Frost heave causes the soil to pulverize so that it is easily removed by strong winds. The weakening of the surface leads to the development of large gullies with shingle fans forming at the lower levels, covering large areas of former forest land. In

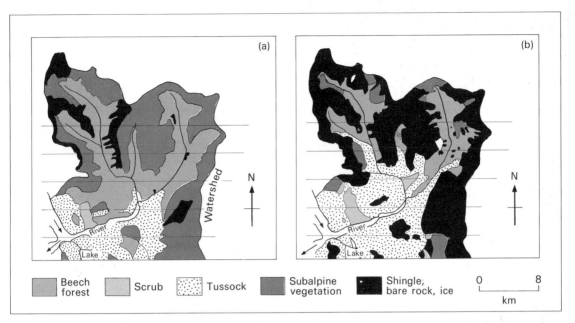

Beech forest · Scrub · Tussock · Subalpine vegetation · Shingle, bare rock, ice

0 — 8
km

Figure 3.7 Lake Coleridge catchment in the foothills of the Southern Alps, South Island, New Zealand. (a) Natural vegetation. (b) Present vegetation. (After W. P. Packard, 1947)

Plate 41 The man-made desert of Central Otago, New Zealand. Gullies are developing in the schist hillsides, which are now almost bare of vegetation. Scabweed can be seen in the foreground. There is water in the Lindis River valley with some vegetation and tree growth, while snow can be seen on the Dunstan Mountains in the distance.

Table 3.4 Lake Coleridge, New Zealand—land surface changes

Transects across Figures 3.7a and b

| | Frequencies | | | | | |
	1	2	3	4	5 and 6	Totals
Natural cover	17	5	8	18	2	50
Present cover	6	6	14	4	20	50
Totals	23	11	22	22	22	100

1 = Beech forest
2 = Scrub
3 = Tussock grass
4 = Subalpine vegetation
5 = Shingle and rock
6 = Bare rock, ice and snow

	Natural surface cover (% area)	Present surface cover (% area)
1 Beech forest	20·2	11·1
2 Scrub	1·7	2·2
3 Tussock grass	42·7	46·0
4 Subalpine vegetation	28·9	5·7
5 Shingle and rock	4·4	31·2
6 Rock, ice, snow	2·1	2·1
7 Plantation	0·0	0·2
8 English grassland	0·0	1·5

Soil erosion	%
Little or none	5·7
Some—not serious	13·8
Obvious	29·9
Considerable	36·6
Excessive	11·9
Rock, ice and snow	2·1

small areas the land has been turned into a man-made desert, supporting only scabweed, which indicates a very poor soil, as this is the first colonizer of bare shingle of river beds.

The deterioration has gone on to an even greater extent in Central Otago, which naturally supported a fair tussock grassland. It has a much lower precipitation, owing to its sheltered position in a rain shadow area. Overgrazing has reduced large areas to a man-made desert, supporting only scabweed and showing evidence of gullying. (See Plate 41.)

In the subalpine zone the snowgrass vegetation has suffered an 80 per cent reduction of area from 18,494 ha to 3,925 ha, the change being to bare shingle-covered slopes. This grey talus area is quite unstable and supports practically no vegetation now. Burning, summer grazing by merino sheep and depredation by deer, chamois and hares is responsible for the change. The dark green beech forest appears to be effective over large areas, but in this zone deer are preventing the reseeding of the trees and so in time the forest will deteriorate, where it has not already been swept away by the shingle slides. The forest area has been halved by fires and its quality elsewhere has deteriorated by destruction of the ground cover, thus reducing its water-holding capacity. Deer are largely responsible for this deterioration. (See Plate 40.)

Exercise 3.7. Assessment of vegetation change

The maps in Figure 3.7 illustrate the change, which is also indicated in Table 3.4. Test the significance of the changes statistically by taking transects across the maps showing natural cover and present surface cover. Use the same lines on each map and note the type of vegetation at regular intervals along the transects. List the frequencies for five different types of vegetation. Use the chi square test to examine the null hypothesis that there is no difference between the natural cover and the present cover. This example illustrates clearly the dangers of interfering with a naturally balanced ecosystem by the introduction of alien species and by burning the natural vegetation in a geomorphologically active terrain of steep slopes and easily eroded rocks in a climate that induces active weathering and mass movement. Once introduced it is very difficult to eliminate the harmful species, in this instance deer, rabbits and hares.

Deserts

The creation of a man-made desert in central Otago was referred to in the last section. It was caused by the deterioration of the natural vegetation through the introduction of species out of adjustment with the natural ecosystem. Natural deserts have developed a type of life that can exist under extremely harsh, arid conditions by means of various adaptations.

Desert soils are dominated by evaporation and chemical processes. A crust often forms on the surface as a result of desiccation, formed of calcium carbonate or gypsum. Calcium carbonate is often dominant in the top layers of desert soils, rendering them highly alkaline. Wind-scoured pebbles may form a surface layer protecting the finer particles beneath. Organic matter is lacking, rendering the soils infertile. Excess alkali and salt are also often problems in desert soils. Irrigation may intensify the problems, unless water is used for washing the soil and drainage is provided to prevent excessive surface evaporation. The temperatures are suitable for cultivation in many desert areas, and they prove fertile if successful irrigation can be supplied. The area of irrigation is about 13 per cent of the arable area of the world, producing 25 per cent of the total food supply. (See Plates 42, 43.)

The plants adapted to live naturally under very dry conditions fall into three groups: (a) ephemeral annuals; (b) succulent perennials; and (c) non-succulent perennials. The ephemeral annuals include 50 to 60 per cent of desert floras. They can complete their life cycle in 6 to 8 weeks, growing, flowering and reproducing within the short period after the occasional rains. Their seeds then lie dormant until the next rain—they avoid the drought rather than withstanding it. The plants are usually small, have shallow roots, germinate speedily, flower soon and mature quickly. Their response depends on the conditions— they remain small if the rain is light, but flower more profusely if the downpour is heavy.

Succulents can store water in enlarged

Plate 42 Bare hills and sand deposits in the Chilean desert between Copiapo and Caldera. Note the stony ground where it is too windy for sand to accumulate.

Plate 43 The River Loa in the Atacama Desert of Chile. Water derived from the Andes penetrates across the desert coastal plain and is used for irrigation and settlement. Fruit and crops can be grown.

tissues to enable them to withstand drought. Some of the cacti and euphorbias only transpire at night when evaporation is at a minimum.

Non-succulents include woody herbs, grasses, shrubs and trees. These plants rely on a large root system, and only grow to large dimensions after the root system is established. Deep tap roots may grow to reach the deeper water table, the root system being very large in comparison with the above-ground shoots. Alfalfa roots, for example, have been recorded to reach a depth of 39·3 m below the soil surface, while tamarisk roots were found growing down to a depth of 45·7 m when the Suez canal was being dug. The transpiring surface is reduced as far as possible by small leaves, no leaves, shedding of leaves and rolling of leaves. The leaf surface is thick and rendered waterproof by wax, hair or resinous covering. Woody structures prevent wilting in dry conditions and so prevent collapse. The stands are open and there is little competition between plants, only between the individual plants and their environment. (See Plate 44.)

The image of a palm fringed oasis is not natural. Palms have been imported into the Saharan oases, which would naturally have been colonized by tamarisk, oleander and other shrubs. However, the date palm is more useful, providing food, fruit, fuel, thatch, fibre and building material, as well as camel fodder in the form of crushed stones of the fruit.

Desert animals must also become highly adapted to their harsh environment, and they are few in number and species. They must adapt to dry conditions and in tropical deserts to very high temperatures as well as large temperature ranges. They resist high temperature by dormancy during the day, by insulation or by an internal mechanism that renders the high body temperature harmless. Most of the animals are small, but the best known, the camel, is large. The large dromodary can keep cool by evaporation of sweat from its large surface area. In small animals the large surface area relative to their size would soon cause them to lose too much body moisture. The camel has a coat that allows sweating from the skin which provides the maximum benefit and cooling effect, while the hair provides insulation. Fat is stored in the

Plate 44 Semi-arid vegetation near the bottom of the Grand Canyon on the Tonto Plateau. The spiny, succulent or thorny character of the plants is apparent. In the distance is a mesa, on the left, and a butte, on the right.

Plate 45 Salton Sea, California. This area of inland drainage dries out to form well developed desiccation cracks in the fine sediments of the lake bottom.

hump and not under the skin, thus further aiding cooling. Physiologically a camel can tolerate the loss of water equal to 25 per cent of its body weight, while most other mammals can only tolerate 12 to 15 per cent. The water is taken from the camel's tissue and gut and not its blood as in other mammals. Camels do not store water, but they can drink very large quantities very quickly, up to 123 litres in 10 minutes, and thus recharge the tissues rapidly when water is available so that they can survive long periods without drinking. The camel can also vary its body temperature by up to 6 °C; this helps to conserve water—by raising the body temperature during the day.

Rodents, reptiles, birds and mammals are usually small, and being small cannot have sweat glands. They tend to remain in burrows during the heat of the day, and forage for food at night, when they may have problems in keeping warm, hence many have fur coats. They only drink when water is available, conserving it by losing little by respiration in the day. They obtain moisture from succulent plants and produce dry faecal pellets. Food is stored in the relatively high humidity of their burrows. Reptiles have a thin, non-fatty skin which allows them to lose heat quickly, and they adapt their position to the conditions, basking in the sun when it is not too hot, and finding shade when necessary. Insects tend to be nocturnal and seek shelter during the day, and some, such as the desert locust, swarm.

Humid tropical forests

The humid tropics differ from other areas in several important respects. This is the only major environment on the earth that has not suffered a major change of climate during the oscillations of the glacial period, which have affected most other areas. The zone may have contracted somewhat during glacial periods but the central core remained under the influence of high temperature and heavy rainfall. These conditions speed up chemical weathering so that the equatorial areas are optimal for this process, which has been able to continue steadily and rapidly for a very long period of time, stretching back well into the geological past. There are also large areas of old land surface that have remained tectonically stable during considerable spans of geological time. Thus it is only in the equatorial areas, on old land surfaces, that soils have had time to reach full maturity, and weathering has been able to penetrate very deeply into the rocks.

The weathering processes of the equatorial forest areas produce weathered material rich in sesquioxides of iron and aluminium. Where the parent rock has few bases the soil has a low pH value and kaolin forms an important element of the soil. On more basic rocks the pH is higher and limonite may form, giving a yellow or red colour to the soil. Soil particles are small, being in the silt and clay grade, but may become linked to form material called pseudo-sand. It is so named because it does not have the properties of clay, such as plasticity and swelling on wetting, although it is formed of clay minerals. The red soils are called latosols or tropical red earth. When the amount of kaolinite is low it is called a *laterite*, although this name is confusing so the modern term is sometimes *ferralite*, when it is in association with basic rocks. There is a rapid turnover of organic material in the soil in the humid forest zones and it continues throughout the year. Humus is nearly colourless.

In general humid tropical forest soils are difficult to manage, and may not be fertile, being deficient in plant nutrients. They may be friable, however, and easy to work. They are usually acid, but liming may not be helpful. Humus forms the best fertilizer, but leaching is very rapid and exposure to the sun may be dangerous, speeding up the breakdown of the organic matter. They require careful management. (See Plate 46.)

The natural vegetation of these areas is forest, and it grows under a climate that has the minimum of seasonal contrasts so that vigorous growth can continue throughout the year. Temperatures are continuously high and

Plate 46 Near Colombo in Sri Lanka. Tropical vegetation is seen alongside the river, with some settlement.

rainfall always exceeds 150 cm, often being more than 200 cm. In the true rain forest it is continuous with no dry season. Temperatures vary between 24 °C and 26 °C. The diurnal range is small, being only 5 to 9 degrees C, and humidity is constantly high. Wind velocities are low within the forest, resulting in the lack of wind-pollinated species in the lower layers.

The most striking characteristic of the rain forest vegetation is the large number of layers of vegetation. The rain forest is not usually the impenetrable jungle that is often imagined. Thick jungle occurs only along the water courses or where large trees have fallen, breaking the canopy. The forest consists of at least five layers, the upper three consisting of trees. The highest layer consists of the crowns of the largest trees reaching up to between 30 m and 50 m, exceptionally to nearly 60 m. The second layer also consists of shorter trees. The trees have very straight trunks lacking

branches below about 18 m, with well-spaced foliage and smooth bark. Their most unusual characteristic is the buttressed nature of the lower trunk, which is not fully explained. They are also shallow rooted, not needing to penetrate deeply for water. The third layer consists of other, mainly woody plants, which are much larger than their temperate counterparts. Violets can grow to the size of apple trees, and grasses up to 18 m high—it is the reverse of the tundra vegetation. The fourth layer includes many lianas and epiphytes, 90 per cent of these climbers are concentrated in the tropical forests. The fifth and lowest stratum of the forest vegetation consists of herbaceous plants, but the cover is not closed.

One feature of the equatorial forests is the very large number of species; 6,000 species of flowering plants occur in west Africa, 20,000 in Malaysia and about 40,000 in Brazil. One tree species rarely dominates any forest stand,

Plate 47 The Amazon tropical forest area. Note the steep slopes characteristic of tropical forest land with trees up to the ridge crests. Clearance is in progress in the foreground.

which reduces its value for exploitation. Diversity, rapid growth and large vertical extent are the major characteristics of this very prolific environment, which produces the greatest amount of organic plant matter.

The fauna of the forest is also adapted to the environment, and many of the animals live in the tree tops in the light zone. Birds are important in the tree tops; in Costa Rica 14 species eat off the ground, 18 in the shrub layer, 59 in the lower tree layer, 67 in the middle one and 69 in the upper tree stratum. Mammals are also mainly arboreal, including the apes and monkeys. There are few large animals of the forest floor, only elephants, buffalo, okapi, wild hogs and leopards in Africa. Rodents and insects are plentiful, especially the latter, and there are also snakes. Of the insects driver ants are perhaps the most numerous, marching in vast armies, destroying all in their path. (See Plates 24, 47.)

Exercise 3.8. Vegetation and precipitation relationships

In order to quantify the relationships discussed in more general terms in the last few pages the data given in Table 3.5 should be used. They suggest the relationship between precipitation and the growth of vegetation. The mean annual value of precipitation is given in cm, and the production of vegetation is given in kg/ha. The values show that there is a very great range, covering four orders of magnitude, in the growth of vegetation between the desert areas and the tropical hardwood forest of Brazil. Use these data to assess the statistical significance of the relationship between precipitation and growth of vegetation. Plot the data on a graph and then correlate the paired values and draw the regression line on the graph. The great range of values makes the data unsuitable for

Table 3.5 Precipitation and weight of vegetation

Location	Type of vegetation	Mean annual precipitation (cm)	Weight of vegetation	
			(kg/ha)	log (kg/ha)
Las Vegas, Nevada	desert shrub	12·7	111·1	2·045
Salt Lake Desert	desert shrub	20·32	444·4	2·648
Clark County, Utah	sagebrush	30·48	989·9	2·996
Fremont Co., Idaho	sagebrush	30·48	1414·4	3·150
Coconino, Arizona	desert grass	38·10	2,095·5	3·321
Burlington, Colorado	grasses	43·18	2,501·1	3·398
Phillipsburg, Kansas	grasses	55·88	3,888·9	3·590
Lincoln, Nebraska	grasses	68·58	4,963·3	3·696
Sandhills, Nebraska	wheat grass	45·72	4444·4	3·648
Lincoln, Nebraska	grasses	68·58	6,915·5	3·840
Fraser Forest, Colorado	lodgepole pine	63·50	47,777·3	4·679
Rocky Mountain states	conifers	71·12	59,999·4	4·778
NE Central states	mixed forest	76·20	71,110·4	4·852
NE states	hardwood forest	106·68	61,105·0	4·786
SE states	mixed forest	129·54	53,333·3	4·727
Pacific states	conifers	162·56	166,666·5	5·222
Brazil	hardwood forest	304·8	966,665·7	5·985

plotting as they stand on ordinary graph paper. They can be plotted either by using semi-logarithmic paper, or by taking the logarithms of the values and plotting these on ordinary graph paper. Try both methods, and use the logarithmic values for the regression and correlation analysis. Account for the findings by reference to the characteristics of the different zones already described.

3.5 Oceanography

Introduction

The 70 per cent of the earth's surface covered by the oceans will be briefly discussed in this section from a variety of regional points of view. Continuing the theme of biogeography, the example of the living ecosystem of the North Sea will be briefly referred to, as it illustrates certain fundamental facts of oceanic life. It is also an important area from the point of view of exploitation of the organic riches of the sea.

The discussion of certain land environments showed that a wide range of productivity occurs on land, from the tundra at one extreme to the equatorial forest at the other. The oceans also possess zones of fertility and there are marine deserts. Some examples will be cited and the reasons for the variations in fertility explained briefly. In the final section the regional physical geography of an area of ocean of approximately continental size will be considered, illustrating that regional physical geography can apply to an ocean as well as a continent.

Each ocean has its own individual character that depends on many variables. The type of ocean water is one important aspect, the water being defined mainly by its temperature and salinity. These two properties determine its density and hence its stability. Temperature and salinity characterize water masses, in a similar way to the air masses of the atmosphere, so that the oceans can be described in terms of their water masses. The oxygen content is also significant, because it is related to the age of the ocean water, which in turn depends on the circulation pattern and is related to the formation of the water masses. Most water masses form by the sinking of dense water. The density is increased by cooling and/or increase of salinity, because water density increases as the temperature falls and as the salinity increases.

From the point of view of biological oceanography the fertility of the water is very important. This depends on the supply of nutrients, just as the fertility of the soil depends on the nutrients in it and their availability to the vegetation. Plants alone can create organic matter out of inorganic chemical nutrients by the process of photosynthesis in the presence of light. The living organisms in the ocean are also related to the pelagic sediment on the ocean floor, and the study of ocean sediments can tell a great deal about past conditions when they have accumulated slowly over long periods of time. The distribution of pelagic sediment is related both to surface conditions and to depth of water.

Some aspects of the biogeography of the North Sea

The process of photosynthesis whereby nutrients are converted into organic substances can only take place in light, and thus it is limited to the depth to which light can penetrate, which depends on the clarity of the water. Light penetrates to between 100 m and 200 m in clear tropical and subtropical seas, but may be reduced to between 25 m and 50 m in the more turbid waters of middle latitudes, and is even less in the very murky waters of the North Sea, where the limit may be at 10 m to 20 m. It is also necessary to have nutrients in the upper layers, where the light is sufficient for photosynthesis. Thus ocean fertility depends on the presence of both light and nutrients in the upper layers; suitable temperatures and salinities are also required. Productivity tends to be higher nearer the surface.

Diatom production in the North Sea is closely linked with the life cycles of some fishes of commercial importance. These cycles will be examined briefly by reference to three important species, each of which has a different environment. The plaice is a good example of a demersal fish, which is one that spends most of its life on the bottom of the sea. The plaice has adapted to a flat form. The cod provides an example of a round demersal fish, and the herring is a pelagic fish, swimming mainly in the upper layers of the water.

The plaice is very common in the North Sea and very important commercially. The bottom-living plaice lay eggs which float with the plankton. A large female plaice may lay up to half a million eggs at one spawning. Most of the eggs are laid in the Flemish Bight in the southern North Sea in midwinter. About 60 million plaice may have come to this area to spawn, and there are also other smaller spawning grounds in the northern North Sea. The eggs laid in the Flemish Bight drift with the plankton northeastwards towards the Dutch coast at about $2\frac{1}{2}$ to 5 km/day. This drift period with the plankton is vital to the successful survival of the year brood, because they are dependent on the plankton for their sustenance during the spring flowering. If this should fail to occur in the right place at the right time nearly all the young plaice will perish. They take to the bottom as tiny flat fish when they are 4 to $6\frac{1}{2}$ weeks old and only 2 cm long. They must drift into the right area during their period in the plankton to reach the nursery grounds off the coast of Holland. They stay in the nursery grounds for two years, and as they grow larger, thereafter, they gradually migrate towards the deeper water to the northwest. By the time they are 4 to 6 years old they are 40 to 44 cm long. Their migration is shown in Figure 3.8.

Plaice diet consists of polychaete worms in winter and bivalve molluscs in summer. Their

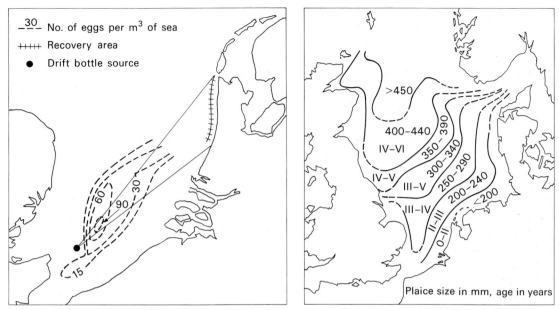

Figure 3.8 The figure on the left shows the plaice spawning area in the southern North Sea and the pattern of drift bottle recovery on the Dutch coast. The figure on the right shows the northward spread of plaice from the nursery grounds off the Dutch and Danish coasts. (After R. S. Wimpenny, 1953)

growth rate depends on the availability of food—they grow much larger and faster when food is plentiful—and being so prolific they require a large food source if they are to grow fast. It was estimated about 20 years ago that there were probably about 2,000 million plaice in the North Sea. However, they can suffer from depletion by overfishing and stocks require careful management to retain the optimum catch without overfishing. Experiments have shown that moving plaice from their overcrowded nursery grounds off the Dutch coast to the Dogger Bank, where plaice do not normally go in their early years, allows them to benefit from a greater stock of food and their growth rate is much faster. The transported plaice had grown 13 to 14 cm, while those remaining had only grown about 4 cm. This suggests one way in which fish stocks could be improved by making better use of the available food supply for the plaice in areas they do not normally frequent.

The cod is an example of a round demersal fish that has a wide distribution around the North Atlantic and of which there is an important stock in the North Sea. Like plaice, cod spend their first months in the plankton, floating passively and feeding on the phytoplankton in their early stages. They are also vulnerable to the drift at this stage. They are spawned on the banks of the North Sea, including the Great Fisher Bank, the Forties and Ling Banks. A large female cod can spawn up to 4 million eggs and they drift for $2\frac{1}{2}$ months with the plankton within the North Sea, taking to the bottom in rough ground when they are 2 cm long. They grow to 30 to 35 cm in $1\frac{1}{2}$ to 2 years and can breed when they are about 5 years old and 70 cm long. When mature cod feed on herring and smaller fish, so that they have a longer food chain than their prey the herring, which is another important North Sea fish.

The herring is a pelagic fish that spends its time in the upper layers of the open sea. However, they do move diurnally, migrating down during the day and rising to the surface in 'the swim' at night. This movement may be related to that of their favourite food, the copepod *Calanus*, which also migrates vertically with a similar diurnal rhythm. The herring has another significant habit, which is useful for the fishermen out to catch them. They live in compact shoals, especially during the swim when they rise to the surface at night. They can be caught relatively easily in drift nets, especially now that the shoals can be located at depth by radar, which enables the nets to be shot at the right position to catch them as they rise towards the surface. As a result herring have been severely overfished in the North Sea, especially by the Danes who catch even the immature herring for fish meal.

Herrings lay their eggs on the bottom of the North Sea, the spawning area moving gradually southwards with the season, so that they used to be caught when spawning in large shoals off Scotland in August and September, off northern England in September and October, and East Anglia in late autumn. The spawn stays on the bottom, where it is preyed on by haddock, another demersal fish. When the eggs hatch the young fish join the plankton, and are very dependent on one special species of zooplankton for their food supply. If this species is lacking then the whole year class may perish. They are very irregular in numbers from year to year; some years when conditions are good large numbers survive, but in other years few mature. The North Sea herring are smaller and live only about 11 years compared with the larger Norwegian ones, which can live twice as long. In the relatively shallow North Sea herrings descend to the bottom during the day, but in deeper water they go down to about 140 m. Herring are preyed on by many creatures, including cod, sea gulls, seals and man. It is probably for this reason that they have developed the shoaling habit, as the safest place is in the middle of the shoal. They all try to reach this position, resulting in a tightly packed shoal. This habit does not save them from man, and they have been very seriously overexploited in the North Sea as a result. Conservation measures are urgently needed to allow stocks to regenerate.

The basis of the marine life pyramid is the phytoplankton, the plant element of the floating and drifting population of minute organisms that live in the upper layers of the sea. These minute plants are many in number but very small in size. Their small size ensures that they have a large surface area compared with their volume. This helps them to remain afloat in the fertile upper layers where they can photosynthesize. Their small size is also useful as they absorb nutrients through their relatively large surface area. They form the bottom of the food web, providing sustenance for the next element, the zooplankton, the floating element of the smallest animals of the sea.

Zooplankton consists of many different types of creatures, including the eggs, larval and young stages of many of the important food fishes, as well as small jelly fish, copepods, worms and other species. The zooplankton in turn form the prey of larger organisms, including the benthos or bottom living forms, the fishes and marine mammals.

Some of the interrelationships between these different oceanic organisms are illustrated by reference to the North Sea. This area lies in temperate latitudes and thus has a strong seasonal rhythm in its organic cycle. The seasonal element is lacking from the lower latitudes in the ocean as on land. In fact seasonal rhythm increases in importance at sea as on land as the latitude increases.

The seasonal rhythm is well shown in the pattern of phytoplankton production in the North Sea. The particular creatures that are important are the diatoms, which are minute plants secreting silica cases. They reproduce by splitting in two, so that their increase is geometrical in rate when conditions are favourable for their growth and reproduction. The diatoms cannot reproduce during the winter months because of the low temperatures so that the supply of nutrients builds up during this season, because they are not being used owing to the low number of diatoms. The nutrients become well distributed throughout the water owing to the stirring activity of the winter gales and large waves of the winter season in the shallow waters of the North Sea, which can be stirred to the bottom. The low light intensity of midwinter also limits productivity.

About early March the light intensity rises above the threshold necessary for effective photosynthesis, at the same time the temperature is beginning to rise from the minimum value reached in February. These conditions allow the rapid reproduction of phytoplankton, which increase very rapidly in numbers and quickly use up the available nutrient in the surface layer. This rapid increase in the number of diatoms is called the spring flowering. After the increase of diatoms has taken place their numbers are severely curtailed by the grazing of the zooplankton, which at this season contains the larval and young stages of many important food fish. The renewal of the diatoms is now restricted by the lack of nutrients. Having used up the available nutrients in their spring growth, they cannot obtain more, because, during the late spring and summer, rising surface temperature and calmer conditions make the water stratification stable. The cooler layers below, in which nutrients remain, cannot rise to the surface owing to the less dense water on the surface; thus growth is inhibited by lack of nutrients, although temperature and light conditions are favourable.

Towards the autumn, however, before the temperature falls too low and the light intensity is too reduced, the autumnal gales stir the waters bringing nutrients once more to the surface, allowing a smaller production at this time of year, called the autumn flowering. This occurs in late August and early September, before the light intensity again becomes deficient by October. Production is then very low until the spring flowering the following year.

Exercise 3.9. North Sea biological activity.

Draw a diagram to illustrate the annual cycle of diatom production in the North Sea based on the variation throughout the year of the following controls:

(1) Light—reaches a maximum value in June and July, and a minimum in December and January.
(2) Temperature—reaches a maximum in August and September, and a minimum in February and March.
(3) Nutrients—values are high from December to March, low in April to August, rise again in August to a subsidiary high, falling again in September, finally rising to December.
(4) Stability of water—the water is stratified and stable from May to September, and stirred and unstable from October to March.
(5) Diatoms—values are highest in April, with a subsidiary peak in September, and low from May to August and October to March.

Account for the variations shown and explain the interrelationships of the variables involved.

Zones of fertility—Peruvian coastal waters

The waters off Peru are renowned for their fertility, even surpassing the North Sea. The coastal waters off Peru are adjacent to a desert coast so that all the fertility of the area is concentrated in the sea. The reason for the fertility is the continuous supply of nutrients that reach the upper layers by upwelling of water from shallow depths in this zone. The upwelling is associated with the pattern of currents and winds. The relatively cold Humboldt or Peruvian current flows north along the coast of Peru, while the wind tends to blow slightly offshore from the south. The upwelling is associated with the effect of the rotation of the earth, the Coriolis effect, on the wind and moving water. Peru lies in the southern hemisphere where the Coriolis effect diverts moving air and water to the left of its path. Thus where the water and wind are

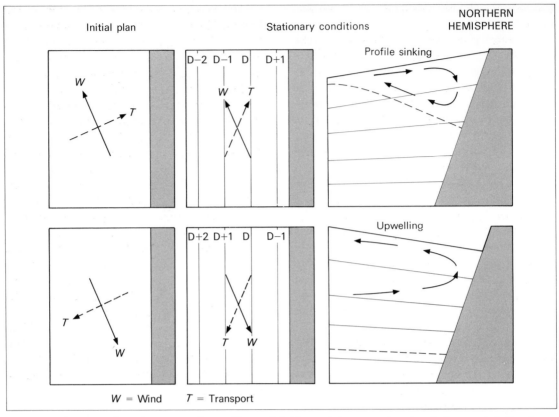

Initial plan Stationary conditions

Profile sinking

D−2 D−1 D D+1

W T

D+2 D+1 D D−1

Upwelling

T W

W = Wind T = Transport

Figure 3.9 Diagram to show the pattern of upwelling and sinking and the water slope in relation to wind direction (W) and water transport (T) in the northern hemisphere. D refers to the height of the water surface. (After H. U. Sverdrup, M. W. Johnson and R. H. Fleming, 1946)

moving north the diversion is to the left in an offshore direction. Figure 3.9 illustrates the situation in the northern hemisphere.

Exercise 3.10. Upwelling

Draw the situation for the southern hemisphere. The water surface slopes up offshore as the contours show, and the transport of water is offshore, to compensate water upwells from below to feed the offshore transport. The upwelling takes place from a few hundred metres depth and is most intense in spring and summer, although off Peru north of 30 °S the maximum occurs in winter.

Peru is not the only place to have fertile waters through upwelling. It is a common feature of the eastern boundary currents that flow equatorward along the eastern sides of the oceans. Thus similar fertile areas occur off Oregon and California in the north Pacific. The Canaries and Benguela currents have similar effects along the coasts of North and South Africa respectively.

The Peru current carries 10 to 15 million $m^3 s^{-1}$ to the north and extends up to 900 km from the shore. Upwelling occurs from 40 m to 360 m depth and is concentrated in four zones between 3 °S and 33 °S. It is in these zones that the fertility and the very prolific life on which it depends is concentrated. Productivity is very high and the plentiful phytoplankton provides food for abundant zooplankton, which in turn is preyed upon by the large shoals of anchovies. These small fish form the basis of the extremely rich Peruvian fishing industry, which until recently was the world's single largest fishery.

However, there are occasions when the upwelling fails and the surface waters become much warmer than normal. This warmth is associated with the southward extension of warm, relatively less saline water from the north. The warm water is called El Niño. Its southward extension is very variable, but it does have serious consequences. It brings wet weather to the normally desert coast and the wind changes from southerly to northerly. The El Niño occurs roughly every 7 years, and was notable in 1911, 1918, 1925, 1932, 1939 and 1941. Particularly severe events took place in 1891, 1925 and 1953. When the El Niño is strong the anchovies either die in large numbers or migrate offshore or southwards. The sea birds that feed upon the anchovies also die or migrate. The fishing industry suffered very severely during the last El Niño of 1973 to 1975, which was particularly severe. The warm water may be due either to the southern extension of tropical water, or it may be the result of warming of the coastal water when upwelling processes fail. The warmth can extend as far as 14 °S. Another problem associated with the warm water is the so-called *red water*, which is the result of excessive phytoplankton growth in the warm water. This uses up the available oxygen and leads to large scale mortality of fish and other creatures.

Only fairly recently have the anchovies been exploited by Peruvian fishermen, but from 1965 to 1970 Peru produced the largest volume of fish of any nation. The fishery was based almost entirely on the anchovies, which are small fish only 15 cm long, belonging to the herring family. Their food consists of 99 per cent phytoplankton and 1 per cent zooplankton, and thus they have a short food chain, like their larger relative the herring that feed on zooplankton. Anchovies occur off central Chile and Peru from 37 °S to 4 °S, extending 48 km offshore in summer and 190 km in winter. They are preyed upon by many creatures, including thousands of sea birds, mainly pelicans, the guano birds, and bonito. The fishery catch reached a volume of 10,520,300 metric tons (tonnes) in 1968, falling to a little over 9 million tonnes in 1969. Since this time the catch has fallen disastrously, after reaching a peak of 12·3 million tonnes in 1970. The 1971 catch was only just over 10 million tonnes and fell to 4·5 million tonnes in 1972. It is estimated that the total bulk of fish is 15 to 20 million tonnes at the height of the annual cycle. The El Niño is partly responsible for the loss of both fish and birds, although over-fishing has also played a part. The 1957 El Niño was severe, reducing the guano bird numbers from 27 to 5½ million, although they had recovered to 17 million by the time of the 1964 El Niño. The severest El Niño on record occurred in 1972, with a temperature rise from its normal of 22 °C to 30·3 °C in February. The guano birds were reduced to 1 million, and the 1973 fish catch was estimated at only 3 million tonnes. The numbers caught in 1975, 1976 and 1977 were 3·5, 4·4 and 4·8 million tonnes respectively, indicating a modest revival of the fishing industry.

Oceanic deserts—the Sargasso Sea

It has already been shown that the upwelling of water makes the Peruvian coastal waters exceptionally fertile, because the nutrients are constantly being replenished in the upper layers in which the sunlight allows photosynthesis to take place. Similarly in the North Sea the fertility increased when stirring of the water allowed nutrients to reach the surface. The Sargasso Sea, on the other hand, exemplifies a marine desert, in which nutrients cannot readily be replenished. It lies in the tropical north Atlantic Ocean in the centre of the subtropical high pressure zone. It forms the central quiet hub of the main oceanic current system of the north Atlantic, with currents flowing clockwise around it.

The Sargasso Sea becomes warm on the surface through the heat of the sun in the belt of continuous high pressure. The surface layers become less dense than those below as the high temperature more than compensates for the increase in salinity brought about by excess evaporation in the prevailing high pressure system. The water is thus stably stratified and once the nutrients in the surface layer are used up they cannot readily be replenished from below. Despite the favourable temperature conditions the Sargasso Sea produces very little organic material. Fertility based only on the supply of available nutrients would be expected to be low in the tropics throughout the year. At 40 degrees latitude it is generally low in summer but higher in winter, with a strong seasonal difference due to greater winter cooling. Higher rates of fertility would be expected at 60 degrees latitude all through the year, because of greater turbulence in the westerly wind storm belt. Values recorded in the Sargasso Sea reach a maximum of $0 \cdot 89$ gC m^{-2} day^{-1} when the water is least stable, but are only $0 \cdot 1$ to $0 \cdot 2$ gC m^{-2} day^{-1} when the stability is high, falling in some places as low as $0 \cdot 05$. An annual rate of 72 gC m^{-2} was measured in the Sargasso Sea compared with 120 gC m^{-2} recorded in temperate and sub-polar seas.

The most fertile zones of the oceans are the coastal waters, particularly along the eastern sides of the oceans where upwelling can take place. Another strip of fertile water extends along the equator, especially in the Pacific Ocean. This fertility is also due to the Coriolis effect, causing upwelling from relatively shallow depths in an area where temperatures are favourable. (See p. 215.)

Physical geography of the Southern Ocean

One of the most fertile oceans of all, despite its high latitude, is the Southern Ocean. This stretch of water is unique from several points of view. It is the only stretch of water that is continuous around the globe and it has distinct and unusual boundaries. On the south it abuts against the ice of the Antarctic continent. Its northern boundary is oceanographic and is marked by the Antarctic Convergence, which is a zone beyond which temperature increases rapidly to the north, as shown on Figure 3.10. The convergence separates two strongly contrasting areas. To the south of the line it is cold, as on South Georgia where the snow line is as low as the tree line on Staten Island off Tierra del Fuego. The former island is south of the convergence and the latter is north of it, although they are both in the same latitude of 54°S.

The convergence lies between about 52°S and 56°S and it marks the boundary between different water masses. To the north the sub-Antarctic water mass is fairly warm compared with the circumpolar water mass to the south. The latter water mass is unique in that it is the only water mass to form by rising rather than sinking. The water consists of various other water masses that have mixed together and reached a density that allows the water to rise to the surface over a wide zone south of the Antarctic convergence. Thus this water mass brings a good supply of nutrients to the surface. It helps to account for the high productivity and prolific nature of the life of these high southern latitude waters. The waters of the ocean also include the coldest waters of the earth. The Antarctic bottom water mass forms in the highest southern latitudes and sinks along the continental shelf in the Weddell Sea to fill the greatest depths of the oceans, because it is very dense. It is very cold and has a fairly high salinity. The Antarctic bottom water is formed by cooling of water in contact with ice, and freezing of the surface layers in winter around the Antarctic continent. Sea ice is relatively fresh and salt from freezing water becomes concentrated in the water below, thus increasing its density, so that it can sink to the bottom. It is partly to compensate for this sinking that water has to rise elsewhere and so the Antarctic circumpolar water partially balances the sinking as it rises to the surface as indicated in Figure 3.11.

The prolific fertility of the Southern Ocean is related also to its bottom sediment, which

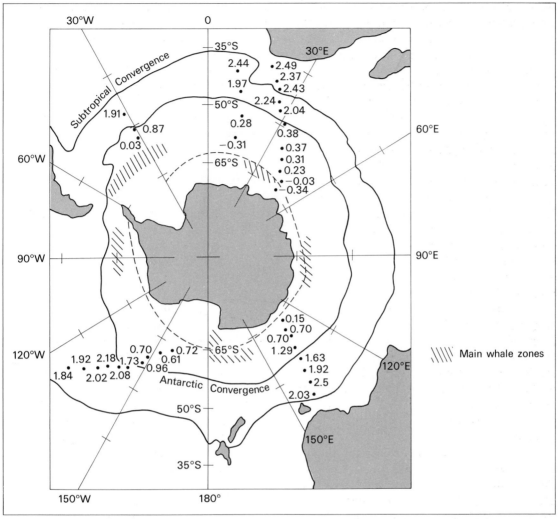

Figure 3.10 Map to show the Antarctic and Subtropical convergences around the Southern Ocean. The dashed line indicates the boundary between easterly and westerly drifts around the Antarctic continent. Figures give the temperature in °C at 2500 m depth. The main whaling zones are indicated by shading. (Mainly after G. E. R Deacon, 1959)

consists mainly of diatom ooze. It is only here that enough diatoms survive, die and sink to the bottom to produce a thick layer of diatom ooze on the ocean floor. (See Plate 48.)

The life of the Southern Ocean is also unique and thrives on the fertile waters in which the diatoms form the basis of the food web. The next member of the ecosystem that supports the higher members is also planktonic. This consists of the krill, a tiny shrimp-like organism that provides food for the largest

animal ever to exist on earth—the marine mammal, the blue whale. The krill is a red crustacean, *Euphausia superba*, which dominates the water south of the Antarctic convergence. It supports many individuals belonging to relatively few species that live in this extensive area of ocean. These creatures include fish, gulls, penguins, seals and whales. The relationships between them illustrate the losses that occur at each step in the food chain. It takes 100 units of diatoms to feed 10 units of

Plate 48 Pack ice off Kulusuk Island, East Greenland. The ice is formed by the freezing of the sea and layering can be seen, with snow lying on the ice surface. The ice is breaking up in the fjord around a Danish fishing boat.

krill, which in turn only support 1 unit of whale.

The whale is one of the most interesting, and certainly the largest, animal on earth at present. The blue whale is the largest of the whalebone whales, which include the fin, sei and humpback. They are all warm-blooded mammals and so need to be fairly large to maintain their temperature in the cold waters in which they live. Whalebone whales have huge mouths with fringes of baleen or whale bone, through which they pass large volumes of water to filter out the krill on which they feed. Therefore, they have a short food chain and make good use of their environment. Whales only reproduce once every two years, moving to warmer waters for the birth of their calves, and returning to the rich krill-filled colder waters to feed in the southern summer. They then take advantage of the spring flowering of the diatoms and the increase in krill at this season.

Unlike fish, whales reproduce slowly and once their numbers are reduced there is a real danger of their extinction. This would be great folly on the part of the whalers who have already hunted several species to the point of extinction despite continuous warnings over a long time from marine scientists of overkilling of the stock. Thus short-sighted greed has replaced common sense in controlling whaling at a level at which a large sustainable yield could be obtained permanently. It would indeed be sad if this largest, probably highly intelligent, mammal, were to become extinct. Blue whales are now completely protected, and their overexploitation has been followed in turn by that of the smaller species, the fin, sei and humpback, in order of size. The blue whale can exceed 30 m in length, while the

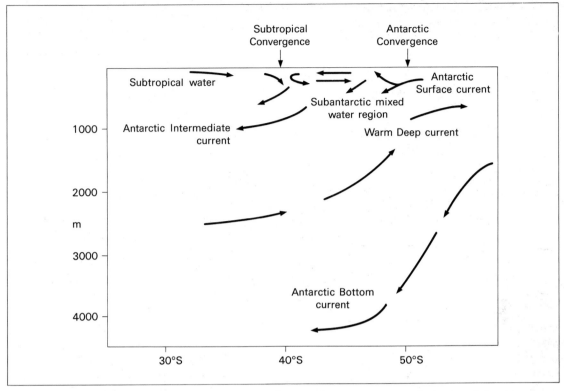

Figure 3.11 Profile through the south Atlantic ocean to illustrate the water movement in depth and the main water masses. (After G. E. R. Deacon, 1959)

other whales are 26 m and 15 m long. The other main animals of this area spend part of their lives on land, mainly for breeding purposes; they include Emperor and Adelie penguins, seals and many sea birds.

Exercise 3.11. Trend analysis of whale catch in Antarctic waters

Table 3.6 provides data for the whale catch in the Southern Ocean for a period of 15 years. Use these figures to calculate the trends of the whale catch over the period. Plot the values on a graph and draw the trend lines from the calculated trend equation. The trend equation is $Y = a + bX$, where Y is the whale catch and X is the year counted outwards from the central year, in this example it is 1957–58. $a = \Sigma Y/n$, and $b = \Sigma XY/\Sigma X^2$; in this example n, the number of years, is 15. The data for the

blue whales illustrate the method. Calculate the trend equations for the other types of whales and for the total whale catch. The figures are given in thousands of tons of biomass.

3.6 Geomorphology

Introduction

On the local scale geomorphologists concentrate on the details of the processes that operate to change the form of the landscape. On the regional scale the form itself often becomes one of the main aspects of study, while on the global scale interest centres on major patterns and world-wide processes. The local scale processes are mainly exogenetic, i.e. acting on the surface of the earth as subaerial, subglacial or submarine processes. On the

155

Table 3.6 Biomass of whales derived from Antarctic pelagic whaling

Year	Fin	Humpback	Sei	Sperm	Blue	Per cent	Total
1949–50	903,050	56,312	2,328	88,665	526,280	33·4	1,576,635
1950–51	838,752	44,010	7,938	163,599	571,212	35·1	1,625,511
1951–52	1,067,040	40,660	734	184,368	420,168	24·5	1,712,970
1952–53	1,059,850	25,853	2,901	75,382	305,414	20·9	1,469,400
1953–54	1,224,314	15,503	5,274	77,750	220,088	14·3	1,542,929
1954–55	1,242,144	13,558	2,675	182,656	173,040	10·7	1,614,073
1955–56	1,213,872	36,659	6,181	213,311	127,269	8·0	1,597,292
1956–57	1,259,300	19,012	15,844	134,695	117,390	7·6	1,546,341
1957–58	1,210,656	11,128	52,900	182,990	131,352	8·3	1,588,026
1958–59	1,216,013	60,134	31,852	154,954	98,853	6·1	1,611,806
1959–60	1,255,972	36,260	72,442	120,002	92,976	5·9	1,577,652
1960–61	1,315,507	20,104	94,820	131,115	119,683	7·1	1,680,229
1961–62	1,251,723	8,652	104,953	128,535	69,563	4·5	1,556,019
1962–63	868,674	7,641	118,865	130,248	62,155	5·2	1,187,583
1963–64	666,548	51	177,320	168,270	6,270	0·6	1,028,459
1964–65	329,667	0	419,341	108,644	1,069	0·1	858,721

global scale the processes are largely endogenetic, i.e. acting within the earth, but apparent on the surface in the distribution of land and sea, and the major patterns of the earth's surface features.

On the regional and continental scales, with which this chapter is concerned, the form resulting both from endogenetic and exogenetic processes will be considered, although more weight will be given to the effects of external processes. Emphasis will be placed on the relationships between geomorphology and other branches of physical geography, in dealing with the exogenetic processes. The first example, however, is concerned with the relationship between form and material.

On the medium scale of both time and space the landscape can become adjusted to the material of which it is formed. The more resistant rocks stand out as eminences and the structural pattern can be appreciated in the relief. In the scarplands of eastern England, for instance, hard rocks form the higher ground. Gentle dip slopes contrast with the steeper scarp slopes, which overlook the lower vales, formed on the less resistant rocks, which are often clays. (See Plate 49.)

The second example deals with the re-

lationship between form and process by examining the pattern of some glacial features. The pattern is dependent on the controlling influence of the climate in association with the morphology and aspect of the landscape. The pattern of corries illustrates these relationships well, especially in the areas near the limit of glaciation during the last ice advance. This reached its culmination about 20,000 years ago in Britain and in many other parts of the northern hemisphere.

The third example considers relationships between processes and climatic control. The field of climatic geomorphology is relevant to the wider field of physical geography, as it links most aspects of physical geography. The processes are directly relevant to geomorphology, but in turn they depend very substantially on the climate, which is in turn dependent to a certain extent on the major aspects of relief. The smaller features affect the microclimate. There are feedback connections at all scales. The climate, relief and rock type control the soil, and both together control the vegetation. Thus both climatology and biogeography are intimately related to the climatic elements of geomorphology through geomorphological process. The aspect that will

Plate 49 Dip and scarp relief in eastern England. Tabular hills near Scarborough, Yorkshire.

be considered is the relationship between rates of erosion and climate. The influence of man on erosion rates has already been considered in the small scale, while on the medium scale the larger range of climatic variation causes important differences from one major region and climatic belt to another.

Relation between form and material—Great Britain

Great Britain is a good area in which to consider the relationship between rock type and landform because a very wide range of rock types of varying ages outcrop within a small area where the climate is fairly similar across the country. The geological time scale is well represented in Great Britain, with the general pattern of decreasing age of rocks from northwest to southeast. The Precambrian rocks are represented in the northwest Highlands of Scotland in the very old Lewisian gneiss, some of which was metamorphosed about 2,400 million years ago. It must already have been deposited and indurated by then. Precambrian Torridonian sandstone outcrops

in northern Scotland. The Cambrian period started at least 600 million years ago, and by then life was well established on the earth. Organisms living before the Cambrian are not well preserved, partly because they often had no hard parts to survive. In the period just before the Cambrian it has been suggested that the earth underwent major glaciation, and the environmental problems and hard conditions, associated with the greater differentiation of climate, may have been partly responsible for the wide divergence of animal life at this time. An environmental challenge of this type often seems to produce increased differentiation and development of higher life forms. The emergence of the human race may also have been partially the result of the Pleistocene ice age, which produced more demanding conditions that needed active response and cooperation in the development of clothing, housing and agriculture. (See Plate 64.)

Figure 3.12 illustrates the major structural regions of Britain (b) and the geological column of the rocks (a). The rocks of the lower Palaeozoic are well represented in Wales, southern Scotland and the Lake District. Vol-

157

Figure 3.12a The geological column of the rocks.

Era	Period	Beginning of period (million years ago)	General lithology
Cainozoic (Tertiary)	Pleistocene	1	Alluvium, gravel, till and minor littoral deposits
	Pliocene	10	Thin littoral deposits in S.E. England
	Miocene	25	Absent—alpine folding in S.E. England
	Oligocene	40	Largely lacustrine clays, sandstones and limestone in Hampshire Basin and Devon
	Eocene	70	Predominantly clay, with sand and flint gravel in the London and Hampshire basins. Basalt lavas in N. Ireland and W. Scotland
Mesozoic	Cretaceous	130	Upper: chalk mainly in S.E. and E. England and N. Ireland Lower: clays and sand in S.E. England
	Jurassic	180	Limestones and clays, sandier in N.E. Yorks. Ironstones, sands, shales, limestone and thin coal locally
	Triassic	230	Sandstones, conglomerates and mudstones
Upper Palaeozoic	Permian	270	Dolomitic limestone and sandstones in E. England and Ulster Sandstones and breccias in N.W. and S.W. England Variscan folding, granite intrusion in S.W. England and Scotland
	Carboniferous	350	Upper: sandstones, shales and coal Lower: massive limestone, shales, sandstone and coal in north basic lavas in Derbyshire and Scotland
	Devonian	400	Mainly sandstones in north and west. Marine sandstones, shales and limestones in S. W. England Main Caledonian folding in L. Devonian. Lavas in Scotland with granite intrusion
Lower Palaeozoic	Silurian	440	Shales, sandstones and greywackes in N.W. Britain. Argillaceous limestone in western Midlands
	Ordovician	500	Shales and sandstones in N.W. Britain. Lavas and ash in Lake District, N. Wales and Ireland
	Cambrian	600	Shales, sandstones and quartzite. Slate and flagstones in Isle of Man. Dolomitic limestone in N.W. Scotland
Pre-Cambrian			Mainly schists, gneisses and quartzites, marmorized limestone and igneous rocks in N.W. Britain
	Torridonian		Sandstones with shales, arkoses and conglomerates
	Dalradian		Schists, slates and quartzites, arkoses and conglomerates
	Moinian		Schists and granulites mainly
	Lewisian		Gneisses in two groups separated by dolerite dykes

canic rocks are important in the Ordovician in the Lake District, but elsewhere the rocks are typical of sedimentation in subsiding troughs, called geosynclines, in which very great thicknesses of sediment can accumulate. The sediments are often fine grained and later become indurated to form mudstone or greywacke, or metamorphosed to form slate. The pressure of metamorphism develops a cleavage in the slates making it possible to split them into thin sheets, and so providing a useful roofing material. (See Plate 50.)

The lower Palaeozoic consists of the Cambrian, Ordovician and Silurian groups, which are called after Wales and the ancient tribes of the Welsh borderland, the Ordovices and

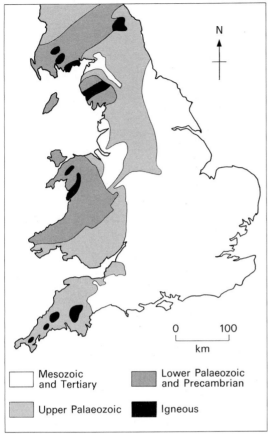

Figure 3.12b Map to show the major geological elements of England, Wales and south Scotland.

Silures, because these rocks were first described in this area. The divisions of the upper Palaeozoic include the Devonian, Carboniferous and Permian. The first gets its name from Devon, where the rocks of this age are marine in type. Elsewhere in Britain Devonian rocks are largely desert sandstones, indicating a dry climate over much of Britain at this time. This period covers the Caledonian orogeny, when the lower Palaeozoic sediments of the ancient geosyncline were folded, uplifted and eroded. The Carboniferous is so called because it contains most of the important coal seams of Britain. However, the series starts with limestone formed of organic remains of calcium carbonate accumulating in clear, warm seas. Muds then washed into the sea to form shales, while gradual shallowing brought sand as well to form the Millstone Grit above the shales and rhythmic Yoredale strata. Finally swampy conditions allowed thick vegetation to flourish that eventually became transfor-

Plate 50 Looking up the Nant Ffrancon Pass towards Llyn Ogwen, with Tryfan in the background. The ice scoured landscape is apparent, with roches moutonnées, the rougher sides of which are facing the viewer. The stream is falling over a rock step in the waterfall in the middle distance.

med into the coal seams that interdigitate with layers of deltaic sandstone. The Permian that overlies the coal measures is partly marine, the Magnesian limestone of eastern England, and partly desert sandstone, the New Red Sandstone, to differentiate it from the Devonian, the Old Red Sandstone. Both Red Sandstones indicate desert conditions and consist of wind-blown sand and pebble beds laid down by desert flash floods. (See Plate 51.)

The New Red Sandstone runs on into the Triassic, which is the first member of the Mesozoic, or middle life group of the geological column. The other members of this group are the Jurassic and Cretaceous. These groups cover a wide range of sedimentary deposits, ranging from clays to estuarine sandstones, and limestones, for example in the Cotswolds. The Cretaceous gets its name from the dominance of chalk in the upper part of the series. The rocks outcrop mainly in the south and east of England where they dip gently eastwards so that progressively younger rocks outcrop to the east. This is also the area where the dip and scarp relationship between rock and relief is most clearly developed. In the Weald and

Wessex conditions are rather more complex, owing to the tectonic effects of the Alpine earth movements. There are steeper folds in the Weald, which is mainly anticlinal in structure, producing the inward facing scarps of the North and South Downs and the sandstone ridges of the central Weald. Further west in Wessex many of the folds have been inverted so that the weaker rocks outcrop in the cores of the anticlines to form anticlinal vales, with synclinal uplands around them.

The youngest rocks belong to the Tertiary period that started about 70 million years ago. It is one of the geological periods that is relatively less well represented in Britain, but there are some early Tertiary strata in the London and Hampshire basins, where they consist of colourful sands and other sediments. The Eocene includes the London Clay as well as the pebbly Reading beds. The Miocene is one of the few geological periods represented in Britain only by volcanic activity in the northwest, but there are Pliocene strata on the eastern edge of East Anglia. Certain pocket deposits probably date from the Neogene, or later part of the Tertiary, for example on the

Plate 51 Mam Tor and Lose Hill in the Peak district of Derbyshire. The thinly bedding Carboniferous rocks of Mam Tor are very unstable and frequent landsliding occurs. The landslide relief is seen in the lower right of the view.

limestone uplands of the southern Peak District, where they are preserved in large sink holes into which they have gradually been let down as solution proceeded. Although erosion was taking place over much of Britain during the Tertiary the North Sea was subsiding rapidly during this period and up to 6,000 m of sediments, including thick coal seams, accumulated in the central trough. This trough subsided as the sediments accumulated as most of them are shallow water in type. This great thickness of material has interesting implications for erosion in Britain, as much of the material was derived from the west.

In travelling from northwest Wales across the Wash to the east coast of Norfolk it is possible to cross a great many of the geological series just mentioned, and this route provides a very good indication of the range of rocks outcropping in Britain. Records of rock type, age and relief can be used to study the relationship between these variables on this traverse. (See Plates 52, 53.)

Table 3.7 Rock and relief transect between Anglesey and East Anglia

	Rock type	Rock age	Relief
West			
1	6	4	4
2	10	1	9
3	9	2	6
4	9	2	6
5	6	4	6
6	8	6	2
7	1	6	3
8	6	4	5
9	7	4	7
10	6	4	6
11	6	4	3
12	2	6	2
13	1	7	2
14	4	7	3
15	1	7	1
16	1	7	1
17	1	7	1
18	5	8	4
19	5	8	4
20	3	10	1
21	3	10	1
East			

Exercise 3.12. Relief–rock type relationships

Table 3.7 gives the results of such an investigation. The figures 1 to 21 refer to points sampled every 32 km across Wales and England, from Anglesey in the west to East Anglia in the east. The variables noted at each point were firstly, the rock type, rated in order of resistance to erosion, as indicated on the table. The second column gives the age of the rocks and the third gives the relief. The data were derived from the Atlas of Britain, which gives maps of geological type as well as age of rocks and relief. An examination of the figures suggests that in general the rocks tend to get less resistant from west to east, with the exception of rather more resistant rocks in East Anglia than in the Fenland area, which is crossed before reaching East Anglia. The age

Rock type	Rock age	Relief
1 Clay	Precambrian	under 25 feet
2 Clay and sand	Lower Palaeozoic	25–100 feet
3 Clay, sand and shell	Devonian	100–200 feet
4 Clay and limestone	Carboniferous	200–400 feet
5 Chalk	Permian	400–600 feet
6 Sand and shale	Triassic	600–1,000 feet
7 Limestone	Jurassic	1,000–1,500 feet
8 Sandstone	Cretaceous	1,500–2,000 feet
9 Slates and greywackes	Tertiary	2,000–3,000 feet
10 Igneous	Quaternary	over 3,000 feet

Plate 52 Wootton Fen, Norfolk, showing reclaimed fenland on the left and open mature marshland on the right, separated by the reclamation bank in the foreground. The mature marsh is still liable to flooding during very high tides.

shows an even more consistent variation, with older rocks to the west becoming progressively younger eastwards. The same general pattern applies to the relief, which also falls from the high ground of Snowdonia in the west to the low Fenland and East Anglia in the east. In order to test this relationship statistically the data should be ranked. As the values were in any case only ordinal little precision is lost in this process. It is not feasible to rate these variables more quantitatively as even the relief refers to a height range in the general vicinity rather than to a particular point, and the rock type and age scales are relative. When data are ranked in this way they can be compared by means of the Spearman rank correlation technique. Test the three pairs of values by calculating the Spearman's rho value.

There are many examples in Britain of the relationship between rock type, structure and relief. In general Britain is often thought to consist of a highland zone to the northwest and a lowland one in the southeast. This pattern is reflected in the geological map, with older, harder rocks outcropping in the northwest and younger, softer ones in the southeast. On the medium scale there is a close relationship between structure, rock type and relief in Britain. This pattern is not so clear everywhere, however, and the South Island of New Zealand offers an interesting contrast.

In New Zealand the rocks are more homogeneous. Greywackes outcrop extensively throughout the South Island and they form the high mountains of the Southern Alps. They also outcrop in the lower foothills and much of the block and basin country to the northeast and southeast of the main Alpine chain. The contrast between the two countries is probably mainly the result of their tectonic histories. Britain has not been greatly disturbed tectonically, except in the southeast, during the Tertiary. There has, however, been a general fall of base level that has increased the relief, probably accompanied by some gentle warping of the land to compensate for subsidence in the marginal basins, such as the North Sea, Cardigan Bay and the northern Irish Sea. This gentle upwarping and increase

Plate 53 The subdued relief of the marshland of Norfolk is shown, with marsh drainage in the foreground. The silt forming the low marshland is exposed in the bank of the drainage channel.

of relief has given the eroding agents a chance to pick out the soft rocks and leave the harder ones upstanding, producing the relationships between relief and rock type that have been demonstrated in the transect across north Wales and central England.

In New Zealand the tectonic activity has been much more recent and intensive. The whole of the Alpine chain has been formed and uplifted by over 3,000 m, the great Alpine fault has moved the western part of the country laterally relative to that to the east. Earthquakes still cause visible earth dislocation, and minor earthquake fault scarps of recent age are still visible. The tectonic activity has also created the basin and ranges and disturbed the drainage of parts of Otago and Nelson provinces to the east of the Alps. During these earth movements active erosion has stripped off much of the younger covering strata to expose the Mesozoic greywackes that now form the fault blocks. Much of the relief is directly tectonic in type, rather than being the result of differential erosion. Structural control is thus strong in the landscape of most new

and tectonically active areas such as New Zealand, while in less active areas differential erosion has had more opportunity to exploit differences in rock type. The relief has become adjusted to the type of rock and its attitude, on the medium scale. This important aspect of geomorphology is sometimes not adequately stressed.

The association of rock type with elevation cannot be considered to be a relationship that applies everywhere owing to differences in the tectonic history of different areas and to the variations in response of rocks to weathering in different climates. Thus granite may be a relatively harder rock in temperate climates, but suffer differential lowering in relation to other rocks in hot, wet climates in which chemical weathering is rapid. In Britain it so happens that the harder rocks outcrop to the west where uplift had also been greatest. This pattern is partly due to the long-continued downwarping of the North Sea since at least Mesozoic times. This downwarping can be demonstrated by means of an altimetric analysis across the Yorkshire Wolds.

163

Exercise 3.13. Altimetric analysis

The figures in Table 3.8 give the frequencies of occurrence of heights measured from the OS 1:25,000 maps of the chalk Wolds of Yorkshire. The values are taken as the highest points of each 1 km grid square of the National Grid using three 10 by 10 km squares arranged in a west to east sequence. Calculate the grouped mean elevations for each large square and plot the values as histograms of frequency. Deduce the nature of the surface and the gradient of warping.

Relationships between form and process— distribution of corries

A well-developed corrie is one of the most easily recognized of glacial landforms, with its steep back wall and hollowed floor, sometimes containing a small lake. It is nevertheless difficult to define verbally in terms that cannot lead to exceptions. In fact corries form a continuum of features that range through the gentle hollow of a nivation niche, never occupied by moving ice, to the well-defined corries with lakes, terminal rock bars and surrounding arêtes, such as Red Tarn below Helvelyn in the Lake District and Llyn Llydaw in Snowdonia. As erosion continues a more open form develops, such as some of the Alpine and Antarctic corries. (See Plate 54.)

The best developed corrie forms often occur near the margin of glaciation where the glaciers were active, with a large gross budget, but where the ice thickness was not excessive. The nature of the terrain also plays a part as corries develop best in mountain groups having steep slopes and sharp peaks, rather than on flatter plateau surfaces on which mountain ice caps with outlet glaciers tend to form. (See Plate 55.)

The nature of ice movement in a corrie glacier has been investigated in Norway by digging a tunnel through the corrie glacier. The results showed that the ice within the corrie moved mainly by rotating about a

Table 3.8 *Altimetric data for the Yorkshire Wolds*

Heights (ft)	Frequencies		
	87–97E 97–07E 07–17E 65–75N Grid		
0– 49	0	0	0
50– 99	0	0	0
100–149	0	0	3
150–199	0	2	13
200–249	2	4	20
250–299	0	3	32
300–349	3	16	21
350–399	7	21	7
400–449	12	18	4
450–499	13	16	0
500–549	23	20	0
550–599	27	0	0
600–649	11	0	0
650–699	2	0	0
700–749	0	0	0
Totals	100	100	100

circular arc, as shown in Figure 3.13. The length of the arc is related to the form of the corrie, the longitudinal profile of which approximates a circular arc. The ice moves down in the accumulation zone, over the bed at the base of the corrie where the ice is thickest at the equilibrium line, and then it has an upward component in the ablation zone. There is some distortion as well as the circular rotation, but in general the ice slides over its bed where it is thickest, scouring and smoothing the bed of the corrie. Where the ice is thinner, near the steep head wall, melt water can penetrate below the surface and when it freezes it helps in the process of plucking, whereby the steep head wall is created, and the blocks broken off in this way provide the tools for the scouring action further down the corrie where the ice slides over the bed.

Useful results can be obtained by measuring various dimensions of corries on maps and comparing the values for different regions. The values that can be measured fairly accurately include the length, the height and the breadth, as indicated on Figure 3.14. The ratio of length to height, length to breadth both illustrate interesting variations. The volume can also be estimated by multiplying the three values, length, breadth and height and dividing by 2.

These values can be measured for a number of corries of different mountain groups and the results correlated and compared. Thus for the Lake District the correlation of length and height gives a value of $r = 0.558$, for Scotland a value of $r = 0.645$ was given, while the Rocky Mountain National Park corries gave $r = 0.562$, all similar values. The positive correlation suggests a similarity of form in each group.

The ratio of length to height (L/H) is also of interest. The smaller the ratio the steeper and higher the back wall is relative to the length, thus when the ratio is small the corrie is more protected from insolation and provides better shelter for snow accumulation. It can be argued that in areas where the gross budget of the glacier is large ablation will be rapid and flow fast, then the ratio is likely to be small, which could apply to marginal areas and well-fed corrie glaciers, while polar regions of cold glaciers would tend to have a larger ratio.

The actual size of corries also varies from area to area. They are exceptionally large in Alaska, which is a high altitude, heavy precipitation zone, where glaciers are active. In the San Juan Range of the southern Rocky Mountains, on the other hand, the corries are exceptionally small. They vary by two orders

Plate 54 Red Tarn and Helvellyn, English Lake district. The view shows the arête of Striding Edge leading up to the flattish top of Helvellyn, below which the steep back-wall of the cirque drops to the lake. The glacial lake of Thirlmere is seen in the distance. The forestry around its shores is related to its use for water supply.

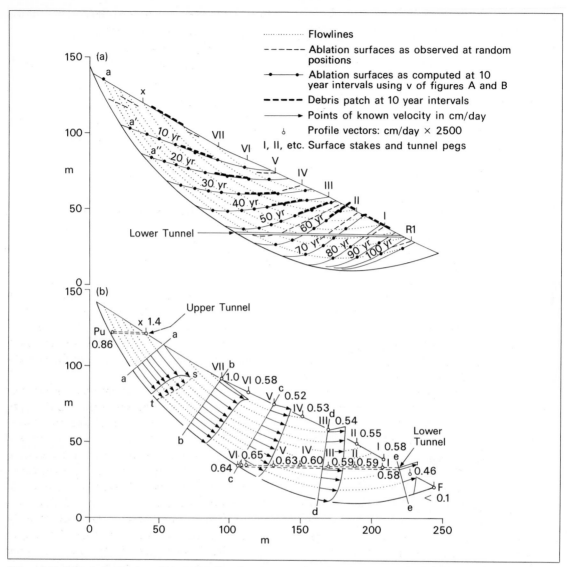

Figure 3.13 (a) Longitudinal section of Vesl-Skautbreen in Norway showing flow lines and ablation surfaces. (b) Velocity vectors at different positions normal to the glacier bed. Velocities are given in cm/day. (After J. G. McCall, 1960)

of magnitude. The San Juan corries are only between 200 m and 800 m long. They also have an exceptional length to height correlation, given by $r = -0.126$. This value indicates a greater height than length, although the lower value shows that the relationship is not statistically significant. This type of study is an example of morphometry, which is the quantitative study of form.

Another example of morphometry is the measurement of the slope of the flattest and steepest part of a corrie long profile, each measured over a vertical distance of 60 m. The steepest slope gives an indication of the character of the back wall, while the flattest slope is a measure of the degree of reversal or flattening of the gradient on the corrie floor. Values measured in corries in the mountains of New England, Montana and Utah, USA, gave the following figures: 4·70, 11·20 and 28·94

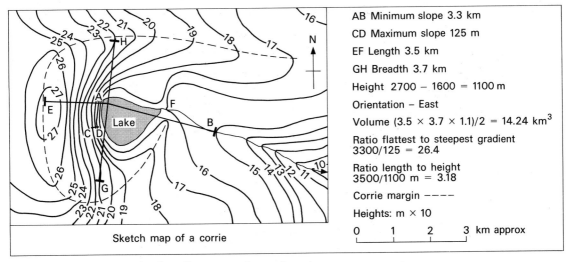

Figure 3.14 Diagrammatic sketch to illustrate corrie morphometry.

respectively. The *t*-test showed that these results are significant statistically at the 99 per cent level for all pairs showing that the different corrie groups can be distinguished in terms of this ratio. They also had significantly different areas. The values for the three groups were 1·89 km², 1·22 km² and 2·80 km² respectively. In all three areas the relief varied between 457 m and 497 m. Similar measurements can be made of corries in other areas.

The orientation and elevation pattern of corries can also be examined quantitatively. There is a clear pattern in North Wales, where the various mountain groups run from southwest to northeast. In marginal areas aspect and elevation play an important part in the pattern of corries. The pattern can be studied in relation to exposure to snow-bearing winds, and shelter in which snow can accumulate. The elevation of the corrie floors bears a relationship to the level of the snow line when the corries were being eroded. In North Wales the Snowdonian group of corries lies furthest southwest and is lowest, the Glyders lie to the northeast and the Carnedds are still further in this direction. Table 3.9a gives details of the mean elevation and orientation of the corries. The heights refer to the level of the corrie lip. The data show that for any one of the three orientations in which corries face their elev-

ation decreases towards the northeast, while the mean values for all directions indicate that the mean elevation rises from 427 m on Snowdon to 655 m in the Carnedds. For each mountain group the lowest corries face northeast.

The reason for these variations is related to the microclimate. The westerly winds were the snow-bearing winds that fed the corrie glaciers so that the hills most exposed to these winds would receive the most snow, as they now have the heaviest rainfall. Thus the snow line would be lower on Snowdonia than on those hill groups to the northeast, and partially in the rain shadow. The northeasterly aspect has the double advantage of being in the shelter from the strongest winds so that the snow could accumulate here more effectively. This aspect is also the most shaded from the midday and evening sun, which is the most effective in melting snow in these northern latitudes.

Exercise 3.14. Analysis of corrie pattern

Analyse the data in Table 3.9a by two-way analysis of variance to assess the effect of aspect on corrie elevation and to test whether the mountain groups are significantly dif-

167

Table 3.9a North Wales corrie data

| Aspect | Altitude with respect to group and aspect | | |
	Northeast	Northwest	Southeast
Group			
Snowdon	1,200	1,700	1,300
Glyders	1,500	2,000	2,000
Carnedds	2,000	2,100	2,200

Table 3.9b Lake District corrie data

Aspect	Northeast	Northwest	Southeast	Southwest
Group				
1 and 2	1,600	1,600	1,700	1,800
3 and 4	2,100	2,000	1,900	1,400
5	1,800	1,800	1,700	1,900

1 Hills south of Buttermere
2 Pillar and Great Gable
3 Scafell hills
4 Langdale fells
5 Coniston fells

ferent with respect to the elevation of the corries. The pattern for the Lake District, given in Table 3.9b can be analysed in a similar way, noting that the groups in this area are arranged from northwest to southeast.

The frequency of corries with respect to orientation can also be studied by statistical tests. The data in Table 3.10, showing orientation frequency for the five groups, can be used to apply the chi square test to establish whether there is a preferred orientation of these corries. Explain the results by reference to the snow-bearing winds and shelter.

Exercise 3.15. Corrie orientation—vector analysis

The mean orientation of a group of corries can be calculated quantitatively and also illustrated graphically. The graphical method is done by drawing a line proportional to the number of corries in the sector 0 to 30 degrees in a direction 15 degrees, and from the end of this line drawing another line proportional to the number of corries in the next sector 30 to 60 degrees in a direction of 45 degrees. The process is continued until all 12 lines have been drawn. An example is shown in Figure 3.15, using the data in Table 3.11. The final point is joined to the origin to give the preferred orientation. Different frequency groups can be used if required. The results can be checked and evaluated to a greater degree of precision by trigonometrical methods as follows: use the data set out in Table 3.11, which gives for each 30 degree sector the frequency of corries and the values of the sine and cosine for the central orientation of each sector. Multiply the sine and cosine values by the frequencies, noting the signs. Sum the results to give $\Sigma F \sin \alpha$ and $\Sigma f \cos \alpha$. The angle of the preferred orientation is then given by

$$\tan \alpha = \frac{\Sigma F \sin \alpha}{\Sigma F \cos \alpha},$$

the tangent can then be converted into

Table 3.10 Frequency corrie data for Lake District

Aspect	Northeast	Northwest	Southeast	Southwest
Group				
1	6	2	0	0
2	11	13	2	2
3	4	2	5	0
4	10	1	3	1
5	7	0	3	1

Table 3.11 North Wales corrie orientation data

Frequency					
000–030	19	sin 015	0·2588	cos 015	0·9659
030–060	22	sin 045	0·7071	cos 045	0·7071
060–090	13	sin 075	0·9659	cos 075	0·2588
090–120	12	sin 105	0·9659	cos 105	−0·2588
120–150	4	sin 135	0·7071	cos 135	−0·7071
150–180	1	sin 165	0·2588	cos 165	−0·9659
180–210	3	sin 195	−0·2588	cos 195	−0·9659
210–240	1	sin 225	−0·7071	cos 225	−0·7071
240–270	0	sin 255	−0·9659	cos 255	−0·2588
270–300	2	sin 285	−0·9659	cos 285	0·2588
300–330	1	sin 315	−0·7071	cos 315	0·7071
330–360	6	sin 345	−0·2588	cos 345	0·9659

degrees. The result should check with the graphical method in both direction and distance, in that the ratio of the X value divided by the Y value should be proportional to the tangent value. The significance of the orientation can be found by using the relationship $L\% = \sqrt{[(\Sigma F \sin \alpha)^2 + (\Sigma F \cos \alpha)^2]}$. The value of $L\%$ can be checked in tables. The length of the vector is an indication of the strength of the orientation pattern.

Rates of erosion in different climates

The importance of climate in the pattern of corries has been considered, although in this instance it is the relatively small scale regional variations that are important in determining the height and orientation of the corries. On a larger scale the climate determines the gross

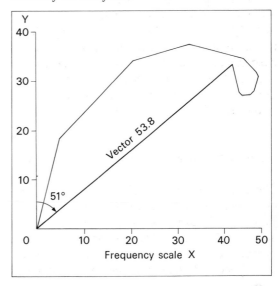

Figure 3.15 Diagram to illustrate corrie vector analysis. Each line is proportional to the corrie frequency in the direction indicated. The vector of mean orientation is shown by the heavy line.

budget of the cirque glaciers and hence their activity and capacity to erode. An approximate idea of the rate of corrie erosion can be determined by measuring the volume of the corries. The variation in size between corries in Alaska and San Juan Range has already been noted. If the time that they have been occupied by corrie glaciers is known it is possible to calculate a rough estimate of the erosion rate. Results suggest that rates vary between 50 m and 1,000 m/1,000 years. Thus even one major geomorphological process can cause very different erosion rates according to the environmental circumstances. The cool oceanic climate of Alaska allows active glaciers to be much more effective as erosion agents than the small glaciers of Colorado in a continental climate.

There is naturally much greater variation when totally different types of erosion are

Plate 55 Glaciers in south Baffin Island, N.W.T., Canada. The vertical view shows a number of features associated with valley glaciers. 1 indicates pyramidal peaks and horns, often linked by arêtes, 2 indicates medial moraines, 3 shows cirques, 4 shows terminal moraines, 5 shows lateral moraines. Note that the moraines in the lower parts of the glaciers form high ridges and this is due to their ice core. 6 and 7 refer to crevasse patterns, 6 being characteristic of compressing flow and 7 of extending flow, where the glacier bed is concave and convex respectively. 8 indicates avalanche chutes leading down to avalanche boulder tongues. 9 indicates nunataks.

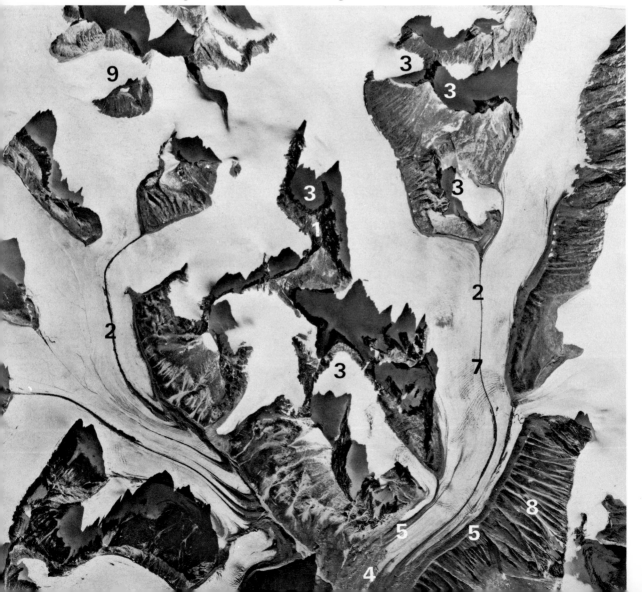

compared. Human interference adds further complications in considering erosion rates in those areas where natural conditions have been disturbed. One difficulty in assessing the rate of glacier erosion is the wide variations that occur. Glaciers can under some conditions be one of the most effective agents of erosion. Where active glaciers are constrained to flow fast through narrow troughs they can erode very effectively, particularly when they are not frozen to their beds. Erosion amounting to more than 2,000 m below sea level has taken place in some of the Antarctic fjords, while the Sogne Fjord in Norway is over 1,300 m deep

and similar depths are reached in some of the fjords of south Chile. Both the last two areas are very wet and have steep slopes into open water, allowing ice to flow fast. (See Pls 56, 58.)

On the other hand, ice can lie passively on a plateau surface, protecting the underlying rock from erosion. This contrast is seen in close proximity in eastern Scotland in the Cairngorms, where deep troughs, eroded by through-flowing glaciers, lie between flattish plateaux that were covered by passive ice.

Another method of estimating the rate of glacial erosion is by measuring the amount of sediment carried from the glacier in its turbid

Plate 56 Irtirbilung Fjord, eastern Baffin Island, Canada. The straight, steep sides of the fjord are shown, with U-shaped valleys entering above the fjord bed. A diffluent col and hanging cirques are also shown.

meltwater streams. Values derived by this means vary from 1,620 mm to 3,110 mm/1,000 years for the Hoffellsjökull in Iceland to 59 mm/1,000 years for the Decade glacier in east Baffin Island. Larger east Baffin Island meltwater streams suggest values of nearly 3,000 mm/1,000 years in a polar climate with cold glaciers. Intermediate values are given for Austrian glaciers and those of the Karakoram. One of the problems of this method is to measure the bedload, which may well be a considerable part, up to a half, of the total load. (See Plate 57.)

The effectiveness of glaciers is illustrated by some comparative figures. Glacial erosion probably lies within the range of 1,000 to 5,000 $m^3 km^{-2} yr^{-1}$ for large active glaciers. Erosion in the Mississippi basin is about 50 $m^3 km^{-2} yr^{-1}$, the Colorado above the Grand Canyon is 230 $m^3 km^{-2} yr^{-1}$ and the Hwang Ho 1,000 $m^3 km^{-2} yr^{-1}$, so that of these large rivers it alone comes anywhere near the value for active glaciers. Under favourable conditions glaciers are probably one of the most effective agents of erosion. Climates that will support active glaciers can

Plate 57 Skeiðara in Iceland after the jökulhlaup of 1954. The icebergs in the view were carried out of the subglacial meltwater channels by the flood water and left stranded on the sandur as the flood subsided.

be considered as one of those associated with really efficient erosion. The time scale must also be considered, however, because glaciers are probably more effective over a short time span when they are modifying a landscape not adapted to their flow than over a longer time span, when they have already removed the weathered material, and adjusted their form to their motion. At this stage positive feedback, enhancing erosion, comes to its limit. There is, for example, a limit beyond which a glacier cannot deepen rock basins along its profile, the limit being set by the curvature of the bed in response to flow. (See Plate 58.)

In non-glacial climates erosion is much more dependent on the type of vegetation, which in turn is controlled by the soil, climate and relief, as well as man's activity. All these variables can influence the vegetation. Thus once again the whole range of physical geography is involved in estimating the rates of erosion to climate and process.

Exercise 3.16. Erosion rates in different climates

Table 3.12 gives values for erosion in $m^3 \times 10^6$/yr for different climates according to Corbel (1964). The rates for arid and humid,

Table 3.12 World climate erosion rates (from Corbel, 1964)

	Arid	Humid	Total
Warm,			
equatorial	1·0	52·5	353·5
intertropical	3·7	20·0	248·7
total	19·0	80·0	459·0
Temperate	175·0	325·0	1,550·0
Cold			
Extra polar	70·0	90·5	710·0
Polar	77·5	75·0	502·5
Total unglaciated	346·2	642·5	
Glaciated	550·0	200·0	4,750·0

Arid: less than 200 m/yr precipitation
Normal: 200–1,500 mm/yr
Humid: more than 1,500 mm/yr

and the totals are given. Work out the rates for normal precipitation regions, and the totals for the different precipitation types. The results can be partly explained by reference to the effectiveness of the different processes of weathering and denudation considered later and also be related to Schumm's and Douglas's analyses illustrated in Figure 3.16.

Plate 58 The Lauterbrunnen valley, Switzerland. Alpine glaciers have eroded a deep, U-shaped valley in several stages. A hanging valley is clearly shown, with its steep lower course, and on the left of the valley a high level Alp with a small hamlet and cleared fields is visible. Note the zig-zag road pattern at the valley head as the road negotiates the steep step in the valley head.

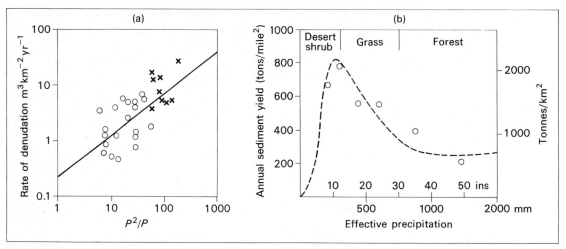

Figure 3.16 Graphs to illustrate denudation in relation to precipitation. (a) Rate of denudation plotted against p²/P. (After I. Douglas, 1967). (b) Annual sediment yield plotted against effective precipitation. (After W. B. Langbein and S. A. Schumm, 1958).

A useful method of measuring the rate of erosion is to record the sediment yield of streams. A number of observations in different areas and under different conditions of vegetation and relief have been made by different people. For example, in France on experimental plots rates of $10 \cdot 5$ m³ km⁻² yr⁻¹ were recorded with values of between $1 \cdot 0$ and $40 \cdot 0$ being recorded from areas of natural vegetation. A forested catchment in Oregon gave a value of $5 \cdot 5$ to $9 \cdot 9$. These values are high for areas under natural conditions. The values increase as the size of the catchment increases, but from the present point of view the relationship between sediment yield and erosion rates and the climatic factors is more relevant. A large range of values has been recorded from Queensland in north Australia to New South Wales in the more equable part of the continent. In Queensland the rainfall is seasonal and very heavy when it does fall, while in New South Wales it is much more evenly distributed, both throughout the year and in any one storm it is less intense. This difference is revealed in the strong contrast in the rates of erosion, values for Queensland range between 2 and 40 m³ km⁻² yr⁻¹ while in New South Wales they vary from about $0 \cdot 6$ to well under 10 m³ km⁻² yr⁻¹.

In order to relate these differences to the climate the value of p^2/P is useful. This ratio, called the Fournier ratio after the person who first used it, relates the precipitation of the month of maximum precipitation, p, to that of the whole year, P. The value of the ratio increases both with the seasonality of the rainfall and with the total annual precipitation. The importance of this ratio is its relationship with vegetation. Dense vegetation cannot grow where rainfall is very seasonal and falls in heavy storms at one season of the year. Heavy rainfall is very effective in eroding the relatively bare soil, hence providing rapid erosion.

Normally the sediment yield is greater in steeper country, which is usually in the upland part of a catchment, but there are exceptions. In southeast Asia, for example, the Irrawaddy has a silt load of $75 \cdot 5$ m³ km⁻² yr⁻¹ at Mandalay, but below this town it increases to 347 as the Chindwin joins the river, and still more sediment raises the value to 507 when the intermittent streams from the dry zone enter the system. The total sediment removal then amounts to 310 m³ km⁻² yr⁻¹ in the combined streams.

In catchments seriously affected by human intervention the values for sediment removal

Plate 59 Clutha River, New Zealand.

can amount to 2,500 $m^3 km^{-2} yr^{-1}$ in the humid tropics, for example in a catchment in Java. It is often difficult to distinguish the human factor from natural variations in sediment yield, because so many areas have been affected by human activity in the form of cultivation on a large scale and locally urbanization has a marked effect.

The effectiveness of precipitation in terms of sediment removal depends not only on the seasonality of the precipitation, but also on the temperature. When the air is hot evaporation is much greater and hence the rainfall less effective, while at low temperatures precipitation may fall as snow and runoff is delayed. Precipitation can be adjusted in relation to temperature to provide a value of effective precipitation, based on a mean temperature of 10 °C. The graph in Figure 3.16 relates the annual sediment yield to the effective precipitation and reveals an interesting variation. The values rise to a peak of

about 2,070 tonnes/km^2 for an effective precipitation of 25 cm, and then fall to values of about 520 tonnes/km^2 for an effective precipitation of 125 cm. The maximum value lies at the boundary between scrub vegetation of the desert areas and grassland, while the lower rate applies in the forest zone, which starts when the effective precipitation exceeds about 75 cm. The vegetation plays a major part in accounting for this pattern. There are several complicating factors that disturb the pattern in some areas. This is related to the fact that over the recent past many areas have been affected by considerable changes in climate and vegetation.

Exercise 3.17. Sediment yield

The values given in Table 3.13 refer to a number of large drainage basins. The size, total average annual suspended sediment load,

175

Table 3.13 *Discharge and sediment load. (From Holeman, J. N. (1968)* Water Reserves Research **4**, 737–47)

River	Drainage basin area ($\times 10^3$ km^2)	Average annual suspended sediment load ($\times 10^6$ tonne)	(tonne km^{-2})	(m^3 s^{-1})	Average discharge (m^3 s^{-1} km^{-2})	(cm^3 s^{-1} km^{-2})
Yellow	673	1,887	2,804	1·501	0·00223	2·23
Ganges	956	1,451	1,518	11·723	0·01226	12·26
Bramaputra	666	726	1,090	12·178	0·01829	18·29
Yangtze	1,942	499	257	21·806	0·01123	11·23
Indus	969	435	449	5·551	0·00573	5·73
Ching	57	408	7,158	0·057	0·00100	1·00
Amazon	5,776	363	63	181·248	0·03138	31·38
Mississippi	3,222	312	97	17·842	0·00554	5·54
Irrawaddy	430	299	695	13·565	0·03155	31·55
Missouri	1,370	218	159	1·954	0·00143	1·43
Lo	26	190	7,308	—	—	—
Kosi	62	172	2,774	1·812	0·02923	29·23
Mekong	795	170	214	11·045	0·01389	13·89
Colorado	637	135	212	0·156	0·000245	0·245
Red	119	130	1,092	3·908	0·03284	32·84
Nile	2,978	111	37	2·832	0·000951	0·951

load/km^2 and average discharge are given. Use Spearman's rank correlation to assess whether these variables are significantly related. The values are already ranked in descending order of total average annual load. Explain the values given in the table for the average suspended sediment load/km^2 as far as possible by reference to the environment in which these large drainage basins lie by reference to their relief, climate and vegetation.

Figure 3.17 illustrates Peltier's morphogenetic regions, which are based on temperature and precipitation data. He also attempts to relate the type of process that is dominant under different climatic regimes to the morphogenetic regions. The problem is complicated, however, because of the recent rapid shifts of the climatic belts in association with the waxing and waning of the ice sheets. Most areas now show a polygenetic landscape that has been affected by a range of climatic types. These are areas that have been subjected to a variety of climatic influences in the recent geological past. Examples include the whole of

the British Isles, where the major part of the country north of the glacial limit has been shaped under glacial as well as periglacial conditions as well as the more recent temperate conditions. In the southern part of Britain, south of the limits of the ice sheet, the landscape was strongly affected by periglacial processes for a considerable time. Between the limits of the older and newer ice advances over the Midlands the area was first affected by glacial modification, then temperate conditions in the interglacial, followed by periglacial conditions during the newer ice advance, with a final return to temperate conditions. Thus many landscapes are not now in adjustment with the processes operating upon them. The slopes of southern Britain are probably gentler as a result of the periglacial processes. This helps to account for the relative lack of serious soil erosion, despite long-continued agricultural activity, on these gentle, and therefore, relatively stable slopes. Further north the oversteepened glaciated slopes have been modified by scree formation and a reduction of gradient. (See Plates 61–64.)

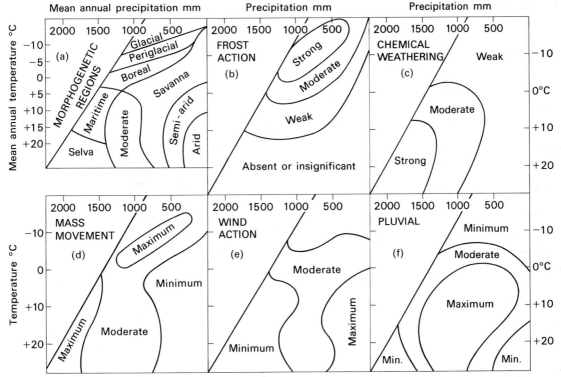

Figure 3.17 (a) Morphogenetic regions. (b) Frost action related to temperature and precipitation. (c) Chemical weathering related to temperature and precipitation. (d) Mass movement related to temperature and precipitation. (e) Wind action related to temperature and precipitation. (f) Pluvial action related to temperature and precipitation. (After L. C. Peltier, 1950)

Exercise 3.18. Morphogenetic regions and processes

Figure 3.17a shows the morphogenetic regions proposed by Peltier, while Figures 3.17b, c, d, e and f, illustrate the climatic conditions under which the five different processes operate most effectively, and those in which they are relatively inefficient. Use the process elements to describe the most effective processes operating in each of the morphogenetic regions. Some processes that are not included in the diagrams should also be mentioned so include these in the analysis. A diagram of weathering regions can be constructed on a similar basis, using the temperature and precipitation controls. One of the weaknesses of this method is the use of mean annual data that can hide a wide range of variation at different seasons. Nevertheless some interest-ing conclusions can be drawn. The results will help to provide relevant information for the previous exercise.

3.7 The time element

Introduction

All aspects of physical geography vary over time and change is fundamental. The vari-ations are sometimes cyclic and regular, such as the seasonal and diurnal changes caused by the earth's rotation round the sun. The tidal fluctuations are more dependent upon the moon's movement with respect to the earth. Other variations follow less regular or more complex patterns of cycles, such as the sun spot variations and the El Niño outbreaks in the waters off Peru. Yet others appear to be

177

Plate 60 A solifluction lobe in west Baffin Island, showing sorting, with fine material in the centre and large stones arranged with their long axes parallel to the lobe at the margin. An altiplanation terrace is seen in the background.

Plate 61 Small scale periglacial sorting in Baffin Island, showing fine material in the centre of the polygons and coarser material around the edges. Sparse vegetation facilitates the development of these patterns.

random in their occurrence, such as major volcanic eruptions and the sequence of wet and dry summers.

Various methods of studying variations over time will be mentioned briefly by reference to medium scale fluctuations. The first example is in the field of climatology. It is concerned with the changes of climate during the glacial and postglacial period, and the advances and retreats of the ice sheets and the associated climatic variations. This time scale covers a period of several million years. It is much shorter than the time interval that separates major ice ages, or orogenic periods, which can be measured in hundreds of million years. The second example is from the field of biogeography. It is concerned with vegetation changes during the postglacial period, which covers the last 10,000 years approximately. The third example, which is related to the first, and also associated with oceanography and hydrology, is in the geomorphological field. It is the change of sea level during the last 10,000 years. This period covers part of the Flandrian transgression, when sea level rose rapidly as the ice sheets melted quickly

away. Changes of sea level have a marked effect on many things, not least the operation of geomorphological processes related to base level. Climate is also involved because this factor was essential in causing the fluctuations of the glaciers, which were the most important element in this particular change of sea level.

There are statistical methods by which a series of observations made over time can be analysed. The observations are called a time series and it is possible to calculate a trend in the values, while the cyclical element can also be isolated. There are also methods whereby one time series can be related to another to ascertain their degree of correlation.

Variation of climate over time—glacial and postglacial

The major events due to climatic change over the last few million years have been the growth and decay of the large ice sheets, of which only the Antarctic and Greenland ones now survive. Not until about 1840 was it realized that ice had once spread over much of

Plate 62 Frost shattering in west Baffin Island, breaking the rocks into angular blocks, note figure for scale. The rocks are those of the Canadian Shield basement of granite, gneissic type.

Plate 63 An ice wedge formed in limestone on Foley Island in Foxe Basin. The ground ice has caused the shattered limestone to heave up along the line of the wedge.

northern Europe and North America, and the evidence that convinced scientists of that time was largely geomorphological. Recent work on the fluctuations of climate that gave rise to glacial and interglacial conditions has been based to a considerable extent on oceanic evidence, where the record is more complete and changes in fauna and other evidence give information concerning past temperatures.

Dating techniques are essential in the study of variations over time and several have been developed. There are the relative methods, such as stratigraphic position, by which the lower deposit is known to be older than the upper one, except in rare circumstances. The degree of weathering and soil formation on till sheets provides evidence of relative age. Palynology, pollen analysis, has been used a great deal for relative dating, because pollen is extremely resistant and very widespread and readily preserved in suitable materials. Pollen studies give information concerning both climatic change and vegetation change. Rates of sedimentation can be used in some circumstances particularly in oceanic data where time horizons are available. These stratigraphic markers include ash layers from volcanic eruptions. The study of ash layers is called *tephrochronology*.

Absolute methods of dating are of greater value. Some methods depend on biogeography, including tree ring analysis, called *dendrochronology*, in which the annual bands can be counted where trees are affected by a strong seasonal growth pattern. Slow growing lichens provide a time scale in suitable areas, and where a growth rate curve can be established absolute dates can be obtained covering several hundred years. The annual sediment layering characteristic of some glacial lakes produces varves or rhythmites that can be used to establish an absolute chronology under favourable conditions. Radiocarbon and isotope methods have provided the most useful information. Radiocarbon dating has a range of up to about 40,000 years, as the half-life of ^{14}C is 5,570 years. The term BP used in relation to ^{14}C dates refers to the years before present based on 1950. Potassium–argon dating is being increasingly used for the earlier part of the Quaternary. It can be used to date volcanic materials, while ^{14}C dating requires organic remains, including wood, peat and shells.

Another useful occurrence that can be used for dating in some situations is the reversal of the earth's magnetic field. At present compasses point to the north, but in the past they have at intervals pointed south. These reversals can be recognized in volcanic rocks, and where they have been dated by an isotope method, a time scale can be established.

From studies all over the world it has been found that major climatic changes were world-wide in their occurrence, although in detail the correlations may not be exact. The general pattern of glacial events suggests four major ice advances, although each was complex and had several interstadials, if not full ice advances within them. The total time involved in the Pleistocene is of the order of $1\frac{1}{2}$ to 2 million years. The Pliocene–Pleistocene boundary may not occur everywhere at the same time, as it is fixed by faunal change, as are most geological boundaries. The Pleistocene does not mark the beginning of the Ice Age. Ice has existed in the Antarctic since the middle Tertiary at least, a period of up to 50 million years, and ice was also present in the northern hemisphere at least 3 to 5 million years ago.

Exercise 3.19. Glacial Fluctuations

The curves shown in Figure 3.18a, b and c provide data on a variety of different time scales and derived from a wide range of data. Use the curves shown in Figure 3.18a, b and c (iv) to compare the variations of temperature, insolation and ice volume. Note the major high points and low points on the three curves going back to 400,000 years ago, and assess how far they agree. Is there any evidence for a double curve of temperature fluctuations and insolation compared with the evidence provided by variations in ice volume? How could you account for this?

The initiation of the whole Ice Age was probably a combination of events, of which the drifting of Antarctica to high southern latitudes was important, as well as the oro-genic upheaval that created the high ranges of the Alps, the Himalayas, and the Rocky Mountain system, as well as the uplift of older mountain ranges, such as the Caledonides of northwest Europe. All these events would help to cause greater snowfall, and once initiated the ice sheets were partly self-generating by positive feedback.

The oscillations of the ice sheets must have another cause or causes as they operate on a shorter time scale and are reversible. Five suggestions made are: (a) variations in insolation, (b) the effect of an open Arctic Ocean, (c) inherent instability of ice sheets, (d) isostatic adjustments, and (e) the closure of the Panama isthmus. This unique event could have diverted warm water to high northern latitudes and increased the precipitation as snow in the north and thus acted as a trigger to start glaciation.

Variations in insolation have been associated with the three cycles studied by Milankovich. The cycles he studied in developing his theories of climatic change are some of the most important regular orbital cycles associated with the movement of the earth around the sun. One of the cycles he used is the variation of the tilt of the earth's axis, which has a period of about 40,000 years. Its average value is 23·5 degrees, but it can vary between 24·5 degrees and 21·1 degrees. The variation in tilt, or the obliquity of the ecliptic as it is called, is most important at high latitudes, where it can affect the amount of incoming solar radiation by 4·17 per cent during the summer half year. When the tilt is high the northern hemisphere gets more insolation. Another cycle is the eccentricity of the earth's orbit. At present the earth is nearest to the sun at perihelion in January. When perihelion occurs in July and tilts are high the situation favours interglacial conditions. This change is called the precession of the equinoxes and it has a period of about 21,000 years. A third cycle is the ellipticity of the earth's orbit, with a period of 96,000 years. These three curves of different periods can be combined in the same way as the tidal curves

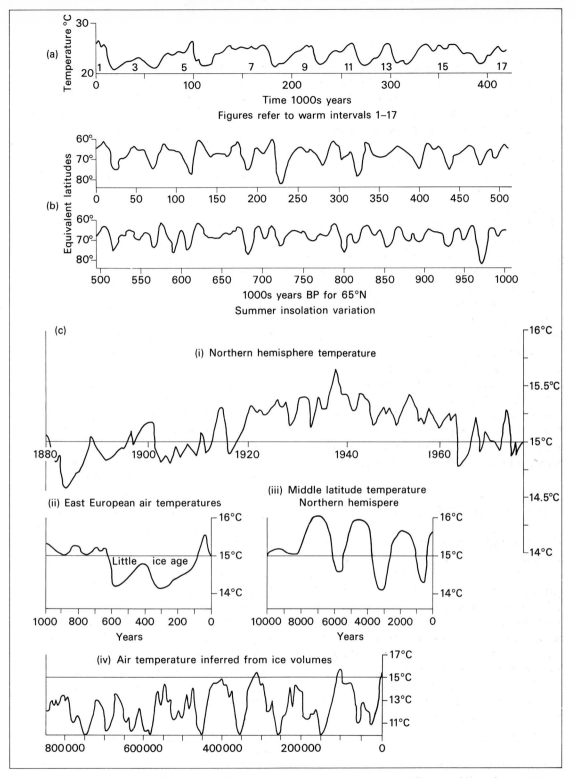

Figure 3.18 (a) Curves to show the relationship between palaeotemperatures, derived from Caribbean deep sea cores, and time. Odd numbers refer to warm intervals. (b) Curve of summer insolation variation in terms of latitude equivalent of 65 °N. (After C. Emiliani, 1966) (c) Curves to show temperature fluctuations on different time scales.

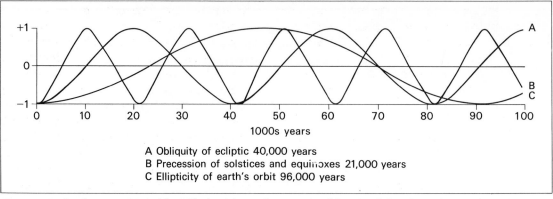

A Obliquity of ecliptic 40,000 years
B Precession of solstices and equinoxes 21,000 years
C Ellipticity of earth's orbit 96,000 years

Figure 3.19 The three curves used by Milankovich are shown. A = obliquity of the ecliptic (40,000 years) B = Precession of the equinoxes (21,000 years) C = ellipticity of the earth's orbit (96,000 years). Timescale is in 1000s years.

to provide a more complex curve.

Exercise 3.20. Summation of curves—Milankovich cycles

Sum the curves shown in Figure 3.19 for the three cycles used by Milankovich to illustrate how they could interact. The effects are different in different latitudes and are of varying dimensions over the earth's surface. The actual changes induced directly by the astronomical factors are probably small, but they can trigger bigger changes by positive feedback mechanisms. They may, therefore, help to induce glacial and interglacial periods in combination with many other variables. In fact the peaks of the Milankovich curve correlate well with climatic events in some areas as shown in Figure 3.18.

One criticism of Milankovich's theory is that it suggests that insolation variations are related to glacial events, whereas it is more likely that precipitation changes are more important in terms of glacier budgets. If, however, the solar radiation cycles are associated in a complex way with earth climate, they must have some effect on the advance and retreat of ice masses. Some of the other suggested causes of climatic fluctuations may also exert an influence and they are not mutually exclusive. Ice sheets, especially the large land-based ones over northern Europe and North America, appear to be inherently unstable, and once initiated

they tend to grow rapidly to a limit, when they bring about their own decay. They melt even faster than they grow, until a threshold is again passed and once again they may advance rapidly, setting up their own climate as they grow. Glaciers respond slowly to a cooling of the climate, lagging behind vegetational changes, and growing more slowly than they decay. The build up may take about 20,000 years, although a quicker time of 1,000 to 5,000 years has been suggested. During the last major glacial advance two glaciations probably took place before about 50,000 years ago. A warm interstadial then intervened until about 30,000 years ago, when the last major ice advance started. It reached a maximum about 20,000 to 25,000 years ago and was still massive about 15,000 years ago. The ice sheets and glaciers had, however, largely disappeared by 7,000 years ago, in North America and north Europe. The Scandinavian ice sheet retreated most rapidly about 11,000 to 10,000 years ago, followed by the Laurentide ice sheet at about 9,000 to 8,000 years ago.

The date by which the Pleistocene ended and the Holocene started is fixed at 10,200 years ago, based on evidence of sudden warming about 11,000 years ago. Small glaciers existed in Britain between these dates in the Highlands of Scotland and other British hills. There is a general similarity of pattern in Europe and North America during the last major ice advance. The Wisconsin in North

America is the equivalent of the Wurm of the Alps and the Weichsel of north Europe and the Devensian or Newer Drift of Britain. The zones following the last glacial are based essentially on vegetational changes and pollen analysis.

Vegetation change over time—postglacial zones

The ice sheets started to withdraw about 15,000 years ago and sea level was very low at this time so plants and animals could migrate across the channel from continental Europe to Britain on dry land. Moraines were exposed in the North Sea and peat developed from vegetation growing on the North Sea banks. Between 13,500 and 14,000 BP there was a park-like landscape with some trees in England and Denmark, the Bølling interstadial, after which ice sheets again thickened and advanced in the zone I, early Dryas zone, which is characterized by the arctic plant, *Dryas octopetala*, which now grows well in Iceland and Baffin Island. Zone II was another warm spell, the Allerød of Europe and the Two Creeks of North America. This period was warm enough for birch trees to grow and lasted from 11,000 to 10,800 BP. After that zone III, called the younger Dryas zone, marks a recurrence of colder conditions and renewed glacier advance. In zone III small glaciers occupied the corries of the British uplands and valley glaciers grew in the Scottish Highlands for the last time. The climate deteriorated over a period of 50 to 80 years, completely killing large forests.

Following this cold phase of the late glacial, climatic amelioration set in quickly. The Scandinavian ice sheet broke up rapidly and the climatic optimum was established. The zones of the Holocene are as follows:

Pre-Boreal	10,200–9,650 BP
Boreal I	9,650–9,140
Boreal II	9,140–8,450
Atlantic I–III	8,450–5,980
Atlantic IV	5,980–4,680
Sub-Boreal	4,680–2,890
Sub-Atlantic	2,890–1,690

The Boreal was generally drier and the Atlantic wetter. The temperature during the warmer spells was 2 °C to 3 °C higher than the present mean values, on the evidence of plants. The natural forests of Britain during the climatic optimum included warmth-loving trees such as lime, and the tree line was high. Trees also grew in the northern islands of Scotland, the Faroes and Iceland. The climate was probably less windy and more anti-cyclonic and Arctic sea ice probably almost disappeared. The climate started to deteriorate from the optimum between 5,000 and 2,500 BP, although better spells, such as the later part of the warm sub-Boreal from 3,200 to 3,000 BP occurred. By 2,900 BP the climate was getting worse and summers became cooler and wetter in the sub-Atlantic period, and Alpine glaciers grew again in size. Temperature declined by about 2 °C and forests receded from the high ground, although effects of human interference are difficult to differentiate from natural effects.

The climate recovered in the early Middle Ages. Between 1,200 and 1,000 BP the Vikings established colonies in Iceland and Greenland and reached North America, but after this the conditions deteriorated again. The thirteenth century was particularly stormy, and the Little Ice Age set in, culminating between AD 1550 and AD 1850. Glaciers reached their postglacial maxima and vegetation again deteriorated. Vines would no longer grow in England, the Viking colonies in Greenland failed, and harvests throughout Europe were bad. Only since AD 1700 have instrumental temperature records been taken. Rainfall figures are fairly reliable from 1727. Britain has the oldest reliable records and longest runs.

Once quantitative data are available it is possible to apply statistical methods to the figures. The rainfall averages for England and Wales averaged for months provide a useful set of values to establish a trend. The figures hide a great range of values because once a mean is used the extremes are hidden. The figures given in Table 3.14 are means calcu-

Table 3.14 Weather data

Rainfall by decades (mm)	West winds No. days	Winter index

Number of days westerly type over British Isles			
1873	110	1919	123
1874	117	1920	149
1875	85	1921	167
1876	74	1922	154
1877	124	1923	182
1878	78	1924	140
1879	75	1925	126
1880	86	1926	145
1881	85	1927	127
1882	110	1928	145
1883	105	1929	129
1884	99	1930	148
1885	87	1931	129
1886	98	1932	142
1887	96	1933	119
1888	89	1934	166
1889	97	1935	139
1890	121	1936	124
1891	101	1937	121
1892	96	1938	138
1893	110	1939	80
1894	107	1940	93
1895	77	1941	94
1896	95	1942	116
1897	84	1943	138
1898	128	1944	114
1899	100	1945	107
1900	107	1946	122
1901	93	1947	95
1902	120	1948	117
1903	137	1949	131
1904	129	1950	113
1905	135	1951	104
1906	147	1952	95
1907	130	1953	105
1908	118	1954	149
1909	99	1955	80
1910	100	1956	97
1911	134	1957	107
1912	122	1958	80
1913	130	1959	75
1914	153	1960	74
1915	96	1961	110
1916	135	1962	98
1917	134	1963	80
1918	134		

Year	Rainfall by decades (mm)	West winds No. days	Winter index
1730	88		−9
1740	83		−4
1750	90		0
1760	97		−6
1770	99		−1
1780	89		−4
1790	96		−2
1800	89		−2
1810	97		−2
1820	99		0
1830	97		−3
1840	98		−3
1850	80		−1
1860	98		+8
1870	105		0
1880	100	95	−2
1890	90	100	−3
1900	93	120	+2
1910	103	134	+5
1920	102	142	+8
1930	101	130	+4
1940	93	112	+1
1950	100	100	0

lated over decades, still further reducing the range.

Exercise 3.21. Rainfall analysis—trend of time series

Use the data in Table 3.14 to establish the trend for rainfall over the period from 1730 to 1950, covering 23 decades. Plot the data and draw the trend line. The instrumental records show that the climate is continually fluctuating. These fluctuations involve a complex series of meteorological elements so that their combined effects are rarely similar, giving rise to the very complex pattern revealed in the longer-term trends already considered. These complex interactions account for the difficulties of weather and climate predictions at all scales of both space and time. It would be very useful if it were possible to predict future climatic oscillations and to know whether a new major ice advance is imminent, as suggested by some authorities. To make this possible eventually a useful approach is to analyse the recorded and deduced variations of conditions, to attempt to establish trends and cycles, and to correlate time series. These exercises provide an introduction into some of the methods available for this type of analysis.

Plot the mildness/severity index for the period 1730 to 1950 and draw the trend line. Information concerning the frequency of west winds for 1880 to 1960 in decadal values can be correlated with the mildness/severity index over this period. Three sets of regression and correlation values result from this analysis. Explain the results in terms of the planetary wind systems to be considered in more detail in the next chapter.

Exercise 3.22. Calculation of running mean for climatic data

Use the data in Table 3.15 to establish the 10-year running mean for the winter mildness/severity index and the frequency of westerly winds. The 10-year running mean is calculated by adding each 10-year set of values and dividing by ten, placing the result against the central year. The curve derived in this way is much smoother than the year to year values. Plot both annual values and the 10-year running mean for Britain. Figure 3.20 shows the time series for Germany and Russia.

Exercise 3.23. Correlation of time series—climatic data

Use the data in Table 3.15 to correlate the variations in winter mildness/severity index for Germany, Russia and Britain. Each pair can be correlated. Count the number of times the p value is greater or less than the $q - 1$ value, e.g. $p_i - q_{i+1}$ sign and $q_i - p_{i+1}$ sign, and calculate the product of each pair of signs, thus $- \times - = +$, $+ \times + = +$, and $- \times + = -$.

Count the pluses and minuses and calculate chi square on the assumption that there are an equal number of pluses and minuses in the set. The method is a non-parametric one, based on increases and decreases of the two sets of data to be correlated. The calculations are simple and provide a useful means of establishing the relationship between two time series. It is not possible to establish which one causes the other, or whether they both depend on a third or more variables. In this instance it is likely that the westerly type of weather is the cause of the greater rainfall in England and Wales.

Vegetational evidence has played an important part in establishing the climatic changes during the postglacial period. Pollen analysis has provided the zones already mentioned, which were originally based on macroplant remains collected a long time ago by Blytt and Senander. Forest dominated the British Isles during the postglacial apart from the cold phases of zones I and III. Botanical evidence includes the level of the tree line, the limit of vineyards and average dates for grape picking. Vineyards were successfully established in England during the warmer periods and extended as far north as York. The vineyards were particularly successful between 1100 and 1300, when they competed seriously with French ones. Vines can withstand some frost, but are killed by severe frost. They particularly need summer sunshine and not too much rain. The local terrain was important in locating vineyards in these marginal areas. Lowering of the tree line and the upper limit of vineyards in south Germany heralded the Little Ice Age during the period after AD 1300. The tree line fell by between 100 m and 200 m between AD 1300 and AD 1500 in the Vosges and Black Forest.

Plants and animals spread very quickly over an area so they provide good evidence of climatic changes and their remains can be found in fossil form. The study of beetles has added much information to changing conditions during the interglacial and postglacial periods. Biological evidence in the

Table 3.15 Winter mildness/severity index for Europe near 50°N. (The values refer to the number of unmistakably mild months (Dec., Jan. and Feb.) minus the numbers of unmistakably severe months (Dec., Jan. and Feb.) per decade. Britain about 0°E, Germany about 12°E, Russia about 35°E)

Decade	Britain	Germany	Russia	Decade	Britain	Germany	Russia
1100s	− 1	− 2	0	1530s	− 3	+ 6	− 6
1110s	− 4	− 5	0	1540s	− 2	− 1	+ 6
1120s	− 3	− 3	− 6	1550s	− 1	− 3	−12
1130s	0	− 3	− 6	1560s	− 7	− 3	+ 4
1140s	− 3	− 9	+ 2	1570s	− 3	− 1	0
1150s	− 6	− 3	0	1580s	− 7	− 5	+12
1160s	0	− 2	+ 6	1590s	− 5	− 5	0
1170s	+ 2	+ 2	+ 4	1600s	−10	0	− 6
1180s	+ 6	+ 5	− 6	1610s	+ 2	− 3	−10
1190s	+ 7	+ 5	+ 6	1620s	− 7	− 3	0
1200s	+ 1	− 3	− 6	1630s	− 5	+ 2	−10
1210s	− 2	− 5	−12	1640s	− 3	+ 5	−36
1220s	+ 1	+ 6	+ 2	1650s	− 9	− 7	−24
1230s	+ 6	+ 1	0	1660s	− 7	− 3	−14
1240s	+ 1	+ 3	− 6	1670s	0	− 6	−24
1250s	+ 3	− 5	0	1680s	− 2	− 7	0
1260s	− 3	− 1	0	1690s	− 7	− 5	− 6
1270s	+ 7	0	+ 6	1700s	0	+ 1	− 6
1280s	+ 3	+ 3	+18	1710s	− 2	+ 1	0
1290s	+ 2	+ 5	0	1720s	− 4	− 1	−12
1300s	− 1	+ 1	+ 6	1730s	+ 9	+ 3	+ 6
1310s	− 1	−10	− 6	1740s	− 4	− 1	− 6
1320s	0	− 1	0	1750s	0	− 5	+10
1330s	− 4	− 1	0	1760s	− 6	0	−12
1340s	+ 2	+ 4	0	1770s	− 1	+ 4	− 6
1350s	− 3	− 4	0	1780s	− 4	− 4	−20
1360s	+ 4	− 2	− 6	1790s	− 2	− 2	0
1370s	− 3	− 5	−12	1800s	− 2	− 1	+ 1
1380s	+ 2	+11	− 8	1810s	− 2	− 3	− 1
1390s	− 2	+ 2	− 6	1820s	0	0	0
1400s	− 3	− 2	−22	1830s	− 3	− 2	− 1
1410s	+ 2	− 1	−10	1840s	− 3	− 2	+ 1
1420s	+ 4	+ 7	− 6	1850s	− 1	− 2	+ 3
1430s	−12	−18	−10	1860s	+ 8	+ 6	+ 1
1440s	− 5	− 2	−28	1870s	0	0	− 3
1450s	− 1	− 3	+ 8	1880s	− 2	0	+ 1
1460s	− 9	− 6	− 6	1890s	− 3	− 1	+ 1
1470s	+ 3	+ 1	− 6	1900s	+ 2	0	+ 2
1480s	− 6	+ 1	−10	1910s	+ 5	+ 6	+ 1
1490s	− 5	−12	− 8	1920s	+ 8	+ 6	+ 2
1500s	+ 7	− 3	− 6	1930s	+ 4	+ 4	+ 5
1510s	− 2	+ 2	0	1940s	+ 1	− 1	− 1
1520s	+ 2	+11	− 6	1950s	0	+ 1	+ 2

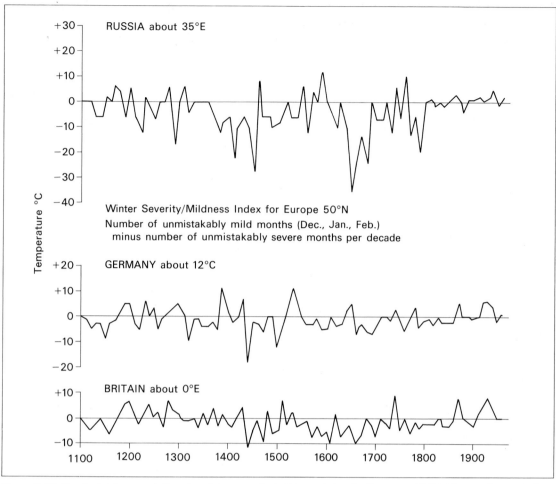

Figure 3.20 Graphs to show variations in winter severity/mildness index for Europe at 50°N. The graphs cover the period from 1100 to 1950 and show the hard winters of the Little Ice Age. Data are from H. H. Lamb.

oceans is also useful. Deep cores from areas undergoing continuous sedimentation of deep sea oozes provide valuable data. Some species have developed the habit of changing their coiling direction in response to changes in temperature. When the deep sea cores can be dated by absolute methods they also provide a valuable time scale. The results show a cyclic change of colder and warmer conditions, with at least ten cold stages during the Cainozoic. A close correlation with Milankovich's cycles has been established.

Changes of sea level—Flandrian transgression

The change of sea level during the last 15,000

years has been dominated by the rapid melting of the large land-based Laurentide and Scandinavian ice sheets, which were near their maxima at this time, but had almost completely melted away by 7,000 years ago. Different factors affect sea level and should be distinguished as far as possible. There are very few stable areas, where a complete record of the eustatic changes of sea level can be recorded. Eustatic changes are caused by variations in the capacity of the oceans basins and the amount of water in the oceans. Glacioeustatism has been by far the most important aspect of the recent eustatic changes, and sea level has oscillated extremely rapidly during

187

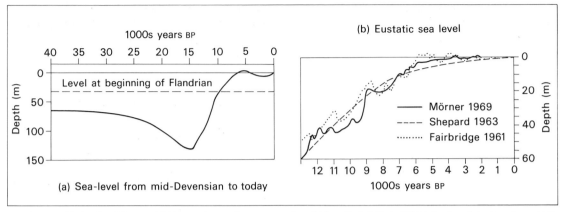

Figure 3.21 (a) Sea level from mid-Devensian to the present day. (After G. F. Mitchell) (b) Sea level during the middle and later part of the Flandrian transgression, showing three different curves.

the glacial fluctuations.

The major pattern is of a steadily falling level superimposed on the much more rapid oscillations associated with the waxing and waning of the Laurentide and Scandinavian ice sheets. The general fall of the later Tertiary was probably associated with the growth of the Antarctic ice sheet, which is much more stable than the northern hemisphere ones. Another important process causing a fall of sea level in the Tertiary was the orogenic upheaval that gave rise to the Alpine mountain chains and the compensating deep sea basin formation.

Recent curves suggest that sea level was 70 m below its present height about 15,000 years ago. It rose very rapidly after this time until about 4,000 years ago, when the level stood about 2 m below the present. Three estimates of the eustatic rise over the last 13,000 years are shown in Figure 3.21. In the last 4,000 years the level has risen slowly to its present height. As the figure shows there is some disagreement in detail both from place to place and with different workers, partly because of the many uncertainties involved. These include the relationship between sea level and the surviving effects of marine activity, such as raised beaches, and the problems of establishing sea level relative to these features and to dating the features

accurately. Another problem is the lack of stability of the land. Even the relatively stable areas where the record is most complete, such as along the southeast coast of the United States and Holland, there are considerable variations from place to place, and stability is far from assured. In fact subsidence is taking place quite fast in Holland and along parts of eastern North America.

Some curves show periods of sea level higher than the present during the last few thousand years. Some of these may be local effects, due to storm surges or a local zone of warping, but it does seem likely that sea level was a bit above the present a few thousand years ago. Sea level has risen by about 12 cm–30 cm during the twentieth century as glaciers have continued to recede. Sea level should have fallen by a similar amount during the Little Ice Age, when glaciers expanded. It is, however, difficult to be sure of such small changes with present methods of analysis.

Exercise 3.24. Eustatic sea level changes

Plot the data in Table 3.16, using Shepard's figures. The rapid rise in the early half of the period and the slowing down later suggest a logarithmic form of the curve. Test the

Table 3.16 Eustatic sea level changes (value relative to present)

10³ years BP	Mörner	Shepard
1	0	− 0·5
2	− 0·5	− 1·0
3	0	− 2·0
4	− 1·0	− 3·0
5	− 2·0	− 4·0
6	− 5·0	− 7·0
7	−10·0	−10·0
8	−20·0	−16·0
9	−20·0	−22·0
10	−40·0	−31·0
11	−45·0	−40·0
12	−43·0	−48·0
13	−60·0	−58·0

accuracy of this suggestion by plotting the logarithmic values on ordinary graph paper or the values given on semi-logarithmic paper. What can you deduce from the result?

Exercise 3.25. Uplift curves—isostatic recovery

Eustatic sea level affects all the coasts of the world, but sea level changes are locally complicated by isostatic recovery of the land from beneath a weight of ice as it melts. The uplift is related to the ice thickness, which is itself related to the position of the area relative to the ice margin. The time of deglaciation is also involved as the ice retreats past the point under consideration. The uplift curve is a double exponential curve. First it rises slowly as the ice thins, then accelerates to a maximum value at the inflexion point, as the ice retreats past the place under consideration. The rate then slowly decelerates until the uplift is

Plate 64 Raised beaches and degraded cliff lines in Torridonian Sandstone, Gruinard Bay, Wester Ross, Scotland. In the distance the Lewisian Gneiss forms the knobbly hillsides. The raised beaches result from postglacial isostatic uplift.

Plate 65 Raised beach on Magee Island, near Larne, Northern Ireland. The raised wave-cut platform is well developed and the abandoned cliff is as yet little degraded. Note the two sets of waves in the background and the short-crested waves in the foreground. These waves are steep and have breaking crests in places, indicating active generation at the time.

complete. In the example given the ice retreated from the end of the Henry Kater Peninsula in east Baffin Island 10,000 years ago. Andrews (1975) has calculated the percentage uplift for each thousand years since uplift started after deglaciation. In this area the marine limit is at 28 m, and the eustatic effect is 31 m rise, giving a total uplift of 59 m. Draw an uplift and emergence curve using Shepard's eustatic correction given in Table 3.16, and the following figures, which gave the percentage uplift for each thousand years since deglaciation. (See Plates 64, 65.)

10,000 yr	100·0 m	5,000 yr	86·5 m
9,000	97·4	4,000	80·2
8,000	94·6	3,000	70·4
7,000	93·4	2,000	56·0
6,000	91·1	1,000	33·0

3.8 Conclusion

The wide ranging discussion of regional scale physical geography covered in this chapter should have shown the close interrelations of the different aspects of physical geography. It is impossible to understand one aspect without reference to most of the others. The connections are mostly two-way in character, each variable affecting the other, and in turn being affected by it. Feedback relationships are, therefore, very important. The physical make-up of the land, the field of geomorphology, is fundamental in providing the land or sea floor on which the other elements operate. The land and sea floor, however, are affected by the external forces acting upon them and modified by them. Thus the climatic elements are involved, as it has been shown that climate affects the processes of erosion and deposition. Climate is in its turn influenced by the relief on all scales, and strongly so in the medium scale.

Climate in turn plays a vital role, together with rock type and relief, on the formation of the soil, as has been exemplified in the discussion of soils, vegetation and life in different major climatic environments. The soil and climate control the vegetation, and from the geographical point of view, man's use of the land must also be considered. Relief must not be forgotten because both the absolute height and the slope are important variables in both natural vegetation and man's use of the land. Fertility of the soil is vital in the situation of rapidly growing population and the need for more food from the land. Fertility is not only a property of the land surface, but it also applies to the sea, where examples of marine deserts and zones of exceptional fertility have been examined.

The oceans are of considerable and growing importance in physical geography. They cover 70 per cent of the earth's total surface areas and play a vital role in many fields. They maintain the reservoir of water without which life on land would be impossible. Both on the regional and global scales the interaction of the oceans and atmosphere is complex but essential. The latter aspect will be further considered in the next chapter. Evaporation takes place over the ocean surface, carrying moisture up into the air where the wind can carry it over the land and precipitation provides the life-giving moisture that maintains organic activity on land. This is one of the most significant steps in the hydrological cycle, which involves all aspects of physical geography, and which provides a very important link between the different aspects of the subject.

References and further reading

*Andrews, J. T. (1975) *Glacial systems*. Duxbury Press, N. Scituate, Mass.

*Barry, R. G. and Chorley, R. J. (1976) *Atmosphere, weather and climate*. 3rd Edn, Methuen, London.

Corbel, J. (1964) L'érosion terrestre, étude quantitative (méthodes–techniques–résultats). *Annales Géographiques* **73**, 385–412.

Cotton, C. A. (1945) *Earth beneath*. Whitcombe and Tombs, Christchurch.

Deacon, G. E. R. (1959) The Antarctic Ocean. *Science Progress* **47**, 637–60.

Douglas, I. (1967) Man, vegetation and sediment yield of rivers. *Nature* **215**, 925–8.

Emiliani, C. (1966) Isotopic paleotemperatures. *Science* **154**, 851–7.

Fairbridge, R. W. (1961) Eustatic changes of sea level. *Physics and chemistry of the earth*. Vol. 4, Pergamon, Oxford.

Godwin, H. (1941) Pollen analysis and quaternary geology. *Proceedings of the Geological Association* **52**, 328–36.

Godwin, H. (1956) Quaternary history and the British flora. *Advances in Science* **50**, 1–7.

*Hardy, A. (1956) *The open sea. Vol. I The world of plankton*. Collins, London.

*Hardy, A. (1959) *The open sea. Vol. II Fish and fisheries*. Collins, London.

*Lamb, H. H. (1966) *The changing climate*. Methuen, London.

Langbein, W. B. and Schumm, S. A. (1958) Yield of sediment in relation to mean annual precipitation. *Transactions of the American Geophysical Union* **39**, 1076–84.

Lee, A. J. Marine life in relation to the physical environment. In Lake, P. (1958) *Physical*

Geography. 4th Edn, Ed. J. A. Steers. Cambridge University Press. pp. 420–55.

*Leopold, L. B. (1974) *Water: a primer*. Freeman, San Francisco.

*Manley, G. (1952) *Climate and the British scene*. Collins, London.

*Maunder, W. J. (1970) *The value of the weather*. Methuen, London.

McCall, J. G. (1960) in Lewis, W. V. (Ed.) Norwegian cirque glaciers. *Royal Geographical Society Research Memoirs* 4, 39–62.

Mörner, N. A. (1969) Eustatic and climatological changes during the last 15,000 years. *Geol. Mijnbouw* **48**, 389–99.

Packard, W. P. (1947) Lake Coleridge catchment: a geographical survey of its problems. *New Zealand Geographer* 3, 19–40.

Pardé, M. (1933) *Fleuves et rivières*. A. Colin, Paris.

Peltier, L. C. (1950) The geographical cycle in periglacial regions as it is related to climatic geomorphology. *Annals of the American Association of Geographers* **40**, 214–36.

Østrem, G. (1975) Glaciologiske undersokelser i Norge 1974. *Rep. 5 Norges Vassdragsdirektoratet Hydrologisk Ardeling, Oslo*. 71 pp.

Shepard, F. P. (1963) 35,000 years of sea level. In *Essays in marine geology in honor of K. O. Emery*. University of California Press, Los Angeles, pp. 1–10.

Sverdrup, H. U., Johnson, M. W. and Fleming, R. H. (1946) *The oceans*. Prentice-Hall, Englewood Cliffs.

Warwick, G. T. (1964) in *British Isles: a systematic geography*. Eds J. Wreford Watson and J. B. Sissons. pp. 91–109.

Wimpenny, R. S. (1953) *The plaice*. Buckland lecture for 1949. Arnold, London.

CHAPTER 4

Global scale studies on land, and in the sea and air

4.1 Introduction

On the global scale the world is considered as one unit. In considering the physical geography of the earth as a whole it is essential to remember that it has one continuous closed surface, which is spherical in shape. The surface of the earth is more important to geographers than the interior, but some information concerning the processes at work within the earth is relevant to an understanding of the pattern of land and sea. The earth has a rocky crust overlain over 70 per cent of its surface area by ocean and completely enveloped by air. Half of the mass of the earth's atmosphere lies within the lowest 5·6 km, 90 per cent within 32 km and 99·9 per cent within 80 km of the surface.

The highest mountain, Everest, is about 8,880 m high, the greatest ocean depth is rather greater than this, giving a total vertical extent of nearly 20 km, compared to the radius of the earth, which is about 6,370 km (equatorial radius 6,378 km, polar radius 6,357 km, radius of sphere with the earth's volume

6,371 km). The earth's surface is nearly smooth when its height range is compared with its surface dimensions. The average elevation of the land is 623 m and the average ocean depth is 3·8 km, a ratio of 1:3,353 of the diameter. The ocean is, therefore, a thin film of water relative to the earth's diameter. Nevertheless the thin envelopes of water and air are vital to life on earth. The biosphere exists where these envelopes meet and interact. Life can only be created and flourish in the presence of moisture, sunlight and nutrients, close to the surface of the land and sea.

Some of the essential features of the world-wide pattern of land and sea, and the processes that control this pattern are considered in this chapter. The major relief features of the earth are closely related to the processes that determine the pattern of land and sea. This is the geomorphological aspect of world physical geography. The three-dimensional circulation within the oceans and the atmosphere are vital to life on earth and are next investigated briefly. The three-dimensional pattern must be stressed as the vertical element is most important in the circulation of both ocean and atmosphere. The interaction between the two circulating systems is also very significant. This is explored further in the next section in which the global aspects of the hydrological cycle are discussed. Some elements of this

Plate 66 An aerial view of Mount Ngauruhoe, one of the active volcanoes of the North Island of New Zealand. The volcano is situated near the subduction zone of the destructive plate margin in the southwest Pacific Ocean.

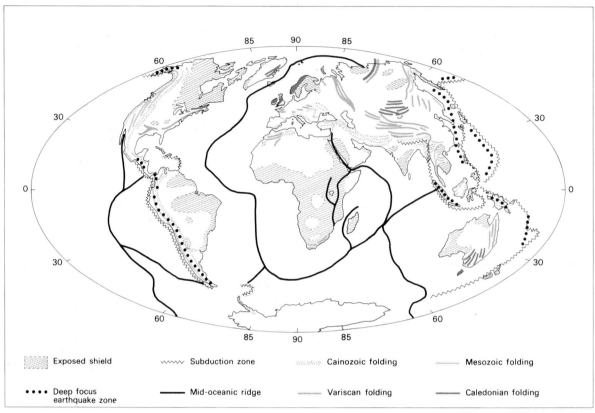

Figure 4.1 World pattern of plates, shields, zones of folding and subduction, oceanic ridges, deep sea trenches and deep focus earthquake zones.

major cycle have already been considered, but the world-wide pattern and processes draw together some important elements of the previous two chapters. Some relevant biogeographical aspects are also mentioned.

4.2 Within the earth—plate tectonics

The pattern of land and sea is a fundamental geographical fact, that exerts its influence throughout all aspects of physical and human geography. Recently a great many lines of evidence have combined to support the modern theory of how the earth's crust behaves and how the pattern of land and sea evolved and is evolving. Most people now accept the theory of plate tectonics, which is closely related to the concept of sea floor

spreading. Figure 4.1 illustrates the pattern of plates and their relative movement. The term *plate* is used to describe these relatively rigid pieces of crust because they are very thin relative to their surface area.

The earth's crust consists of relatively light material resting on the denser material of the mantle, which in turn surrounds the core. The crust is of two distinct types, and the recognition of the complete difference between the continental and oceanic crust of the earth was one of the most important steps in the development of the theory of plate tectonics. The continental crust consists of light, granitic type rocks and is about 35 to 40 km thick, while the oceanic crust consists of denser, basic type rocks, and is only about 6 km thick. The two types of crust can remain in equilibrium because the thickness of the continental crust is compensated by its light

Table 4.1 Crustal types—density and thickness

Oceanic crust			Continental crust		
Density	Thickness	Type	Density	Thickness	Type
1·0	6 km	water	2·65–2·95	36 km	continental crust
2·9	6 km	sediment and ocean crust	3·3	4 km	mantle
3·2–3·3	28 km	mantle		40 km	
	40 km				

Table 4.2 World area—height data

Height (m)	% area		cumulative % area	
	A	B	A	B
more than 5,000	0·1	0·2	0·1	0·2
4,000– 5,000	0·2	0·2	0·3	0·3
3,000– 4,000	0·4	0·4	0·7	0·7
2,000– 3,000	2·7	1·4	3·4	2·1
1,000– 2,000	4·5	4·3	7·9	6·4
500– 1,000			13·5	12·3
300– 500	20·6	22·1	19·0	18·6
0– 300			28·7	28·7
0– −200	8·8		34·7	
−200−−1,000			37·6	
−1,000−−2,000	3·7		41·3	
−2,000−−3,000	4·0		45·3	
−3,000−−4,000	15·3		60·6	
−4,000−−5,000	21·5		82·1	
−5,000−−6,000	15·0		97·1	
−6,000−−7,000	2·8		99·9	
more than −7,000	0·1		100·0	

A = altitude of actual surface (or of ice caps). Maximum height 8·85 km Average height of land 0·8 km
B = altitude of solid crust after isostatic recovery. Maximum depth 11·04 km Average depth of sea 3·7 km

weight, compared with the thinner, denser oceanic crust.

Exercise 4.1. Crustal types and hypsometric curve

Assess the isostatic balance of the crust of the continents and oceans using the values given in Table 4.1. Take the mean value where a range is specified. Plot the world hypsometric curve, using the data given in Table 4.2. Relate the two parts of the exercise by reference to the values calculated from the data in Table 4.1. Comment on the form of the hypsometric curve in terms of the structure of the earth's crust.

As Figure 4.1 shows, all the plates, with the exception of the Pacific plate and some of the minor plates, have a core of continental crust. In most of the continental crustal areas there are relatively large expanses that have been tectonically undisturbed for a very long time. These areas are the continental shields, which are also shown on the figure. They consist of very old rocks that have undergone upheaval in the remote geological past, but have been quiescent for at least the last 700 million years, when the Palaeozoic period started. The shields include the Baltic area, the Canadian shield, large parts of Africa, much of western Australia and eastern South America, as well as parts of northern and eastern Asia. The plates are rigid areas of crust that move around over the earth's surface. Along their margins the internal or endogenetic forces are active and create forms that have a surface expression. The volcanoes and earthquakes are concentrated along these linear belts.

The pattern of land and sea gives a clue to the operation of plate tectonics, first appreciated towards the beginning of the twentieth century by Wegener. He suggested the idea of continental drift. He was a climatologist and was intrigued by the presence of fossil glacial deposits all belonging to the Permo-Carboniferous period in many low latitude areas, including Africa, eastern South America, India and Australia. At this time the coal swamps of the higher northern latitudes, Britain and North America, indicate a much milder climate in these zones. To account for this distribution Wegener suggested that all the land masses were grouped together to form one major continent, which he called Pangaea. The glacial deposits and associated cold climate flora could then conveniently be grouped around the south pole, situated near southern Africa. He then argued that this super-continent subsequently split up and the separate pieces drifted apart to their present locations. The similarity of shape of the two sides of the Atlantic Ocean strongly supported this idea.

Since Wegener first suggested this idea the fit of the two continental margins has been studied quantitatively by computer and the best fit pattern shows that they do in fact fit together remarkably well, too well for the pattern to be fortuitous. For many decades, however, the idea of continental drift was not accepted by the majority of scientists, partly on the grounds that no force sufficient to cause the continents to drift through the oceans could be imagined, let alone proved, as the oceans were considered more rigid than the continents. Even in 1950 the theory of continental drift was far from being generally accepted. The great advances in oceanography since the Second World War have, however, altered the situation very considerably and in the oceans much of the most compelling evidence is found.

During the last 15 years a great deal of evidence in favour of the new global tectonics has been assembled and many aspects of the pattern of land and sea fit well in this scheme. The diagrams (Figure 4.2) show the distribution of land and sea at various stages during the past.

Exercise 4.2. Plate tectonics

Project the movement of the plates into the future by using the directions of plate movement shown on map 4, depicting the present situation, and draw a map for the situation 50 million years hence. The earth is now at one stage in an ever changing pattern, in which the pieces remain recognizable, but their relative positions vary.

The plates are the stable blocks that move about over the earth's surface, and the oceans change most fundamentally through time. The oceans have provided the geological and geophysical evidence that explains much of the global tectonics. The maps in Figure 4.2 show that the Atlantic Ocean is a relatively new feature and only started to open about 150 million years ago, when South Africa split from South America, and North America started to split from northwest Africa. The split gradually spread northwards up the Atlantic, with Greenland separating from northern Europe and the Canadian Islands relatively late in the Tertiary. The Indian Ocean also opened up as India drifted northwards, Antarctica drifted southwards and Australia northeastwards.

The movements probably played an important part in the initiation of the Cenozoic Ice Age, as Antarctica drifted towards the South

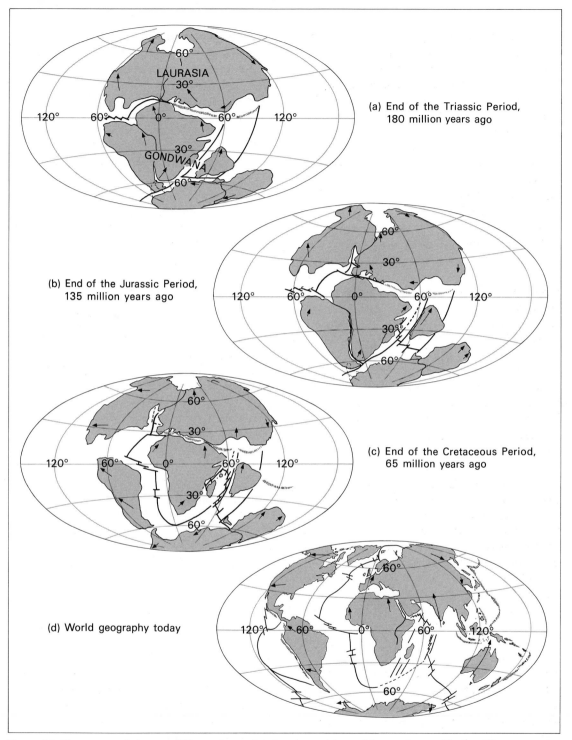

Figure 4.2 The pattern of plates at different times, showing the positions of the continents. (a) 180 million years ago at the end of the Triassic. (b) 135 million years ago at the end of the Jurassic. (c) 65 million years ago at the end of the Cretaceous. (d) The present situation. (From R. S. Dietz and J. C. Holden, 'The breakup of Pangea'. © 1970 Scientific American, Inc. All rights reserved)

Pole and the Arctic Ocean became nearly land-locked. Both movements would encourage cooling and ice formation, first in the Antarctic and later in the northern hemisphere. The process probably started well back in the Tertiary, within the last 50 million years. The major mountain ranges are also related to these movements, with the Alps and Himalayas marking the crumpling at the margins of two colliding plates. The long chain of the Rocky Mountains and Andes all along the western seaboard of North and South America mark the position where the western-moving American plate met minor plates or the Pacific plate, the only oceanic one. The Atlantic and Indian Oceans are now growing at the expense of the Pacific Ocean, which may eventually be eliminated as new oceans develop.

The evidence from the oceans has been one of the most important elements that has led to the general acceptance of the theory of plate tectonics. The theory is closely linked with the process of sea floor spreading that dominates the movement of the plates.

The major oceanic relief feature is the ridge system that runs across the Arctic Ocean, down the middle of the Atlantic, bisecting the distance between South Africa and Antarctica as it passes into the Indian Ocean. One branch turns north as the Carlsberg Ridge, to enter the Red Sea, and the other branches eastwards to pass equally between Australia and Antarctica into the Pacific Ocean. It becomes the East Pacific Rise, which continues across the ocean and passes up the Gulf of California. One branch runs southeastwards, as the Easter Island ridge, towards the coast of Chile, while another runs northwest to form the Christmas Island ridge.

Where the ridge is well developed in the North Atlantic between 17 °N and 54 °N it is 800 km to 1,400 km wide and from 2·5 km to 4 km high. The ridge rises in a series of steps, becoming more rugged as the centre is approached. Along the centre of the ridge there is a central rift valley. This valley is from 1,000–4,000 m deep and 25–50 km wide. It is thought that the oceans are spreading laterally from the rift. A similar feature on land is the Great Rift Valley of Africa, which is connected with the rift of the Red Sea and Gulf of Oman. It is possible the land rift is the beginning of a new ocean. The Red Sea exemplifies a slightly later phase when the ocean has started to spread.

Not all the ocean ridges have a well developed rift valley, which is characteristic of the Atlantic and Indian ocean ridges. It has been suggested that the oceanic ridges go through a series of stages of development. Initially they are smooth, broad rises, as illustrated by the East Pacific Rise. At this stage the split has yet to start and no central rift valley has formed. It is probably imminent, however, because of the very high heat flow on the embryo ridge, indicating upwelling of hot material from below. In time the hot material breaks through, causing a rift to form, as two plates break apart and the ocean begins to spread out. At this stage of maturity the central rift valley becomes a marked feature of the system and spreading is active. In time the ridge becomes dead and spreading ceases. When this happens the ridge tends to be steeper but is no longer a centre of earthquake and volcanic activity. The mid-Pacific Mountains exemplify this stage. Another aseismic ridge is the Lomonosov Ridge that runs across the Arctic Ocean under the north pole. This ridge differs from the tectonically active arctic ridge that lies midway between the Lomonosov Ridge and the north Eurasian coast. The very long Darwin Ridge that runs southeast to northwest across the Pacific is another senile ridge that has sunk and is now marked by many old volcanoes, sea mounts and guyots.

The most important feature of most of the active ridge system is its position central in the ocean. In order for the ridge to maintain its central position it is frequently dislocated by transform faults. The crust moves in opposite directions on either side of the dislocation. The side-stepping of the ridge allows it to maintain its central position as it is traced through the Atlantic Ocean to fit the bulge of South

America and the hollow of Africa. This pattern is necessary to make the moving plates adjust to the spherical surface across which they are moving. The pattern enables the pole from which the plates are separating to be fixed, as the transform faults lie at right angles to the arcs of spreading. The Atlantic appears to be spreading from a point near the south of Greenland.

The theory of sea floor spreading suggests that new ocean crust forms along the centres of the ridges where the rift valley develops. New oceanic crust forms from material welling up at the rift centre and this moves outwards to make room for more material, so in time the ocean crust expands and the ocean becomes wider. This process has created the Atlantic Ocean during the last 150 million years. In the late Cretaceous period there was only a narrow gap between Britain and North America, and Greenland was joined to northern Europe. The northern part of the Atlantic Ocean developed during the early Tertiary period and gradually widened during the middle and late Tertiary.

Exercise 4.3. Spreading rates

Calculate the spreading rates from the following measurements of distance from the centre of the mid-oceanic ridge to magnetic reversals of known age.

(1)	North Atlantic at 30°N	2·1 degrees of longitude
(2)	South Atlantic at 30°S	1·65 degrees of longitude
(3)	South Pacific at 40°S	4·45 degrees of longitude
(4)	South Pacific at 50°S	3·97 degrees of longitude
(5)	Indian Ocean at 30°S	2·25 degrees of longitude
(6)	North Pacific fracture zone at 30°N maximum displacement laterally by faulting	3·05 degrees of longitude

The radius of the earth is 6,371 km. The values refer to the distance away of the 10 million year isochrone from the central ridge. Also calculate the time needed to open the Atlantic at 30°N at the rate calculated, and at 2·5 cm/yr and 3·0 cm/yr.

The rates of spreading have been calculated for a number of different parts of the ocean ridge system. It has been estimated that the Atlantic Ocean is spreading at a rate of about 1·0 to 2·25 cm/yr near the actively spreading island of Iceland. This island, which only developed during the Tertiary, is still actively growing as witnessed by the creation of the new volcanic island of Surtsey that erupted off the south coast of the island during the last decade. The Indian Ocean is spreading at about the same rate as the Atlantic, a slower rate applies to the Red Sea and a faster one to the part of the ridge in the southern Indian Ocean. The fastest rates of spreading have been measured on the East Pacific Rise, where 5·1 cm/yr was recorded at 112°W, 40 to 45°S, and even as high as 6 cm/yr at 17°S, 113°W. These rates apply to the last 3·6 million years, and show that the process is active at present. The average value for the Pacific Ocean during the last 50 million years has been 4 cm/yr, but the rates do not appear to be constant. A change of speed occurred about 25 million years ago.

The spreading rates have important implications concerning the age of the ocean floors. The idea of sea floor spreading suggests that the ocean floors are relatively young. In fact no material on the ocean floor older than Upper Jurassic has been found anywhere. The age of the Atlantic split indicates that the floor of this ocean cannot be older than Cretaceous, and much of it is younger. Much of the eastern Pacific is also of Tertiary age, the ocean floor getting progressively older towards the west. The oldest ocean floor anywhere in the world is probably in the northwest Pacific Ocean, where it is now being destroyed. There appears to be a limit to the age of the ocean floor. Once it was realized that the ocean basins are transient features of the earth's crust, the idea of moving plates could be accepted. The oceans are continually opening and closing in different places.

The character of the plate margins is important, and there are three different types: (a) There are margins where new ocean crust is being formed at the centre of the mid-oceanic rift valleys in the ridges; these are the constructive margins. (b) There are the margins where two plates are colliding and crust is being destroyed; these are the destructive margins. (c) There are the margins where two plates are sliding past each other without either growth or loss of crust; these are the conservative margins. An example of the first type of plate margin is the centre of the Atlantic Ocean between the American and African plates. The Red Sea exemplifies an early stage of this type. An example of the second type occurs where the Pacific plate collides with the Eurasian plate in the western Pacific. The third type is exemplified in the northeast Pacific, where the American plate is sliding past the Pacific plate along the great fault line of the San Andreas rift. Disastrous earthquakes occur from time to time along this fault, such as that which destroyed much of the city of San Francisco in 1906 and which may be repeated at any time.

Exercise 4.4. Plate margins

Use the information given in Figures 4.1 and 4.2 to work out what type of margin bounds the plates shown in Figure 4.1. Draw a map to illustrate the type of margin.

The major landforms associated with the collision margins are well-illustrated where oceanic crust is being consumed, as it sinks down at the junction of two colliding plates, in the western Pacific. Island arcs and deep sea trenches are typical of plate collision. The greatest ocean depths occur in the trenches. The Mariana trench exceeds a depth of 11,022 m at 11 °N, 142–143 °E. The trenches are typically very long, narrow and deep.

They often exceed 2,000 km in length, including the Kurile–Kamchatka trench. The Peru–Chile trench is 5,900 km long, although it is partially filled with sediment, and is not a relief feature for all its length. Most of the trenches form arcuate features, convex to the centre of the Pacific, which is almost ringed by them. They extend in two places into the Atlantic, the Puerto Rico trench of the West Indies and the South Sandwich trench of the south Atlantic. The best developed trenches occur in the western Pacific, where the Pacific meets the Australian and Asian plates. The trenches lie on the seaward side of the island arcs, which often take the form of a line of volcanic islands, although some of the island systems are double, with one volcanic and one sedimentary arc.

The island arcs and deep sea trenches occur where the oceanic crust comes into collision with a thin edge of the continental crust. Where two continental crust margins collide the mountains that result are of larger dimensions, owing to the greater thickness of crust involved. This type of collision will be considered further in the next section. In this section the emphasis is on the oceanographic evidence that has been so important to the development of the theory of plate tectonics and sea floor spreading. There must be a balance between sea floor spreading and the consumption of sea floor crust because the earth has a finite surface area of constant size. The plates cannot increase in size overall, although some may grow at the expense of others. The denser crust will sink under the less dense where plates collide, so that crust is gradually consumed as it reaches lower levels and higher temperatures. The sinking process produces the topographic deep, while the tensions set up by the process allow volcanic activity to generate the island arcs in the active zones.

The pattern of earthquakes and volcanoes is closely associated with the pattern of plates and their movement. In fact there is a danger of circular argument, as the distribution of these phenomena has been used to delineate the plates. There are, however, good reasons for associating particular types of earth disturbance with the different types of plate boundaries. The processes within the earth are made manifest at the surface at the margins of the plates. Most violent earthquakes occur near the surface. Their epicentres lie within about 35 km of the surface. These shallow earthquakes form a circum-Pacific ring and extend along the length of the central oceanic ridge system, as well as through the mountain belts of the Alps and Himalayas. They are associated with stresses set up as the plates pull apart along the central rift valleys, and where the plates meet along the Tertiary mountain belts. (See Plate 67.)

There is another type of earthquake that has a much deeper focus, extending down to a depth of about 700 km. These deep focus earthquakes are restricted in their occurrence to the Pacific Ocean, where they form two belts one on either side of the ocean. One belt occurs in the eastern Pacific and is associated with the very long Peru–Chile deep sea trench. The depth of focus increases away from the Pacific margin along a plane sloping at an angle of about 45 degrees below the continent of South America. The other belt of deep focus earthquakes is found in the western Pacific, where it slopes down under the continent, also along a plane with a 45 degree gradient. Figure 4.3 indicates how these earthquakes can occur deep in the crust. Normally earthquakes can only occur in the upper crustal zone because only in this part have the rocks sufficient strength to give way abruptly to stress, thus setting up earthquake waves. Where the oceanic crust is sinking down at the collision boundaries, however, the colder crust penetrates to greater depths and still retains

Plate 67 The damage done by an earthquake in 1972 that occurred on the San Andreas rift is illustrated. This earthquake zone lies along a conservative plate margin where the American plate is moving past the Pacific plate, with no loss or gain of crust.

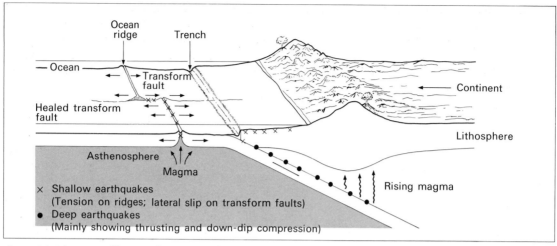

Figure 4.3 Diagram to illustrate the pattern of shallow and deep focus earthquakes on the Benioff zone, where subduction takes place as one oceanic plate sinks beneath another. (After F. Press and R. Siever, 1974, 2nd ed., 1978)

sufficient rigidity to allow earthquakes to take place at these exceptional depths. As shown on the diagram there is a layer of less rigid rock, called the asthenosphere, underlying the rigid lithosphere.

The volcanic action that builds the island arcs derives its material from the hot and relatively mobile asthenosphere. The bending and stresses set up by the subduction of the crust allows the molten material to escape at the surface to form volcanoes. The other major belt of volcanic activity is associated with the mid-oceanic ridge system, with the volcanic activity concentrated along the central rift valley. The volcanic islands of Iceland, Madeira and Tristan da Cunha illustrate the continuing volcanic activity along the ridges. Indeed the whole of Iceland only dates from the Tertiary, forming as the north Atlantic split developed and widened. (See Plate 68.)

Exercise 4.5. Age of volcanoes relative to distance from the ridge

The values given in Table 4.3 give the age of volcanic islands in the Atlantic and Indian Oceans and their distance from the centre of the mid-oceanic ridge. Plot the values against each other, calculate the regression equation and assess the degree of correlation. Use the regression lines to suggest a spreading rate in cm/year. Compare with results of Exercise 4.3.

Table 4.3 Volcanic island age and position in the Atlantic and Indian Oceans

X, age ($\times 10^6$ yr)	Y, distance ($\times 10^3$ km from ridge crest)
120	3·0
120	2·2
120	2·0
83	1·6
60	1·55
50	1·45
50	0·6
50	0·2
35	2·2
35	1·6
30	1·8
20	1·2
20	0·7
20	0·3
17	0·0
10	0·0
2	0·0
1	0·2
0	0·0

The scientific study of earthquake waves, seismology, and other geophysical techniques has provided valuable information concerning the operation of global tectonics and the forces involved. The main methods used to study the earth's crust and its behaviour are: (a) seismology (the study of earthquake waves), (b) the study of the earth's gravity, (c) the study of magnetism, and (d) the study of heat flow through the crust.

Earthquakes set up waves that travel through the crust and which can be recorded on a seismograph. The powerful natural earthquakes can be used, but have the disadvantage that they occur without warn-ing. Artificially triggered waves can be traced more easily but are less powerful. The study of earthquake waves has shown that the earth's interior is layered, and that a sharp discon-tinuity, called the Mohorovic discontinuity, occurs below the crust. It separates the crust from the denser mantle. The mantle extends almost half way to the centre before the core is reached.

The density of the earth's material is studied by recording the variations of gravity over the surface of the earth. The observations can now be made from a ship or aircraft. One of the features of most interest in the gravity surveys was the early discovery that there were long

Plate 68 The inner cone of the volcano Surtsey off southern Iceland. The new volcanic island formed along the constructive plate margin of the mid-Atlantic Ridge. Multiple eruption vents are shown within the caldera wall.

belts of negative anomaly, implying a lesser density, below the deep ocean trenches. This fact can be explained in part by the great depth of water. The subduction of the crust into the depths at these points explains the anomaly satisfactorily, as lighter crustal material penetrates deeper at these points. Visible mountains do not cause an increase of density, suggesting that the visible mountains are compensated by a decreased density at depth, the idea that the mountains have roots. This point is considered further later.

The heat flow studies provided valuable evidence in favour of global tectonics. They showed that, contrary to expectation, the heat flow was as high in the oceans as on the continents. It was thought that the heat flow would be higher in the continental rocks, because they contain more radioactive materials, an important source of heat. The highest of all the heat flow values were found, however, in the central parts of the mid-oceanic ridges, associated with the rift valleys. This finding fits in well with the view that it is along these rifts that hot material is reaching the surface from below. The relatively low heat flow values recorded over the deep sea trenches also supports the idea that here colder crust is being carried below the surface. The generally high value over the oceans reflects the relative youth of the ocean crust compared with the bulk of the continental crust, particularly in the shield areas.

The most persuasive and significant evidence for the global tectonics has come from the study of magnetism, both in the present and the past. Palaeomagnetic studies have provided two main lines of argument. The study of the past magnetic field has shown that continents have not always been in their present positions. It is possible to calculate the latitude from the magnetic dip of particles set in the magnetic field at the time of their formation. The palaeolatitude can be confirmed by evidence of climate associated with latitude, such as glacial deposits, desert conditions and hot–wet conditions. Both lines of evidence indicate considerable wandering of the land masses and of the poles relative to them.

The other field in which magnetism has played an important part is the pattern of magnetic reversals that provides a useful time scale for sea floor spreading. It is known that the magnetic polarity changes periodically and that volcanic rocks record the polarity at the time of their solidification. The discovery that there are belts of rock, magnetized alternately north and south, lying symmetrically on either side of the central rift valleys of the mid-oceanic ridges, has played a very important part in the development of the sea floor spreading theory. This pattern is very striking and provides clear evidence of the reality of the youth of the ocean crust. The magnetic reversal pattern also provides useful dating information for the study of deep sea cores. It covers a period of 30 million years with reasonable accuracy.

All these lines of evidence and the pattern of land and sea, together, confirm the ideas of plate tectonics. The theory that plates move across the earth's surface as new crust is forming at the growing margins and are consumed at the collision margins, is now generally accepted. The force that drives the mechanism is believed to be some form of convection activity, whereby hot material rises under the central ridges and sinks at the deep sea trenches and other collision boundaries when it has cooled. Orogeny takes place at the collision margins. (See Plate 68.)

4.3 On the land—orogenesis and erosion

The mountain building process

Just as oceans are transient features on the earth so are mountains. They are formed deep within the crust, they emerge as visible relief features, and then they are worn down and their roots are eventually exposed by erosion. Sea floor spreading seems to be a fairly regular process, whereas mountain building is more episodic. There have been periods of orogenic upheaval followed by relatively long periods of tectonic quiescence. The earth was probably formed from a collection of cold particles about 4,600 million years ago, the particles heating up as they condensed and consolidated. In the long period before the Cambrian, which started about 700 million years ago, there were probably at least three major periods of mountain building. The oldest dated rocks were formed about 3,500 million years ago, so there has been plenty of time for several cycles of mountain building to have taken place. The rocks now exposed in the Canadian and other shields were formed, folded and eroded during these long periods. The oldest rocks of Britain, the Lewisian gneiss, was metamorphosed in two phases about 2,400 million and 1,600 million years ago, while the rocks of the Canadian shield date from between 1,000 and 3,000 million years ago.

Since the Cambrian dating becomes rather more accurate and the evidence becomes clearer. There was one important period of orogenic activity in the Palaeozoic, culminating in the formation of the Caledonian mountains of northwest Europe, the Appalachians and other ranges. The next major phase of mountain building took place in the Mesozoic period, when the Rocky Mountains were formed and the Hercynian orogeny took place in Europe. The Tertiary period was another important mountain building episode, when most of the major ranges that now form the highest ground were created. These mountains include the Alps of Europe, the Himalayas and Southern Alps of New Zealand, mountains in southeast Asia, Japan and the Sierra Nevada and Cascades in North America. These youngest mountains have not yet been worn down to their roots, although very active erosion during their uplift has exposed sufficient of their structure for the processes of mountain building to be fairly clearly understood.

The distribution of the different periods of mountain building, as shown in Figure 4.1, suggests that each new upheaval is controlled to a certain extent by the last. The belts of orogenic activity in the successively younger periods tend to lie further and further from the rigid cores, formed by the exposed shields.

Exercise 4.6. Orogenic patterns

Using Figure 4.1, locate areas where mountains get progressively older towards the shield centres, and where the pattern has been disrupted by plate movement. Relate shield areas to plate boundaries and movement. Examine the relationships between the shield areas and the orogenies of the period since the pre-Cambrian. Relate the latest orogeny in the Tertiary to the pattern of plates and their movement.

Mountains consist of very large thicknesses of sedimentary rocks, with some igneous intrusions as well as volcanic material. Much of this material is altered to form metamorphic rocks by the processes of orogenesis. A large amount of this material must accumulate first, so that the initial stage of mountain building consists of the development of a negative landform. The deep sea trenches, for example, could form future mountain chains where they lie adjacent to the margin of a plate consisting of continental crust. The first stage of mountain building consists in the formation of a geosyncline, which is a long, relatively narrow belt of subsiding crust, often formed where two continental crust plates are colliding. The northward movement of the African and the Indian plates to collide with the European and Asian parts of the Eurasian plate respectively first created the geosyncline of the Tethys Sea. More recently the European Alps and the Himalayas have developed out of the sediment that accumulated in this sea.

Much of the sediment in geosynclines is shallow water in type, especially in the inner one where two are present, which is a typical pattern. The inner one is the miogeosyncline in which shallow water sandstone and limestone occur, often in great thickness. The outer

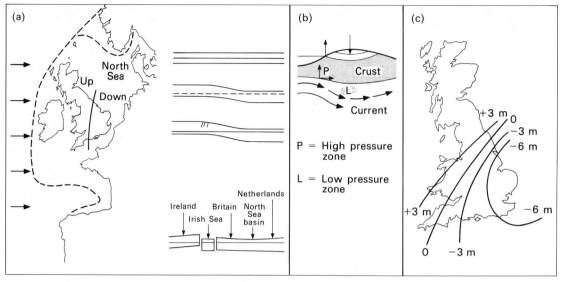

Figure 4.4 (a) Diagram to illustrate the structural effects on Britain and the North Sea of plate movement away from the central Atlantic ocean ridge. Upwarping of the land area and downwarping of the North Sea is shown, and the foundering of the Irish Sea is illustrated in diagrammatic cross sections. (b) Section to show the crustal movements that could occur as convection currents turn down under the thickening continental crust. (After R. L. S. Taylor and I. J. Smalley, 1969). (c) Pattern of sea level change over southern Britain in the last 6500 years. (After D. M. Churchill, 1965)

one is the eugeosyncline in which great thicknesses, up to 10,000 m of deep sea deposits can accumulate. The deep water sediments include turbidites, shales and pelagic limestones with some volcanic material. The inner miogeosynclines appears to form after the outer trough and indeed some of its sediment results from erosion of the material of the first geosyncline, after it has been folded and uplifted. Many geosynclines and orogenic belts show asymmetry as a result of the thrusting of one plate margin against another, so that one plate slides under the other.

Zones of downwarping, full of sediment, are now found along the southern part of the long Peru–Chile trench system, where the small Nazca plate is sliding under the American plate. Another zone of subsidence is the North Sea. This area has been subsiding since the Mesozoic, when the Atlantic first started to open, and great thicknesses of sediment, including much Tertiary and Quaternary material has accumulated in this subsiding area. Sedimentation has kept pace with sub-

sidence and the area has remained a shallow sea. The subsidence in this instance is probably related to the spreading of the Atlantic, as Britain lies on the trailing edge of the moving plate. The convection current that is carrying the plate changes level when it reaches the thicker continental crust. It may be this change of level that causes the subsidence of the North Sea, while the British Isles to the west have been uplifted, as shown in Figure 4.4. The subsidence is of great practical importance, because it has allowed the accumulation of the economically significant oil reserves.

Sediments often accumulate in the geosyncline to a thickness of between 3,000 m and 6,000 m. The material is mainly of shallow water type, although some material, derived from the land, can accumulate in deep water through the action of turbidity currents. The next stage is the compression of this material as the two plates move towards each other. The compression, which operates deep down in the crust, where the material is relatively mobile and plastic, causes complex folds to develop and the material is often altered at this stage by metamorphism. The shales, for example, are altered by pressure to form slates, sandstones are turned into quartzites, and gneisses and schists form where the pressure and heat are most intense, as they show the greatest change. Limestone can turn into marble by metamorphism.

The downwarped crustal rocks are lighter than the mantle into which they have been thrust so their great thickness results in negative gravity anomalies at this stage. When the force holding the rocks down relaxes, the mountains can rise into isostatic equilibrium. This process is similar to the recovery of land masses that have been weighted down by ice sheets. When the ice melts they also rise into isostatic equilibrium resulting in land uplift. When uplift takes place the mountains appear as visible relief features at the late stage of the orogenic process. The whole process from the beginning of the subsidence of the geosyncline to the emergence of the mountains takes up to hundreds of millions of years, although the later stages take place more rapidly than the early ones.

The Caledonian orogeny that affected Scotland, northwest England and Wales took place in the Palaeozoic period. The sediments collected in a geosyncline that trended northeast to southwest, and large thicknesses of fine grained material accumulated. Some of the material was moved by turbidity currents to the bottom of deep basins, as revealed by the sedimentary structures. The material also includes volcanic strata, for example the Borrowdale Volcanic Series of the Ordovician in the Lake District. They lie between shales, slates and mudstones of the Ordovician and Silurian periods. Their different characters are clearly visible in the landscape. Similar mudstones and shales occur in southern Scotland and slate is common in Wales.

The folding and emergence of the mountains did not take place until the Devonian period, by which time Britain had a desert climate. The present mountains of the Scottish Highlands and Southern Uplands, the Lake

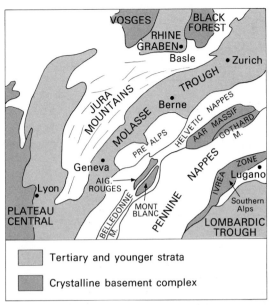

Figure 4.5 Map of the structural zones of the European Alps.

District and Snowdonia are not remnants of the ancient mountains. They were worn down completely and overlain by sedimentary strata before being uplifted anew much more recently during the Tertiary period.

The European Alps provide a good example of a recent mountain range that still has at least some of its original height. The Tethys Sea lay between the African and European plates and large volumes of sediment accumulated in it. Fine sediments accumulated in the deeper water in the centre and coarser material towards the edges. As the two plates moved towards each other the lower sediments became compressed and new geosynclines formed marginally to them. As the early material emerged as mountains it was eroded to provide the coarse deposits of the newer geosynclines, as indicated in Figure 4.5. The accumulation of material took place through the Mesozoic and Tertiary periods, but only near the end of the Tertiary did the mountains finally emerge as relief features.

A transit across the Alps from north to south illustrates the stages of mountain building and the characteristic material pro-

Plate 69 The open folding typical of the Jura Mountains. The view shows anticlines and synclines in the limestones of the Swiss Jura south of Delemont, Switzerland. The conspicuous bedding of the rocks reveals the nature of the folding clearly.

duced at the different stages. The Jura form the northern part of the system. They consist of open folds of material, mainly of Jurassic age. Between these open folds and the Alps proper the Swiss plain is relatively level land. It has been suggested that the lack of folding of the material forming the Swiss plain was the result of the Jura sliding under a wedge of material and folding against it, leaving the plain undisturbed. (See Plate 69.)

Beyond the Swiss plain the Calcareous Alps are built of material laid down on the northern shore of the geosyncline. This area includes the Bernese Oberland to the north of the Rhone valley. To the south of the Rhone valley the Pennine Alps consist mainly of schist lustrés, the metamorphosed fine grain sediments of the deeper central part of the geosyncline. They show intense metamorphism and have been very hot and plastic deep down in the trough

when folding and metamorphism took place. They occur as a series of overfolded nappes, showing extreme distortion. These rocks include the Monte Rosa nappes and about five other complex nappes. There are also granitic intrusions in this part of the Alps, which have been exposed through intense erosion.

To the south of the Pennine Nappes, in northern Italy, is the so-called zone of roots, where the nappes plunge vertically downwards into the depths of the trough. Some folds turn south in contrast to the Pennine nappes that are folded towards the north. The southern direction of folding of the north Italian nappes is thought to be due to subsequent gravity downsliding, as the mountains emerged as relief features.

Coarse sediments mark the beginning of mountain emergence. The rising mountains were steep and erosion was active, producing

210

much coarse grained material, often conglomerate. Continued uplift has maintained the height of the mountains despite continuing vigorous erosion.

Processes of denudation—denuding the land

The uplift of mountain ranges is an example of the effect of the internal, endogenetic forces on the earth's surface. These forces are the positive element creating new land and uplifting existing land masses. On the other hand are the destructive, negative forces of erosion, which are the exogenetic processes. The balance between the two elements determines the morphology of the land surface.

The orogenic cycle has been shown to be long, and to cover a considerable range of processes, from sedimentation, to orogenesis and finally uplift and erosion. The sedimentation, however, implies also erosion, as the material must come from the destruction of already existing land, apart from the volcanic and other igneous material. Thus there is a cycle of erosion, as well as a cycle of mountain building or orogenesis.

The cycle of erosion is often associated with the name of W. M. Davis, who suggested that landforms go through a cyclic development, from a stage of youth through maturity to old age. His model of the cycle of erosion, also called the geographical cycle, is a good example of a theoretical model. He started with an ideal situation of an area raised to an initial elevation without erosion taking place during the process of uplift. The processes of erosion then gradually lower the land surface from its initial plane form. The steep-sided valleys of youth are first created, when much of the original surface still remains as a plateau. The deep valleys of maturity are cut when maximum relief is achieved and most of the original surface is consumed, and rivers have cut down to near base level. The old age stage is one when the relief becomes subdued. The final phase is almost a plane surface again, the peneplain.

The model is highly theoretical, and although accepted in some areas there has always been opposition to it in some quarters, and at present the theoretical model of the cycle of erosion is not fully accepted by most geomorphologists. One of the major problems and a gross oversimplification in the model, is the assumption that no erosion takes place during uplift. The relationship between uplift and erosion was the dominant theme in the work of W. Penck, who was Davis' main early opponent. Penck sought to work out the nature of the earth movements by studying the morphology, particularly the character of the slopes. He appreciated that much erosion takes place during uplift, and argued that if uplift were accelerating the slopes should have a convex profile, steepening downwards, while if it were decelerating the slope should eventually become concave in form. The concavity was held to be the result of slope replacement, the steeper slopes gradually being eliminated from the profile and replaced by flatter ones.

There is plenty of evidence that erosion is very active when mountain ranges first emerge. The coarse Nagelfluh of the European Alps is a conglomerate which shows that erosion must have been very active. The streams flowing down steep gradients must have had a high capacity to carry coarse material. Similar conditions are indicated in New Zealand in the Great Marlborough conglomerate, which is very thick and accumulated quickly, as the fault blocks it partially buries were uplifted.

It has been calculated by D. L. Linton that it takes about 20 million years to lower a mountain range to an area of subdued relief, provided the land remains approximately stable during that period apart from the normal isostatic compensation. The Hercynian mountains of Europe were probably levelled to a large extent during this time period, leaving a relief of about 300 m. If this time scale is roughly correct then it seems that erosion takes place more quickly than orogenesis if the long phase of sedimentation is included in the whole process. It is certainly true that the process of erosion takes place much more rapidly at first, when the relief is considerable and slows down as the relief is reduced. The process thus works on an exponential scale, slowing down to a very low rate of activity as the landscape becomes flatter and the peneplain is approached. The stage of a completely plane surface is never reached in reality.

Just as orogenesis must be considered to operate on a world scale as the plates move across the earth, so the cycle of erosion must be considered as acting on a long time scale and over large areas. This is the continental or global scale of activity. On these scales of both time and space the development of the landscape is time dependent. This time and space scale is called *cyclic time* by Schumm and Lichty. It should be differentiated from *graded time* that operates on the local or regional scale, and is time independent. The state of dynamic equilibrium can exist over a hill slope within a time span of tens of years. In this chapter, however, the scale considered is of cyclic time operating over relatively large areas. Over long periods of time changes occur sequentially and the landscape is gradually worn down and reduced in gradient, provided sea level or base level remain roughly constant and warping of the land does not occur.

The extent to which erosion can reduce the landscape has long been a matter of speculation, and the reality of peneplains has been questioned. One view of the development of landscape is based on the eustatic theory, which considers that landscapes are controlled by world-wide changes of sea level. Assuming that erosion is related to sea level, the proponents of this view consider that it should be possible to correlate surfaces in different parts of the world by altitude. Several attempts have been made along these lines.

One such attempt was made by L. C. King, who recognized two world-wide erosion surfaces of continental extent, the earlier he called the Gondwana surface and the later the African surface in Africa. He called the earlier the Gondwana surface because he argued that it was formed before the supercontinent of that name split up by continental drift. Therefore, it must have been formed during the Mesozoic period before the Atlantic rift opened. The African surface, which also has correlatives in other continents, he considered was eroded during the Tertiary when the splitting was taking place. He has drawn attention to the very flat surfaces characteristic of the more stable parts of Africa, where the ancient shield rocks outcrop and earth movements have not taken place for many hundreds of millions of years.

The flat surfaces he thought were the result of the formation of pediments by scarp retreat, rather than by peneplanation by downwearing. The distinction between the two modes of extensive erosion has aroused considerable argument. There is something to be said for both sides of the argument, and it is probable, as in so many instances, that both are right for different areas. One method operates most effectively under certain climatic controls and the other under different ones. The semi-arid and savannah type climate probably favours pedimentation and backwearing, whereas the humid temperate climate may favour peneplanation and downwearing, because of the well vegetated slopes.

A distinction must be drawn between the stable areas and the unstable ones. The shield areas on the whole are relatively flat, and they have not undergone orogeny for at least 700 million years. They have, therefore, been reduced to areas of low relief. The very old rocks outcrop in the core, around the margins of which zones of younger, little disturbed rocks outcrop. These areas must have been under the sea at times since they became stable, where the younger rocks are of marine origin. There has been plenty of time for these areas to be eroded to subdued relief.

The orogenic belts on the other hand, are much more mobile. The relief is much greater, and erosive activity more powerful and effective. Time has been short, particularly in the Tertiary mountain zones, thus relief has not been subdued. Isostatic uplift, as erosion takes place and mass is reduced, is another factor that delays flattening. It may even increase relief with time, as uplift raises the peaks before these have been worn down below their original level. This may apply, for example, to the Alpine and Himalayan peaks.

It is now generally appreciated that few areas of the earth's crust are stable, except perhaps some of the shield areas. Even the British Isles has probably suffered considerable warping during the Tertiary, and it is still in progress. Many areas in high latitudes are still recovering isostatically from the effects of the ice sheets that weighed them down. The eustatic sea level has been very variable over the last tens of millions of years, since the Antarctic ice sheet started to grow in the Tertiary. The Alpine orogeny also influenced eustatic sea level by creating deep ocean basins. It is, therefore, likely that flat surfaces will be more readily produced by backwearing, which is a process not dependent on a stable base level. Peneplanation is interrupted each time sea level changes because it is related to sea level as the base level of erosion. It is also disturbed by the instability of the land itself. It is not surprising that only in stable areas, such as Africa where climatic conditions are also suitable to pediment formation, that extensive flat surfaces have been developed.

In most areas there is the additional complication of changes in erosive processes during the Tertiary and especially the Quaternary. Glacial and interglacial cycles have resulted in variations of geomorphological processes with time over wide areas. The search for extensive peneplanation surfaces is probably doomed to failure, especially if the surfaces are sought in unstable areas. The correlation of erosion surfaces from area to area is still less likely to succeed, especially if they are correlated by altitude alone.

Nevertheless the fact of reduction of the relief through denudation is amply demonstrated, provided sufficient time is allowed and a large enough area considered. The process of sedimentation and the formation of new rocks could not take place without it. The continuation of the process of denudation in an orderly way through a complete cycle is, however, much more doubtful. The earth has been shown to be much more mobile than was thought at one time. Its crust is in continual movement both horizontally and vertically as the convection currents within the core exert their influence on the surface pattern of land and sea and influence relief. New crust is being created to balance its destruction, while erosion of the surface rocks leads to the formation of new sediments.

Sea level is continually changing because of the very many factors that influence it. The changes have been particularly large and rapid during the last few million years compared with geological time as a whole. Despite all these disturbances large areas of low relief have been created in the past, as the great unconformities beneath the major transgressions demonstrate. They still exist in parts in the earth today, for example in Africa, in the plains of Russia, the prairies of central North America and the deserts of western Australia. These areas, despite their different climates, are in the stable shield zones, and they contrast with the more mobile margins of the plates.

Exercise 4.7. Shields

Use Figure 4.1 to locate the shield areas on an Atlas. Comment on the nature of the relief of the shields. Give characteristic elevations for the main shield areas, including North Canada, Scandinavia, Siberia, Africa, India, Brazil and Australia.

On a world scale the surface of the earth is finite. What is created in one area must be lost elsewhere. One ocean can only grow at the expense of another or of the land surface. Thus the processes that operate on a large scale over a long time must be compensating. Nevertheless the earth was initiated at a point in time,

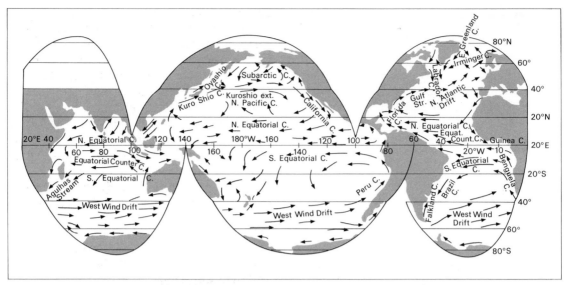

Figure 4.6 The main surface currents of the oceans. (After H. U. Sverdrup, M. W. Johnson and R. H. Fleming, 1946)

and in the very long term, the earth may eventually cease to be an independent body. Over the long time scale, however, there is a delicate balance between the positive and negative elements, and only biological phenomena move seemingly progressively onwards.

4.4 Within the oceans—ocean circulation

Introduction

In this section the world circulation of the oceans will be considered. Both the horizontal pattern and the vertical dimension must be taken into account. This is the world scale view of the oceans in which the global pattern is fundamental, because there is only one world ocean. Water can circulate around the whole earth, just as the air can move over the whole of the earth's surface. There is a very close connection between the circulation of the ocean and that of the atmosphere, and they are only treated separately for convenience. Some of the interrelationships will be considered later.

Ocean currents on the surface

The global wind system is the most important cause of ocean currents on the surface. The wind blowing over the ocean surface moves the water horizontally. The similarity of pattern of the ocean currents and the wind systems confirms this relationship. There are, however, differences in detail between the circulation of the oceans and that of the atmosphere. Other forces also influence the movement of the water.

Just as in the atmosphere the movement of air is the major factor in distributing heat evenly over the surface of the globe, so in the oceans the surface currents help to distribute the excess heat received in lower latitude to higher ones. The distribution of heat is another factor that causes movement of the ocean water. Heat affects density, and where there are differences of density pressure gradients are set up and movement must result. Such movements are, however, more likely to be vertical than horizontal.

The surface winds affect only the upper layers of water down to a depth of about 100 m. The wind stress over the water surface

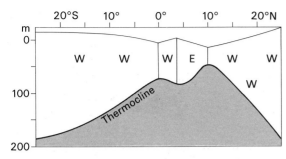

Figure 4.7 The current pattern of the equatorial Atlantic ocean. The direction of the main currents is shown by capital letters and the water slopes on the surface and between the surface and the intermediate water are indicated. (After H. U. Sverdrup, M. W. Johnson and R. H. Fleming, 1946)

creates a gradient sloping up in the direction of the wind. This gradient is of the order of 4×10^{-8} radians, which is very small, but enough to cause downslope currents where the wind is not blowing. The general pattern of surface currents consists of a clockwise circulation in the northern hemisphere and an anticlockwise one in the southern hemisphere. There is a general west to east movement around the southern ocean. There is only one major circulation in each hemisphere in the Atlantic, but the Pacific Ocean is large enough to accommodate two separate circulations as shown in Figure 4.6.

The north Atlantic Ocean contains some of the best known currents, which illustrate many important points. The system is circulatory, one of the major driving forces being the trade winds. These winds blow water westwards across the ocean in low latitudes, forming the north equatorial current, in which water to a depth of about 200 m moves at a rate of about 30 km/day south of 20 °N. On approaching the western side of the ocean the water flows partly into the Gulf of Mexico and some turns northwards. South of the north equatorial current a band of water flows eastwards, the equatorial counter current, which flows downhill in the zone of calms between the two trade wind currents. The south equatorial current spans the equator because the thermal equator lies north of the geographical equator.

This pattern of currents is affected by the rotation of the earth and forms a three-dimensional circulatory system, the main currents are shown in Figure 4.7. The Coriolis effect causes an upslope across the surface of the currents and a subsidiary flow to the right in the northern hemisphere and to the left in the southern. At the lower boundary of the surface current the slope is down to the right in the northern hemisphere and the reverse in the south. The result of the superimposition of these subsidiary flows is to produce a spiral pattern and a series of divergent and convergent zones.

Exercise 4.8. The Coriolis effect

Draw the pattern of subsidiary currents for the equatorial current system and locate the zones of convergence and divergence associated with the current pattern. Figure 4.7 may be used as a base, it shows the main currents and is drawn facing west, so that westerly currents are flowing away and easterly currents towards the viewer. Calculate the vertical exaggeration of the diagram. The subsidiary flows have about $\frac{1}{5}$ velocity of the main flow.

Another interesting equatorial current was discovered in fishing for tuna with a long line at the equator, which is a zone of fertility on account of the upwelling due to the divergence. The line on the surface was carried westward in the south equatorial current, but below the line was carried eastwards, thus revealing the equatorial undercurrent, or Cromwell current after the person who first discovered it. The current is the result of the change of direction of the Coriolis force at the equator, where it is zero. The maximum velocity of the current exceeds 125 cm s^{-1} at a depth of 100 m, and it carries about 25 million m^3 s^{-1} eastwards.

Another important aspect of the equatorial Atlantic circulation, caused by the spread of the south equatorial current across the equator into the northern hemisphere, is the movement of water on the surface from the southern to the northern hemisphere. The south equatorial current divides as it approaches the bulge of Brazil. Most of the water turns south to form the Brazil current, but about 6 million $m^3 s^{-1}$ moves into the northern hemisphere along the north coast of Brazil and meets the north equatorial current in the Gulf of Mexico. This is mainly the result of the land–sea pattern, and does not occur in the Pacific Ocean.

The Gulf Stream system dominates the western north Atlantic Ocean. It is not a stream of warm water passing through the ocean, at least in its southerly portion. It is the fast-flowing margin of the warm water of the Sargasso Sea, separating it from the cooler water over the continental shelf and slope. It has been suggested that if the Gulf Stream intensifies it moves further south, and less warm water can penetrate to ameliorate the climate of northwest Europe. When the Gulf Stream is weaker it moves further north, and more warm water can reach higher northern latitudes. The Gulf Stream system can be divided into three sections. The most southerly is the Florida current, the central section is the Gulf Stream proper, and the northern section is the north Atlantic drift. These names imply a reduction of velocity to the north in the direction of flow.

The fact that at least in its southern part the current flows fast, in a concentrated stream, illustrates a very important difference between the atmospheric and oceanic surface circulation. The oceanic circulation is markedly asymmetrical. There is a strong western intensification of the surface currents. Those on the western side of the ocean are much more concentrated and stronger than those on the eastern side. The western boundary currents all exhibit this characteristic. They include the Gulf Stream in the north Atlantic, the Kuro Shio in the north Pacific, the Brazil current and the east Australian current in the south Atlantic and Pacific respectively, and the Aghulas current in the southern Indian

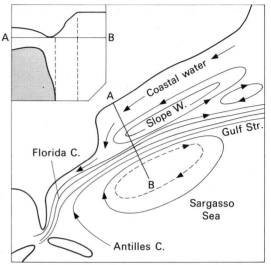

Figure 4.8 Schematic map of water transport in the Gulf Stream system. Each full line represents a transport of about 10 million $m^3 sec^{-1}$. The water slope is shown across the stream. (After H. U. Sverdrup, M. W. Johnson and R. H. Fleming, 1946)

Ocean. The western intensification is also the result of the Coriolis effect, which varies with latitude. It increases away from the equator with the sine of the latitude. In order to maintain equilibrium the western current must become concentrated against the margin of the ocean, increasing the frictional effect, while the currents on the eastern side of the ocean can be more diffuse. This is the Canaries current in the north Atlantic.

The concentration of the Gulf Stream system is most intense where it originates in the Straits of Florida. The Gulf of Mexico stands at a greater elevation than the Atlantic so that water flows out into the Atlantic. The height difference of 19 cm is caused by the water blown into the Gulf of Mexico by the trade winds. About 26 million $m^3 s^{-1}$ pass through into the Atlantic. A further 12 million $m^3 s^{-1}$ joins the current from north of the Antilles (Figure 4.8).

The amount of water increases northwards, reaching a maximum of about 84 million $m^3 s^{-1}$ off Chesapeake Bay, where the current becomes the Gulf Stream proper. It flows in a relatively narrow band extending eastwards from the continental slope. The surface temperature suddenly increases on reaching the current from the shore. Its velocity can reach over 160 km/day over a width of 74 km. A counter-current flows south along the shore in the cooler shelf water, in association with a slope of sea level along the east coast of the United States. Sea level is 35 cm higher in Nova Scotia than Florida, 2,600 km to the south.

The Gulf Stream fluctuates seasonally in volume of transport and its position also varies. Large meanders develop as it moves northwards. These meanders are probably controlled partly by the sea floor relief as they are fairly stable. They occur where the Gulf Stream changes into the north Atlantic drift, which is not a single concentrated current, but a series of separate filaments of moving water. The north Atlantic current does move large amounts of warm water into high latitudes and is mainly responsible for the mild climate of much of northwest Europe, especially the British Isles and Norway. As the water moves northeastwards it bifurcates. Part of the water moves south to form the Canaries current, which in turn feeds the north equatorial current, thus completing the surface circulation. The remaining water turns west to form the Irminger current or north to enter the Arctic Ocean between Norway and Greenland. The Irminger current meets the Labrador current to the south of Greenland, thus forming a secondary circulation in the north Atlantic.

The Pacific Ocean has a current system in many respects similar to that of the Atlantic. The Kuro Shio is the equivalent of the Gulf Stream, but an important difference is the lack of transport across the equator from south to north in the Pacific, and there is no equivalent of the cold Labrador current in the Pacific because the Bering Straits are narrow and shallow. Thus there is no exchange between the Arctic Ocean and the north Pacific as occurs in the Atlantic. The patterns of the southern hemisphere are also broadly similar, and are linked by the west wind drift, which flows between the three oceans, mixing their

waters and creating the world ocean. The Indian Ocean shows changes in current direction related to the monsoons. The pattern of ocean currents changes with the seasons in the north of the ocean.

Water masses

The waters of the oceans can be defined and differentiated on the basis of their characteristic temperatures and salinities. These properties together determine the density of the water. The colder and more saline the water is the greater its density will be. The density of pure water is greatest at a temperature of $4\,^{\circ}C$, but because ocean water is salty its greatest density is reached at a temperature of about $-2\,^{\circ}C$, at which temperature it will begin to freeze. This fact is very important in oceanic circulation because it means that very cold deep water masses can form and sink to the ocean depths. The average density of ocean water is about 1·02575, which is usually expressed as $\sigma_t = 25\cdot75$, calculated by subtracting 1 and multiplying by 1,000.

The temperature and salinity of the water are mainly determined at the surface. Precipitation and evaporation determine the surface salinity, which closely reflects the global pressure patterns. The water is saltier in the high pressure areas where evaporation is rapid and precipitation low. Salinity is low in the low pressure belts of the equatorial zone and the westerlies, where precipitation is high and evaporation lower. Table 4.4 shows the evaporation, precipitation and salinity for different latitudes in the Atlantic and Pacific Oceans.

Exercise 4.9. Salinity analysis

Plot the salinity against the difference between evaporation and precipitation for the Atlantic and Pacific oceans separately. Note whether the observation lies in the northern or southern hemisphere. Calculate regression

Table 4.4 Average salinity (‰), evaporation (cm/yr) and precipitation (cm/yr) for the Atlantic and Pacific Oceans

Latitude	Atlantic Ocean			Pacific Ocean		
	Salinity	Evaporation	Precipitation	Salinity	Evaporation	Precipitation
40°N	35·80	94	76	33·64	94	93
35	36·46	107	64	34·10	106	79
30	36·79	121	54	34·77	116	65
25	36·87	140	42	35·00	127	55
20	36·47	149	40	34·88	130	62
15	35·92	145	62	34·67	128	82
10	35·62	132	101	34·29	123	127
05	34·98	105	144	34·29	102	177
00	35·67	116	96	34·85	116	98
05°S	35·77	141	42	35·11	131	91
10	36·45	143	22	35·38	131	96
15	36·79	138	19	35·57	125	85
20	36·54	132	30	35·70	121	70
25	36·20	124	40	35·62	116	61
30	35·72	116	45	35·40	110	64
35	35·35	99	55	35·00	97	64
40	34·65	81	72	34·61	81	84
45	34·19	64	73	34·32	64	85
50	33·94	43	72	34·16	43	84

equations and correlation coefficients for both sets of data and analyse the results.

Water masses are defined by their temperature and salinity. If the masses were completely homogeneous they could be identified by one temperature and one salinity. Usually, however, they show a range of both values, due to mixing and other causes. The water masses are identified by a T/S curve, devised by Helland Hansen. The T/S curve is derived by plotting salinity against temperature at various depths within the water mass.

Exercise 4.10. T/S curves

Figure 4.9 illustrates a T/S curve from Onslow Bay. Temperature and salinity are first plotted against depth and then against each other to form the T/S curve. Use the data in Table 4.5 to plot curves for both depth and temperature and salinity separately, and then plot these values against each other as a T/S curve.

Density curves can be added to the T/S diagram to indicate the stability of the water. Most water masses form a well defined curve as shown by the examples in Figure 4.10, which shows most of the major water masses of the world ocean. All the surface water masses are much warmer than those in the ocean depths. These warm surface waters only occupy the uppermost few hundred metres, rarely exceeding 200 m, except in the Sargasso Sea and similar areas where they may attain 900 m. The surface currents flow in the surface water masses. There is a reduction of temperature and salinity with depth forming nearly straight lines on the T/S curves.

The central water masses form at the subtropical convergences, where surface water tends to sink as it is cooled in winter, increasing its density. The north Atlantic central water is similar to the south Atlantic except for the lower temperature at about 5 °C at any given salinity. The same pattern applies in the Pacific, which has four different central water masses because of its double circulation

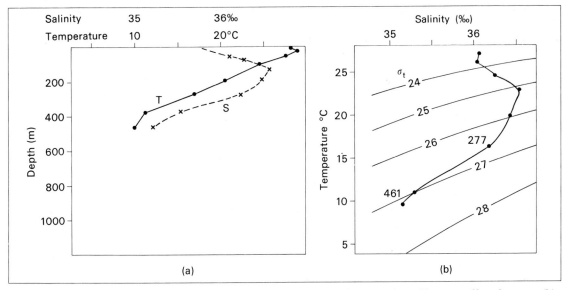

Figure 4.9 (a) The relationship of temperature and salinity to depth is shown in the Gulf Stream off Onslow Bay. (b) The same data plotted as a *T/S* curve on which salinity is plotted against temperature. Density lines are also shown. Two depths in m are shown. (After H. U. Sverdrup, M. W. Johnson and R. H. Fleming, 1946)

system. There are also equatorial water masses in both the central Pacific and north Indian Oceans in which the equatorial counter-currents flow.

Exercise 4.11. Surface water masses

Use the data given in Figure 4.10 to assess the relative density of the different surface central water masses and the equatorial water masses and account for the differences, in terms of the differences in the northern and southern hemispheres resulting from the pattern of land and sea.

To the high latitude side of the subtropical convergence there are cooler surface water masses, the most important of which is the sub-Antarctic water. It lies between the subtropical convergence and the Antarctic convergence. At the latter convergence the most important intermediate water forms. This is the relatively low salinity, cool Antarctic intermediate water which sinks to fairly shallow depths below the surface water

masses and moves northwards. It can be recognized by its low salinity and moderate temperature, as shown in Figure 4.10.

The most important waters from the point of view of volume are the deep and bottom waters that fill the ocean basins at depth.

Table 4.5 Temperature and salinity values in the Gulf Stream off Onslow Bay, Atlantic station 1640

Depth (m)	Temperature (°C)	Salinity (‰)
0	25·2	35·6
25	27·0	36·0
50	26·0	35·9
100	25·2	36·2
150	22·0	36·4
200	20·0	36·6
300	18·5	36·5
400	18·0	36·45
500	17·5	36·4
600	16·0	36·2
700	12·5	35·7
800	9·5	35·2
1,000	5·5	35·1
1,200	4·5	35·0

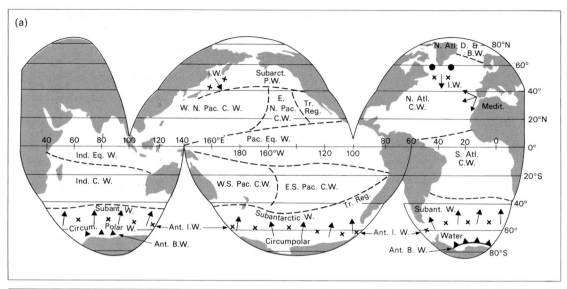

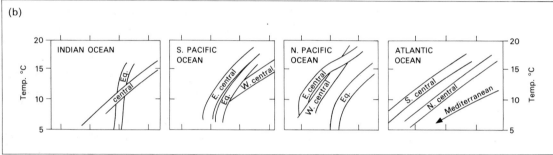

Figure 4.10 (a) Map to show the distribution of surface water masses and the sources of some deep water masses. The crosses indicate areas where Antarctic and Arctic Intermediate water forms. (b) Temperature-salinity relationships of the major surface water masses. (Modified after H. U. Sverdrup, M. W. Johnson and R. H. Fleming, 1946)

These deep waters only form where conditions allow great density to be reached by cooling, or by increase of salinity, or a combination of both. The influence of salinity is best exemplified in the Mediterranean Sea which produces one of the most saline waters. The Mediterranean water flows out of the Straits of Gibraltar as very saline, warm water, with a salinity of 36·5‰ and a temperature of about 12 °C. It attains these characteristics within the Mediterranean Sea. Cooler fresher water flows in on the surface from the Atlantic, but within the sea it is modified by excessive evaporation relative to precipitation and run off and becomes very much more saline and warmer. The amount of water flowing in is con-siderably more than that flowing out to compensate for the loss by net evaporation in the sea. The difference shows that the water of the Mediterranean can be replaced entirely about every 75 years. This warm, very salty water is dense enough to sink to considerable depths as it moves out into the Atlantic.

The Atlantic is also the source of the two most important deep waters which form in high north and south latitudes. At the northern end of the ocean the north Atlantic bottom water forms. This is a mixture of the fairly salty and warm Gulf Stream water, which derived its character when it was in the trade wind belt as the north equatorial current, and the colder, fresher waters of the east Greenland and Labrador currents. These currents bring very cold water south from the Arctic Ocean. The mixture cools in winter to sink to a depth dependent on the degree of cooling. It has been suggested that the bottom water formed in the early years of the nineteenth century when it was exceptionally cold in this area. The oxygen content of the water supports this argument, which also fits the climatic evidence.

The north Atlantic bottom water is rather more saline but less cold than the densest of all water masses, the Antarctic bottom water. This water mass forms by freezing in the Weddell Sea area off Antarctica. The density is increased by the extreme cold of the water and the increase of salinity that results from freezing. Sea ice is relatively fresh and the salt is concentrated in the water beneath. There are no sources of deep water in the Indian and Pacific Oceans.

As water sinks to form the deep and bottom water masses, so it must rise elsewhere to compensate. This partly takes place as a slow and widespread upwards movement of water. Only in one area is the upward movement concentrated enough to form a definite water mass. This is the Antarctic circumpolar water, which rises to the surface south of the Antarctic convergence. It is warmer and more saline than the Antarctic bottom water, which sinks beneath it.

Ocean circulation at depth

There are several ways in which the deep water circulation of the ocean can be investigated. The water masses can be used to trace the movement of the water. As they move their characteristics become diluted by mixing with waters above and below, but they retain their character long enough for their movements to be traced by means of water samples collected both on the surface and at depth. Direct observations of the flow at depth can be obtained by tracing floats weighted to move at different depths. They give acoustic signals that can be received by surface ships and the float position determined. Bottom photographs also reveal movement on the ocean bed. All these methods show that there is a three dimensional circulation within and between the oceans. The direct observations show that the movements are often erratic both in space and time and they can be quite fast. There is also a theoretical method of calculating the volumes of flow involved on a basis of density variations and the constraints of geostrophic flow. The theoretical studies have indicated that there should be western boundary currents at depth, and these have been confirmed by direct observation. A current flows south under the Gulf Stream, continuing all the way to the southern ocean, while western boundary currents theoretically flow north in the other oceans.

The deep water sources are all located within the Atlantic Ocean so that this ocean has a much more vigorous and well defined pattern of deep water movement than the other oceans. The separate water masses of the Atlantic Ocean can be distinguished by their characteristics. The waters form a sandwich with the south-flowing north Atlantic deep water and the Mediterranean water moving south between the north-flowing Antarctic bottom water on its lower side and the Antarctic intermediate water above it. The north Atlantic water can be identified by its salinity maximum, partly due to the Mediterranean water. The water is also warmer than that from southern sources. The pattern is evident from the distribution of temperature and salinity within a section north to south along the length of the Atlantic as shown in Figure 4.11. This figure also shows the main water mass movements and some of the volumes involved.

Exercise 4.12. Water masses

Calculate the amount of north Atlantic deep water moving south at 30 °S and the Antarctic intermediate water and bottom water volumes

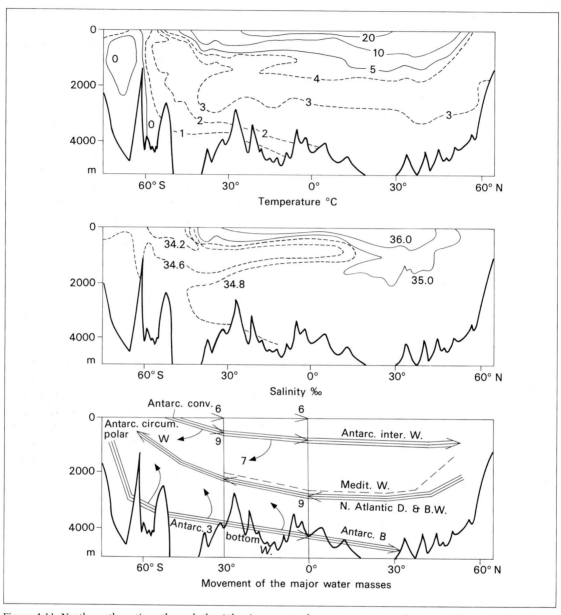

Figure 4.11 North-south sections through the Atlantic ocean to show temperature, salinity and the movement of the water masses in depth. The figures give the volume of transport in million m³ sec⁻¹. (Partly after H. U. Sverdrup, M. W. Johnson and R. H. Fleming, 1946)

at 0°, and the amount of Antarctic bottom water returning south between 0° and 30°S. There must be a balance between the north and south movements at any latitude. The source of the water that makes up the Antarctic circumpolar water on the surface

can now be appreciated. It is a mixture of all the other water masses, which become very homogeneous at depth and so can rise to the surface.

There is a similar sandwich pattern in the

Indian Ocean, with the Red Sea adding salinity to the water at intermediate depths, but the pattern is lacking in the Pacific. The water that eventually reaches the deep Pacific must have been at the surface in the north or south Atlantic or Mediterranean. It then travelled to high southern latitudes and was carried eastwards with the west wind drift that extends right to the ocean bottom. This current carries water right round the world from ocean to ocean until eventually it passes back to the Atlantic. The waters of the Pacific are low in oxygen because of their long journey to reach this ocean, and the waters are much more homogeneous due to extensive mixing.

The world ocean and its water

The estimated volume of the world ocean is $1,370 \times 10^6$ km³. Most of this water has probably been circulating for the bulk of geological time. Some water still enters the hydrological cycle through volcanic eruptions that produce steam that condenses into water. This is called juvenile water, and the rate at which it enters the world ocean is not more than 0.1 km³/yr. During the last 600 million years, since the beginning of the Cambrian, the oceans must have increased in volume by 60×10^6 km³, which is only a small proportion of the total water. By the beginning of the Cambrian the oceans at this rate must have contained more than $1,300 \times 10^6$ km³. The doubtful part of the argument is the amount of volcanic activity that has taken place during this long period, but it seems reasonable to assume that the oceans contained nearly as much water at the beginning of the Cambrian as they do now.

The matter that makes up the oceans, atmosphere and biosphere is not common in the crustal rocks or the deeper mantle material. It seems likely, therefore, that this material which forms the outer layers of the earth had a common origin with the materials that form the crust and core, and was not derived from the latter. The most probable theory of the origin of the ocean water is that it was formed during the first 500 million years of the earth's existence by a process of degassing.

The other major problem is to account for the salt in the sea. Robert Boyle in the 1670s was the first to suggest that the salt was derived from weathering of rocks on land. It was then suggested that the age of the oceans could be calculated if the rate of addition of salt could be determined by noting the salinity of river water entering the sea. The lack of information of river flow was one problem and not until 1899 did J. Joly produce a result. He suggested the age of the ocean water was 90 million years, now known to be far too short.

One fact concerning the salinity of sea water is that the salts are in constant proportions everywhere within the world ocean, although the absolute amounts vary as indicated already. Modern work suggests that the salinity of the oceans has been more or less constant for a long time, at least 200 million years. This lack of change implies that material brought into the oceans by the rivers must be locked up in the newly forming sedimentary rocks, eventually to be carried back onto the land during mountain building or uplift. There is thus a salt cycle within the oceans, as there is a cycle of water movement. The whole system leads to a stable equilibrium state with many complex interactions and feedback processes operating.

The tide

The relationship between the rhythmic rise and fall of the sea and the phases of the moon was recognized long before it was explained. It was not until Newton proposed the theory of gravitation that the control of the moon on the tide was understood.

The tide is one of the few phenomena that are influenced more by the moon than the sun. This is because the force of gravity depends on the masses of the bodies concerned and their distance apart. The much smaller mass of the moon is more than compensated for by its nearness. The moon at its nearest is less than 400,000 km away, although its mass is only $\frac{1}{80}$th that of the earth, while the sun's diameter is 100 times that of the earth. The sun's distance is such that Concorde flying at 2,400 km/h would take about 7 years to reach the sun, but only 7 days to reach the moon. It would take nearly 2 million years to reach the nearest star.

The force that generates the tide is related to the cube of the distance to the tide-producing body, hence the effect of the sun is only 0·47 that of the moon. It is, however, sufficient to make the difference between the neap and spring tides, the former occurring when the sun and moon pull in opposition to one another, and the latter when they combine to produce the much higher spring tides.

Tide-producing forces. The tide-producing forces exerted by the moon will be analysed. Those of the sun are similar but of smaller proportions. The earth rotates around its own axis in one day, while the moon goes round the earth in one lunar month. The lunar day is 50 minutes longer than the solar day so the tide gets later by this amount each day.

Figure 4.12 illustrates the forces exerted by gravity on the earth's surface and the balancing centrifugal force that holds the earth and moon in their respective orbits as they rotate around each other. The gravitational force is everywhere proportional to the distance to the moon, and so is greatest on the side of the earth nearest to the moon and least at the furthest point, and it is directed towards the moon. The centrifugal force, on the other hand, is everywhere the same. The resultant between these two forces is shown on the diagram. It is directed perpendicular to the earth's surface along the meridian through the poles and at points under and opposite the moon on the equator when the moon is overhead at the equator. The tidal resultant between the points where it acts normal to the surface has a force parallel to the earth's surface. This is called the tractive force as it has the power to drag particles over the surface. It is the tractive force that generates the tides. When the moon is overhead at the equator the pattern of its tractive force is shown in Figure 4.12 and its magnitude is equal to $\frac{3}{2}g(M/E)(e^3/r^3)\sin 2C$, where M and E are the masses of the moon and earth respectively, e is the earth's radius, r is the distance apart of the moon and earth, and C is the latitude. The pattern is symmetrical for a declination of zero, but becomes increasingly asymmetrical as the declination increases. The moon's orbit is set at an angle to the earth's axis so the moon's declination varies between $28\frac{1}{2}°$N and $28\frac{1}{2}°$S.

The example shown in Figure 4.12 is for a declination of $15°$N and the tractive force for $30°$N is analysed diagrammatically, in Figure 4.12a. The projection is stereographic so that angles can be measured directly from it to give the complex curve for the tractive force shown in the figure. This complex curve can be simplified by dividing it into its eastern and northern components as shown in Figure 4.12b. These components can be further simplified by dividing them into smooth sine curves. The northerly component is shown in Figure 4.12c.

Exercise 4.13. Tidal analysis

Analyse the easterly component by reference to the data provided in Table 4.6 which gives the hourly values for the northerly and easterly components. Complete the list of values under the columns (a) and (b) and calculate the semi-diurnal value by means of [(a)+(b)]/2, and the diurnal values are given by [(a)−(b)]/2. Plot the semi-diurnal and diurnal curves for both the northerly and easterly components, making the hours 13 to 24 a mirror image of the hours 0 to 12. Add the two curves to give the observed tidal forces. This method separates a complex curve into smooth harmonic constituents which are sine and cosine curves.

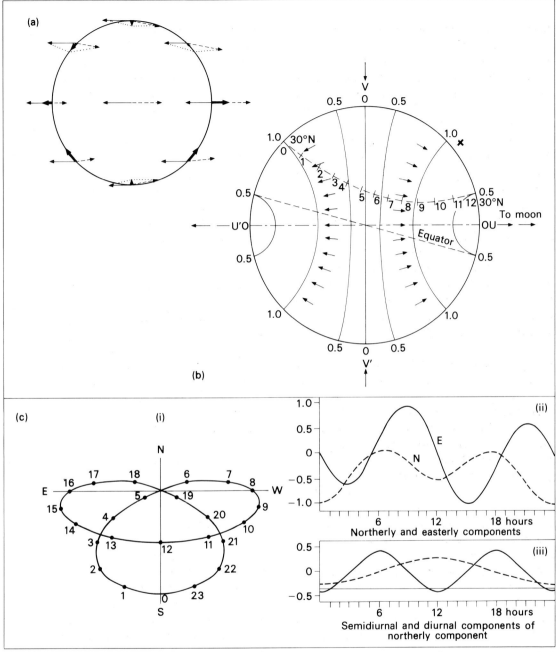

Figure 4.12 (a) Diagram to illustrate the tide producing forces. The thin full line is the centrifugal force, the dashed line the gravitational force and the thick line the resultant force. This tractive force only operates effectively when it is directed parallel to the earth's surface. (b) Diagram to illustrate the lunar tractive force for latitude 30 °N with a lunar declination of 15 °N. The figures on the circumference and the thin lines indicate the relative strength of the tractive force. The direction of the tractive force is shown by arrows. The direction and strength for each hour at latitude 30 °N can be obtained from the diagram, which is on the stereographic projection. (c) (i) The direction and strength of the tractive force at 30 °N for each hour of the lunar day is shown. (ii) The tractive force has been split into its easterly and northerly components for each hour of the lunar day. (iii) The northerly component of the tractive force has been split into its diurnal and semidiurnal components for each hour of the lunar day. (After A. T. Doodson and H. D. Warburg, 1941)

225

Table 4.6 Tidal forces for latitude 30°N with a lunar declination of 15°N

Hour	Intensity	Direction			N component	E component
0	1·00	S			−1·00	0·00
1	1·00	S	20°	W	−0·94	−0·34
2	0·95	S	37	W	−0·77	−0·58
3	0·82	S	50	W	−0·52	−0·63
4	0·55	S	61	W	−0·27	−0·48
5	0·17	S	69	W	−0·06	−0·16
6	0·26	N	77	E	+0·06	+0·25
7	0·65	N	84	E	+0·07	+0·65
8	0·92	S	89	E	−0·02	+0·92
9	1·00	S	80	E	−0·17	+0·99
10	0·89	S	68	E	−0·33	+0·83
11	0·65	S	46	E	−0·46	+0·47
12	0·50	S			−0·50	0·00
13	0·65	S	46	W	−0·46	−0·47
14	0·89	S	68	W	−0·33	−0·83
15	1·00	S	80	W	−0·17	−0·99
16	0·92	S	89	W	−0·02	−0·92
17	0·65	N	84	W	+0·07	−0·65
18	0·26	N	77	W	+0·06	−0·25
19	0·17	S	69	E	−0·06	+0·16
20	0·55	S	61	E	−0·27	+0·48
21	0·12	S	50	E	−0·52	+0·63
22	0·95	S	37	E	−0·77	+0·58
23	1·00	S	20	E	−0·94	+0·34
24	1·00	S			−1·00	0·00

Tidal data analysis: semi-diurnal and diurnal tide-producing forces.
Northerly component; subtract 0·35 from all values given above.

	0–11 hours (a)	12–23 hours (b)	Semi-diurnal 0–11	Diurnal 0–11
Hour 0	−0·65	−0·15	−0·40	−0·25

Easterly component

Hour 0	0·00	0·00	0·00	0·00
Hour 1	−0·34	−0·47	−0·40	+0·06
Hour 2 etc.				

Semi-diurnal force = [(a) + (b)]/2
Diurnal force = [(a) − (b)]/2

It is also possible to analyse the curves for other latitudes and different declinations. The patterns produced can all be broken down into their symmetrical sine components. All these will consist of two elements at least. One of these has two crests and two troughs and the other has only one crest and one trough during the lunar day. These components are called the semidiurnal and diurnal components respectively. The diurnal component increases as the declination of the moon increases. When the declination of the moon is zero, and it is overhead at the equator, the diurnal component disappears. The tide is then symmetrically semidiurnal. This method of analysis of complex curves is useful in providing a simple example of harmonic analysis or spectral analysis, whereby complex curves can be split into symmetrical components. It was on this basis that mechanical tide prediction was carried out before computers were available. The individual elements were summed mechanically to make the complex curve of the tide-producing forces.

Types of waves. The forces that create the tides have been outlined, but there remains the problem of how the oceans respond to these forces. The tide is not simply a bulge of water attracted by the moon and moving round the earth as it rotates about its axis. The tractive forces that drag the water over the surface generate the tide, not the perpendicular bulge, which is too small to be effective. The tides thus involve dynamic movement, so that the nature of the wave action involved must be considered.

If there were a channel of infinite length the wave would be a progressive one. A wave of this type can be defined by its length, the distance from crest to crest, and its period, T, the time it takes to travel one wavelength. The height is the distance from trough to crest. The wave moves at a speed dependent on the depth of water if the wavelength is more than twice the water depth. This relationship is true for tidal waves, as only two crests and two troughs encircle the earth. The current moves in the direction of wave advance at the wave crest, and in the reverse direction at the trough. The velocity of the waveform is much greater than that of the currents. The speed of water flow within the wave is equal to $u/c = y/d$, where u is the velocity of the tidal stream and c is the wave velocity, y is the height of the wave crest above the mean level and d is the water depth. The wave velocity is equal to $\sqrt{(gd)}$, where g is the force of gravity and d is the water depth.

The oceans do not resemble an infinitely long channel, except in high southern latitudes, where water is continuous around the globe. This fact led to the development of an early theory of ocean tides, the progressive wave theory. The theory suggested that a forced tidal wave moved around the southern ocean, kept going by the lunar gravitation. It then sent offshoots up the major oceans as free progressive waves. There is some evidence to support this theory but many facts that it cannot account for, one of the most significant being the disposal of the energy at the northern limits of the oceans. The energy could be lost as heat, but there is no evidence for this.

Amphidromic systems. The actual ocean tides can be best explained by considering what happens if a progressive wave hits a barrier across its path. Figure 4.13 illustrates the situation. The wave energy is reflected from the barrier with little loss, creating a wave of equal dimensions but moving in the opposite direction. These two waves, indicated by the full and dashed lines respectively, can be added together to give the wave shown by the dotted line. The streams now behave in an interesting way, whereby at every half wavelength the streams are always zero, while at intermediate points a nodal line can be placed across the channel where streams are at a maximum but at which the range is zero. This is called an oscillating stationary wave system.

The system can be set up by rocking the water in a basin so that it moves from one end to the other through a nodal line across which

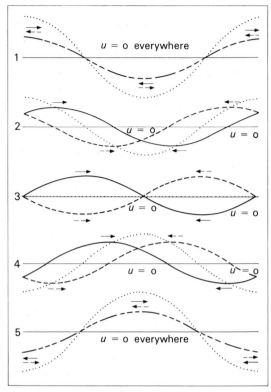

Figure 4.13 Diagram to show the formation of a standing oscillation by the reflection of a progressive wave. The full curve shows the wave moving to the right, and the dashed curve the one moving to the left. The dotted curve is the sum of the two waves. Current directions are shown. (After A. T. Doodson and H. D. Warburg, 1941)

there is no change of height. The basin has a natural period of oscillation that depends on its length and depth according to the relationship $T = 2L/\sqrt{(gd)}$. Each basin thus has its natural period of oscillation. The ocean basins, gulfs and seas can be considered as basins that will respond to the tide producing forces if their natural period of oscillation corresponds to the tide producing forces. This is the principle of resonance, and it was the basis of the next important theory of oceanic tides, the standing wave theory of Harris.

Harris' theory suggested that the ocean basins could be divided up into oscillating systems that responded to the tide producing forces when their natural period of oscillation

coincided with one or other of the tide producing forces. The larger Pacific Ocean could respond to the diurnal forces, while the smaller Atlantic has predominantly semi-diurnal tides. Tidal inequality and diurnal tides are more common in the Pacific than the Atlantic. Some tides follow the sun rather than the moon, such as Tahiti, and this could be explained by the islands being near the nodal line of the lunar tide but far away from that of the solar tide.

This theory was a great step forward, but it lacked one important element, that of the Coriolis effect. The tidal streams are influenced as all other moving bodies are by the rotation of the earth. The simple oscillating system is modified to become an amphidromic system, which is the basis of modern tidal theory. Figure 4.14 illustrates this modification by reference to the tides of the North Sea. As the water moves from one end of the basin to the other subsidiary slopes develop, up to the right of flow in the northern hemisphere. Three hours later when the main flow ceases, and the elevations are at their maxima, a subsidiary current develops.

Exercise 4.14. Amphidromic systems

Combine the information shown on Figure 4.14 to produce an ideal co-tidal diagram for the North Sea, and compare it with Figure 4.15. Draw similar diagrams for the southern hemisphere.

The tide moves round the amphidromic point (the term is derived from the Greek roots, *amphi* = round, as in amphitheatre, and *dromos* = run, as in hippodrome or aerodrome), as a progressive wave in an anticlockwise direction in the northern hemisphere and clockwise in the southern. The tidal streams are at their maxima in the direction of propagation of the wave at high tide, and flow in the opposite direction at low tide, producing at any one point a rotatory system of currents. The tidal range is greatest at the

228

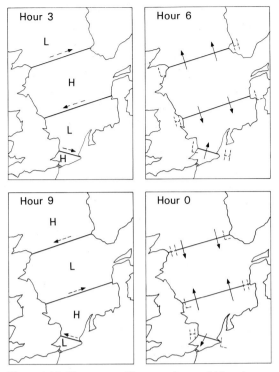

Figure 4.14 Diagrams to illustrate the amphidromic system of the North Sea. The full letters and arrows refer to the primary standing wave and the dashed ones to the modification due to the Coriolis effect. (After A. T. Doodson and H. D. Warburg, 1941)

margins of the basin, zero at the nodal point.

Modern co-tidal charts showing the time of high water along the co-tidal lines are based on a system of amphidromic points. In the north Atlantic, for example, there are three nodal points, one in the centre, one between Iceland and Scotland and one near the Caribbean. Figure 4.15 shows that there are three amphidromic points in the North Sea which interconnect to produce a tidal wave that moves south along the east coast of Britain and north up the coast of continental Europe. The energy for this system is derived from the tidal rise in the open Atlantic to the north, and this energy is transmitted anticlockwise round the North Sea. The actual positions of the amphidromic points, shown in Figure 4.15, lie to the east of the theoretical positions with the exception of the most southerly point. The reason for the easterly shift is friction. The North Sea is shallow and the energy of the tidal wave is reduced by friction as it moves around the sea, thus the tidal range is greater on the British coast and this has the effect of shifting the amphidromic point towards the east. The southern one remains central as some tidal energy gets into this part of the sea from the southwest through the Straits of Dover. Even more energy is lost in the amphidromic systems of the English Channel and Irish Sea so that the amphidromic points have become degenerate and have moved inland, giving low tidal ranges on the central south coast of England and much higher ones along the north coast of France. Tidal ranges are also much greater in the Severn Estuary and off northwest England than off southeast Ireland in the Irish Sea.

Ocean tides. Over most of the ocean the tidal range is fairly small, but where resonance can take place effectively it can produce large tidal ranges. The Bay of Fundy is noted for its very high tidal range of 15·4 m. The bay averages 68·5 m depth and its length is 256 km, which is very close to the critical length of 259 km to respond to the semidiurnal lunar tide-producing forces. Its natural period of oscillation is 6·29 hours, which is almost the semidiurnal lunar tide period.

The tides are one of the most regular and reliable natural phenomena, although at times range can be upset by unusual meteorological conditions, such as hurricanes. In the North Sea deep depressions moving at a critical speed can set up their own system of waves, called surges, that can influence the normal tide. Serious flooding can result from these excessively high tides, such as occurred around the North Sea in 1953. The east coast of North America is influenced by hurricanes in a similar way.

Tidal regularity in some places is such that some fish have developed spawning habits dependent on the tide. The grunion, which is a small fish living off northern California, comes to the beach to lay its eggs at a specific state of the tide near the spring full moon. The beach

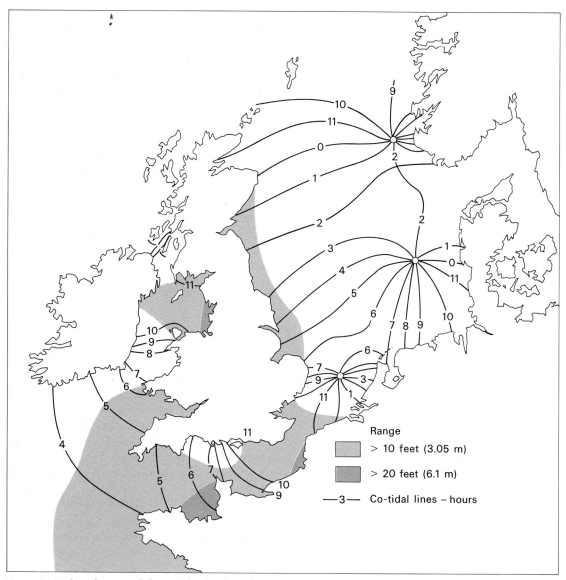

Figure 4.15 The tides around the British Isles. (Based on Admiralty Chart No. 5058)

processes are such that the eggs are then safely buried by sand at that particular state of the tide. The eggs are ready to hatch when the accretion that buried them is replaced by erosion as the tidal cycle changes, and the young fish can then be released to swim safely to the sea as they hatch a fortnight later. The spawning pattern is so regular that its date is advertised in local tide tables for the benefit of fishermen who come to catch the fish as they mass towards the beach to spawn. There are also other examples of the adaptation of different marine creatures to tidal cycles, which implies long continued regularity.

4.5 Within the atmosphere

Introduction

The atmospheric circulation must be considered in three dimensions. There is one important difference between the oceanic and atmospheric circulations. The oceans receive heat from above, while the atmosphere is heated from below. The heat of the sun passes through the atmosphere to heat the land and sea surfaces. The main cause of the atmospheric circulation is the uneven distribution of heat over the earth's surface. Thus the pattern must be considered on a global scale as it is one unified system.

Solar energy—the driving force

The solar energy received by the earth is about 1/2,000,000,000th of the total output. The earth's share is about 1.8×10^{14} kW. The solar constant is the amount of energy received on a surface perpendicular to the sun's rays at its mean distance, and it is 1.95 cal cm^{-2} min^{-1} or 1.36 kW m^{-2} min^{-1} (1 Langley min^{-1} = 1 cal cm^{-2} min^{-1} = 0.6975 kW m^{-2} min^{-1}). The solar constant may vary slightly, which would affect the earth's mean temperature. This is about $10\,°C$, compared with $6,000\,°K$ ($K° - 273 = °C$) at the sun's surface. A change of 2 per cent in the solar constant could change the temperature about $1.2\,°C$ on the earth.

The average solar radiation reaching the top of the earth's atmosphere is about 262 kcal cm^{-2} yr^{-1}. This amount is regarded as 100 units on Figure 4.16. The behaviour of this energy is illustrated in Figure 4.16 and Table 4.7 gives the meaning of the letters.

Exercise 4.15. Atmospheric energy

Complete the table given in the lower part of Table 4.7, using the data provided in Figure 4.16 and Table 4.7.

Some of the incoming energy never reaches the earth as it is absorbed by ozone, by carbon dioxide, dust and water droplets in clouds. Some is reflected back to space from clouds, and some from the surface, while a little is scattered in the atmosphere. The total amount

reflected is called the planetary albedo, which is about 31 per cent. The remaining energy reaches the earth either directly or as diffuse radiation from the clouds or downward scattering. The scattering is most effective at the short-wavelength end of the visible spectrum and hence the sky appears to be blue. The incoming radiation is short wave, while the outgoing is long wave.

The values in the figure and table are only approximate and satellite observations suggest that the earth's albedo may be 29 per cent rather than 31 per cent. There are marked variations over the earth and at different seasons due to the way in which the earth moves around the sun. The earth rotates about its own axis once a day and round the sun once a year. The polar axis is set at an angle of $23\frac{1}{2}$ degrees to its orbit around the sun, providing the differences between the two hemispheres at different seasons.

The energy that reaches the earth's surface from the sun is mainly short-wave radiation with lengths of 0.2 to 4.0 μm, with a strong peak at 0.5 μm (1 μm = 1 micron = 10^{-6} m). The outgoing long-wave radiation is much weaker and has a peak at 10 μm, and a range between 4 and 100 μm. There are various ways in which heat energy can be transferred.

Radiation is effective in transferring energy between two bodies without the necessity of a material medium between them. It consists of electromagnetic waves, and it is in this form that the solar energy travels to the earth. The earth's atmosphere restricts the travel of the electromagnetic waves to certain wavelengths. Ozone absorbs all the ultraviolet radiation below 0.29 μm, while water vapour absorbs radiation to a lesser extent, in certain bands between 0.9 and 2.1 μm. The absorption of incoming energy by the atmosphere is important as it prevents a fall of temperature on the earth's surface of about $40\,°C$, which would make most lifeforms impossible.

Another mechanism of heat transfer is conduction, which requires molecules through which the heat travels. Air is a poor conductor and so this process is not important

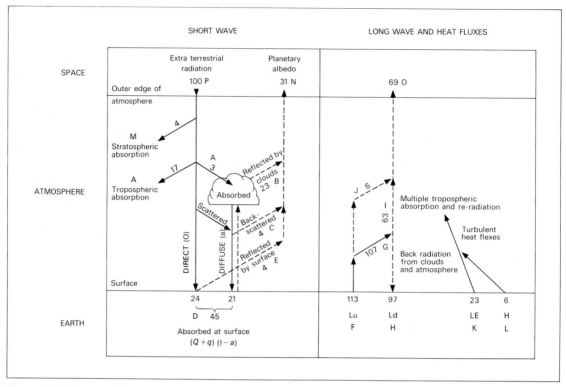

Figure 4.16 Diagram to illustrate incoming and outgoing solar energy, the letters are explained in Table 4.7. Solid lines indicate energy gains by the atmosphere and surface in the left hand diagram, and by the tropopause in the right hand one. The figures are referred to 100 units of incoming solar radiation at the top of the atmosphere, which is equal to 0·5 cal/cm²/min or 263 kcal/cm²/year. (After R. G. Barry and R. J. Chorley, 1976)

in the atmosphere, but it is on the ground.

The third form of heat transfer is by convection, which needs circulation of the fluid, liquid or gas. Air has a low viscosity so the process is very important in the atmosphere. Convection transfers heat in two forms, as sensible heat and as latent heat of vapour condensation. The latent heat of vaporization, L, is given by L cal g^{-1} $\simeq$ $(597 - 0·56T)$, where T is the temperature in °C. The same amount of heat is needed to convert liquid to vapour as is given off by the reverse process. The latent heat of melting ice at 0°C is 80 cal g^{-1}. Sublimation of ice requires 677 cal g^{-1}, i.e. $(597+80)$ cal g^{-1}.

The sun's heat reaches low latitudes in excess of that reaching high latitudes because of the high angle of the incident rays in low latitudes compared with high ones, as shown

in Figure 4.17. The elevation of the sun is never great in high latitudes, even in mid-summer, and in winter the sun lies below the horizon for considerable periods beyond the Arctic and Antarctic circles.

Exercise 4.16. Global energy

Figure 4.17 shows for all latitudes the energy transfer polewards. Draw the curve for total energy transfer by summing the curves for latent heat, sensible heat and ocean currents. The surfeit of energy in low latitudes is in the form of heat energy, but this must be converted into kinetic energy, the energy of motion, before redistribution can take place. The development of kinetic energy must be initiated by upward movement of the air.

Table 4.7 Atmospheric energy disposal

A = Absorbed in the troposphere by ozone, water vapour and water droplets in clouds	I = Re-radiation from the troposphere to space
B = Reflected back into space by clouds	J = Loss through radiation windows of long-wave radiation to space
C = Scattered by air molecules, water drops and dust particles	K = Latent heat received by the atmosphere by evaporation from the earth's surface—convection
D = Reach the earth directly or as diffuse radiation transmitted via clouds or downward scattering	L = Sensible heat received by the atmosphere through turbulent transfer of sensible heat from the earth's surface—conduction
E = Reflected back into space from the earth's surface	M = Short-wave radiation received from space in the stratosphere
F = Loss of long-wave radiation from the earth's surface	N = Planetary albedo, short-wave radiation lost to space
G = Proportion of this long-wave radiation absorbed by the atmosphere	O = Total loss of long-wave radiation to space
H = Re-radiation from the troposphere to the earth's surface	P = Total gain of short-wave radiation from space

Total   Space   N = (B + C + E),   O = (I + J)   P = 100

Atmosphere

Gain from		Loss to	
Short-wave	M + A	Space	I
Long-wave	G	Surface	H
Convection	K		
Conduction	L		

Earth

Gain from		Loss to	
Short wave	D	Atmosphere (long wave)	F
Long wave	H	Convection	K
		Conduction	L

Vertical movement is in turn related to the major horizontal movements that produce the global wind systems. The laws of horizontal motion, which are similar in the atmosphere and ocean, determine the pattern. The Coriolis effect is important in determining the path of the moving air once it has been set in motion.

The fundamental force that causes air to move is the difference in pressure from place to place, and this in turn is due to differences in temperature. The pressure gradient force depends on the differences of density of the air. Hot air is less dense than cold air, and

tends to rise, causing a reduction in pressure, while descending air increases the pressure. The air then moves horizontally from areas of high pressure to ones of low pressure. The stronger the pressure gradient the higher the air velocity. The pressure gradient force is inversely proportional to the density of the air. This is particularly important in considering the upper wind, which plays an important part in the general atmospheric circulation and in surface weather.

The rotation of the earth causes moving air to be deflected to the right in the northern hemisphere and to the left in the southern.

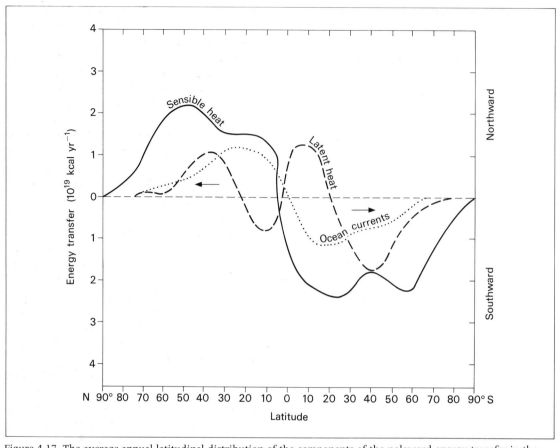

Figure 4.17 The average annual latitudinal distribution of the components of the poleward energy transfer in the earth-atmosphere system. (After R. G. Barry and R. J. Chorley, 1976)

This deflection causes the winds to blow parallel to the isobars, rather than at right angles to them, as they would in the absence of the Coriolis effect. The wind blowing along the isobars is called the geostrophic wind. With one's back to the wind high pressure is on the right in the northern hemisphere. The velocity of the wind is dependent both on the pressure gradient and the latitude. The wind will be stronger for the same pressure gradient in lower latitudes because the Coriolis effect increases with increasing latitude away from the equator.

The ideal pattern rarely occurs naturally as pressure gradients are always changing and the isobars are rarely straight for any great distance. When they are curved another force operates, the centrifugal and centripetal accelerations. These cause the air to move in towards a low pressure centre and away from a high pressure centre. Winds blowing round a low pressure centre have speeds lower than the geostrophic value and those blowing round high pressure centres have higher speeds. The effect is in practice, however, only important when intense low pressure centres develop, such as tornadoes and hurricanes. Anticyclonic pressure gradients are usually slack so the effect in this instance is small. If the flow were purely geostrophic there would be no growth or development of pressure systems, but accelerations associated with vertical movements cause circulations to develop in a three-dimensional pattern.

Exercise 4.17. Geostrophic wind and gradient wind

Draw diagrams to illustrate the effect of the rotation of the earth and the curvature of the isobars on the pressure gradient force to explain the geostrophic and gradient winds for the northern hemisphere.

The global winds

Surface circulation. The major pattern of global winds depends on the surface pressure pattern, and this in turn is linked with the general movement of air in the three-dimensional circulation. The pattern transfers heat from low to high latitudes in order to maintain the temperature of the earth as a whole. Most of the heat transfer is by the air, although the oceans account for between one fifth and one quarter of the total, being more than half in the tropics but only about 10 per cent at 55 °N on average. In the atmosphere two mechanisms achieve the poleward flux of heat. One is the large scale movement of air in a vertical north–south (meridional) circulation, the other is the heat exchange associated with horizontal movement, in the form of cyclones and anticyclones and general meandering of the major wind systems, both at low and high elevations. Convection plays an important part in the first process, assisted by the release of large amounts of latent heat of condensation in the hot tropical latitudes where thick convective clouds develop. The meandering associated with the second process is variable and has important repercussions on the vagaries of the weather and climate over large areas of the earth's surface. The pattern can at times allow warm air to penetrate to higher latitudes or cold air to reach abnormally low latitudes. It can also upset the normal pattern of zonal movement and result in long spells of weather and gives rise to extremes of different types. Recent work on climatic change has shown how important are these variations in the general circulation pattern in accounting for fluctuations in the weather on a wide range of different time scales. The variations tend to be greater in higher latitudes.

One of the most persistent features of the global pressure pattern is the low latitude high pressure cells situated at about 30 degrees north and south. The cells move slightly with the seasons, following the movement of the sun. The large amount of land in the northern hemisphere means that the high pressure cells become stronger in winter when the land becomes much colder, and they tend to weaken in summer as the land heats up. The most stable cells are located over the Bermuda–Azores Atlantic Ocean area, over the south and southwest of the United States, which is liable to seasonal variation, over the east and north Pacific, and over the Sahara desert. This last cell is more prominent in winter. In the southern hemisphere, apart from one over Australia in winter, the cells are oceanic, because of the large extent of ocean in the southern hemisphere.

Close to the equator a belt of low pressure lies between the high pressure cells. It also migrates with the seasons, especially moving north in summer towards the hot lands in the north. On the poleward side of the high pressure cells lies another low pressure zone. This is almost continuous around the southern ocean. In the northern hemisphere it is split into cells, with the main centres over the Iceland area and the Aleutians. In summer the continental land mass also develops lower pressure. There is no permanent high pressure zone over the Arctic, although high pressure does develop over the Canadian Arctic in spring. In winter the high pressure cells are over the cold continents, over Siberia and northwest Canada. High pressure also exists over the elevated east Antarctic plateau. The cells mostly migrate with separate high and low pressure centres moving across the zones. In low latitudes high pressure cells are permanent and extend up into the higher levels of the atmosphere. They are a fundamental feature of the global circulation.

The high pressure cells dominate the most

persistent and important global wind systems, the trade winds and the westerly winds, which blow on their equatorward and polar sides, respectively. The trade winds, so-called because they were an important influence on the development of trade routes in the days of sailing vessels, cover about one half of the earth's surface. They tend to be the strongest during the winter half year, as are the westerly winds. This is partly due to the contraction of the pressure zones as the equatorial heat centre moves with the seasons away from the geographical equator. Both hemisphere trade winds tend to converge towards the equatorial low pressure zone, called the intertropical convergence zone (ITCZ). It is more continuous and marked over the oceans. The zone of light and variable winds, between the trade winds, is known as the doldrums, on account of its dangers in the days of sailing ships, which could become becalmed in this area. Doldrum belts occur in all three oceans, one stretching 16,000 km through the Indian and western Pacific Oceans, particularly in March and April and again from October to December. In some parts of the equatorial zone there is a belt of westerly winds between the trade winds during the summer season. It occurs over Africa and south Asia during the northern summer, but in the Pacific and Atlantic this wind does not develop as the ITCZ does not move far enough from the equator.

The belt of westerly winds on the poleward sides of the high pressure cells are much more persistent, although they are also more variable. These winds are changeable because they are frequently associated with the passage of depressions and anticyclones, which have their own circulations. The cells of high and low pressure themselves travel eastwards in the generally westerly wind system. The pattern is more diverse in the northern hemisphere because of the greater extent of land and the complicating influences this exerts on the circulation. In the southern hemisphere the westerlies are both stronger and more persistent, blowing around the continuous stretch of ocean in high southern latitudes. In the Scilly Islands 46 per cent of winds come from a westerly quarter, compared with 29 per cent from the opposite direction, while at Kerguelen in 49°S 81 per cent of the winds blow from a westerly quarter and 75 per cent blow from this quarter in the Macquarie Islands at 54°S. (See Plate 70.)

The areas poleward of the westerly wind belt are in places affected by the cold polar anticyclone. These systems create a dominant easterly wind, for example in the Indian Ocean sector of the Antarctic and on the northern side of the northern hemisphere depressions in the north Pacific and Atlantic. There is not, however, a permanent or well established system of easterly winds in high northern latitudes. The westerly winds dominate the belt between 30 and 35°N and S and 60 to 65°N and S according to the season.

The global wind system plays an important part in the general energy distribution over the earth's surface. The energy arrives as heat from solar radiation, and is converted into both potential and kinetic energy. The kinetic energy of movement produces the wind system. The rate at which kinetic energy is generated must balance its loss. The rates have been calculated as about 1 per cent of the incoming solar radiation, so the system is a highly inefficient heat engine. Kinetic energy is lost through friction and small scale turbulent eddies. The system is controlled by the earth's angular momentum. As the air moves its angular momentum is proportional to the rate of spin, which is at a maximum at the equator, furthest from the axis of rotation, where it is 465 m s^{-1}. The angular momentum must remain constant if the earth's rotation is uniform; this is called the conservation of angular momentum. As air moves to a different latitude where the rate of spin is different, its angular velocity must change to allow angular momentum to remain constant. As air moves north in the northern hemisphere it develops higher easterly speeds. The mid-latitude westerlies impart westerly relative momentum to the earth by friction. This frictional dissipation would cause them to cease in a little over a week if their momentum was not being continually replenished. On the other hand in low latitudes the trade winds are receiving westerly relative momentum by friction as the earth rotates in the opposite direction to their flow. This excess momentum is transferred poleward at a high level, by the tropical jet streams at about 250 mbar at 30 °N and S. This poleward transfer of momentum is needed to maintain the westerly winds. Thus the three-dimensional pattern becomes essential to the working of the whole system.

Upper air circulation. In the oceans a vertical circulation is set up by differences in the density of the water. In the atmosphere warm air, which is less dense, has a tendency to rise, while colder, denser air tends to sink. On the small scale this effect can, for example, cause sea breezes. On the large scale, there are also important vertical circulations.

Early ideas suggested that hot air rises at the equator and is replaced at ground level by in-blowing trade winds. Above the air moves poleward to complete the circuit by descending in the subtropical high pressure zones.

Plate 70 The satellite photograph shows a well developed depression with a strong air circulation anticlockwise around the low pressure centre. The depression has developed along the polar front in the northeast Atlantic. The cloud pattern defines the positions of the fronts.

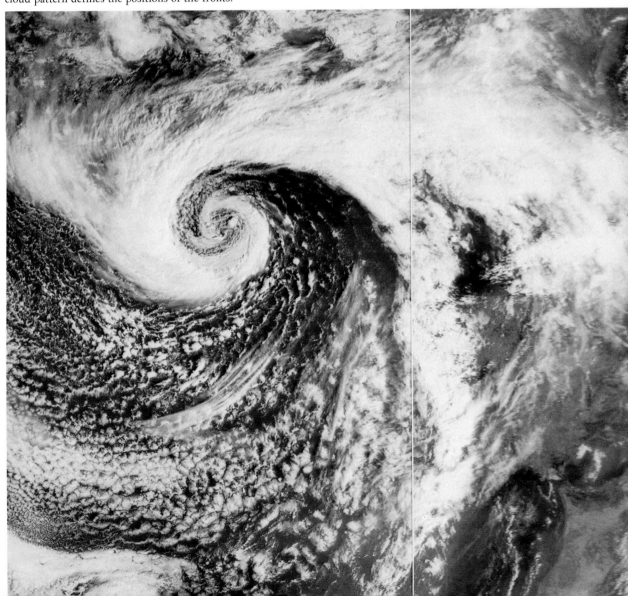

This is the simple, direct Hadley cell, called after G. Hadley who first made this suggestion in 1735. It has been shown recently that this simple pattern is not entirely adequate. According to the earlier ideas the simple, low latitude direct cell is replaced poleward by a cell with poleward movement on the ground, i.e. the westerly wind belt. The air rises at the polar front along the discontinuity where it is undercut by colder polar air. The rising air initiates the depressions characteristic of the polar frontal zone. The air completes the circuit by moving equatorward at upper levels and then descending in the subtropical high pressure zones. A final direct cell was thought to complete the pattern in very high latitudes, with air descending over the polar high pressure area and moving up at the polar front.

A single cell would not be possible because if there were only easterly winds at the surface the angular momentum could not be maintained. Easterly and westerly winds at the surface must balance over the earth as a whole. According to the Hadley pattern there should be upper easterly winds, but in fact observations show that the upper winds are mainly westerly. The upper westerlies flow in concentrated bands called jet streams in the upper troposphere, near the tropopause, just below the stratosphere, at about 13 km height in the subtropics and 10 km in the middle latitudes. The high level westerly winds are related to the temperature pattern in the upper air. Temperature decreases towards the pole in the upper air and upper winds blow with the cold air, which has a smaller thickness, on their left in the northern hemisphere. Thus westerly winds would be expected in the upper air. The high level westerlies are strongest in winter when the temperature gradients are steepest. There are two major jet streams moving from west to east in the upper air. The polar jet is situated at about 45 °N and the subtropical jet at about 25 to 30 °N. The subtropical jet has a velocity of about 45 to 67 m s^{-1}, increasing to 135 m s^{-1} in winter, with the maximum speed concentrated into narrow bands. The polar front jet stream is at a rather lower elevation

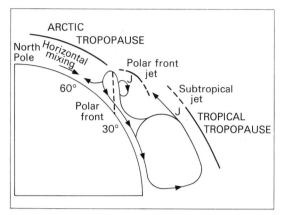

Figure 4.18 Diagram to illustrate the meridional vertical circulation for the northern hemisphere in winter, showing the positions of the jet streams and tropopause. (After G. L. Haltiner, 1957)

and blows above the polar front, where there is a steep temperature gradient, with polar and tropical air in contact. The northern jet is more variable than the southern one and may be discontinuous at times.

A more recent interpretation of the vertical circulation in the northern hemisphere is shown in Figure 4.18. The pattern has been modified from the simpler Hadley pattern, although the low latitude direct Hadley cell still has a place. It is necessary to distribute the heat energy effectively. In higher latitudes the heat energy can be distributed adequately by means of the horizontal circulation, in which travelling depressions and anticyclones can carry heat and energy poleward. The energy of the zonal winds, blowing from west to east, is derived from these travelling waves rather than from a meridional, north to south, circulation with a vertical axis.

The giant cumulo-nimbus clouds of the equatorial belt provide the motive power that carries energy and moisture upwards in the low latitude cell. These clouds which reach great heights are associated with the equatorial trough, which moves from 5 °S of the equator in January to 10 °N of the equator in July. The figure shows the relative position of the two jet streams and their relationship to

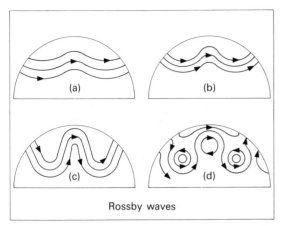

Rossby waves

Figure 4.19 The development of Rossby long waves in the upper westerlies is indicated. (a) indicates a high zonal index and (d) a low one, when blocking anticyclones develop. (After R. G. Barry and R. J. Chorley, 1976)

the tropopause, which varies in elevation with latitude. The vertical transports are complex in the vicinity of the polar front, with horizontal mixing playing a part in the area north of the polar front in the northern hemisphere.

Mean maps can hide many features of the upper air pattern. This differs considerably from year to year causing seasonal contrasts. Hot, dry summers and mild winters are the result of one pattern, while cooler, damper summers and hard winters are related to a different one. Once established the patterns are often persistent for a long time. The high level westerlies frequently develop waves in middle latitudes. Figure 4.19 illustrates the development of the waves. They gradually become larger and may eventually change from well marked ridges and troughs to separate cells. The strength of the westerlies is called the zonal index. It is high when the westerlies are strong, low when they are weak and a well marked cell pattern has developed. The index is also low if the westerlies move far south of their normal position and under these circumstances the southern westerlies may increase in velocity. The waves that develop in the westerlies are called Rossby waves and they play an important part in the generation of depressions in the polar front area. One reason why the waves develop may be related to the variation of the Coriolis effect with latitude. As the air moves northwards the Coriolis effect increases, and cyclonic vorticity tends to decrease. The air then tends to return towards lower latitudes, in which direction the Coriolis effect is decreasing. This in turn results in a swing northwards again, and so the wave pattern can develop. There are usually between three and six Rossby waves around the middle latitudes.

The wave pattern of the westerlies is related in part in the northern hemisphere to the distribution of land and sea, and to the position of high land. A ridge tends to develop over eastern North America and over the Eurasian continent, particularly in winter. The main stationary waves are located at 70°W and 150°E, related to the positions of the Rocky Mountains and the Tibetan Plateau respectively.

The upper level waves control the development of ground level depressions because they result in zones of convergence and divergence, which cause the surface depression to develop and deepen. Once formed the movement of the depression is related to the upper westerly jet stream. They move at an average rate of 32 km h^{-1} in summer and 48 km h^{-1} in winter, when the westerly jet stream is stronger.

During periods when the zonal index is low and the cell pattern well developed, the movement of depressions may be blocked by the formation of high pressure cells. These blocking anticyclones can be very persistent and result in long spells of settled or unsettled weather. The dry summers of 1975 and 1976 in Britain were the result of a blocking anticyclone which lasted a long time. Conditions were persistently wet over central Europe, and Moscow had its wettest spring and early summer for many years. The British drought of 1976 was related to exceptional conditions in the northern Pacific. When the zonal index is high the westerlies are strong and persistent. Precipitation from passing fronts is high in western Europe and winters mild and wet.

Table 4.8 *Wind components in January for the area between London and Copenhagen. X = anticyclonic, Y = cyclonic; some years are omitted owing to lack of data*

Decade	N	E	S	W	X	Y
1550s	2	1	1	9		
1560s	1	5	3	3		
1570s	4	4	1	6		
1580s	5	5	2	3		
1590s	4	3	5	6		
1600s	1	6	–	3		
1610s	1	1	2	8		
1620s	3	4	4	3		
1630s	4	4	5	6		
1640s	2	3	4	7		
1650s	3	4	–	3		
1660s	2	3	3	4		
1670s	3	2	5	8		
1680s	8	2	2	8		
1690s	6	4	4	6		
1700s	5	4	5	6		
1710s	5	4	5	5		
1720s	1	1	9	8		
1730s	3	1	7	9	2	1
1740s	3	3	6	6	6	2
1750s	2	2	5	5	2	–
1760s	1	3	7	3	5	1
1770s	1	1	3	5	1	4
1780s	2	–	5	7	2	–
1790s	–	2	9	7	5	1
1800s	–	3	9	5	2	1
1810s	–	2	6	6	2	1
1820s	2	2	3	5	5	1
1830s	1	2	2	7	5	–
1840s	–	3	8	6	–	–
1850s	1	1	8	7	2	1
1860s	–	–	9	8	2	2
1870s	2	1	7	8	1	–
1880s	1	2	4	6	5	–
1890s	–	1	2	7	4	1
1900s	3	–	3	10	3	–
1910s	–	2	3	5	3	1
1920s	1	1	4	7	1	–
1930s	–	–	7	8	2	3
1940s	2	3	3	6	1	1
1950s	2	–	2	7	3	–

Exercise 4.18. Patterns of weather

Table 4.8 provides data collected by H. H. Lamb on the frequency with which different wind direction dominated the weather in January during the decades from 1550 to 1950 for the area between London and Copenhagen. Plot these values and compare them with the values plotted on Figure 4.20, which gives the values for July. Use the chi square test to establish the probability that the weather during the decades 1550 to 1690 was significantly different from that during the decades 1700 to 1850, each period covering 15 decades, using the January values. The test can also be used for the July figures. Then test whether there is a significant difference between the values for July and January during the periods 1550 to 1690 and 1700 to 1850.

Climatic zones and classification. The contrast between maritime and continental climates is particularly striking, and Exercise 4.19 is concerned with this aspect of climate.

Exercise 4.19. Continentality

Table 4.9 gives a list of places and their latitude and mean annual temperature range. Those places with latitudes greater than 50 degrees have been listed with the distance from the west coast provided. Rank the places in order of temperature range and also in order of distance from the west coast. The latter value provides a measure of the continentality of the place. Use the Spearman rank correlation test to establish whether there is a relationship between the two variables.

Several different types of classification of world climate have been devised. Many are based on the main climatic elements of temperature and precipitation, including the much used one of Köppen. The original classification has been revised by Geiger and Pohl. This classification is based on the relationships between climate and vegetation,

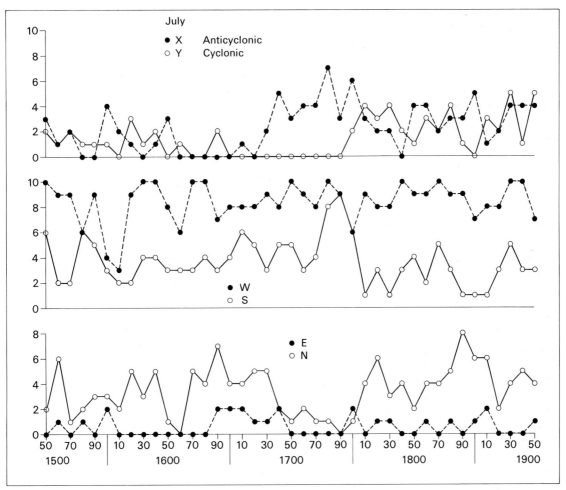

Figure 4.20 Time series of wind frequencies and anticyclonic and cyclonic conditions, for 10-year intervals from 1550 to 1950 for July. (Data from H. H. Lamb, 1966)

the climatic boundaries being related to major changes in vegetation type. It is an empirical classification and is entirely objective. The main division of the climatic zones is based on specific monthly values of temperature and precipitation. They are denoted by capital letters as follows: A = tropical rainy climates, B = dry climates, C = mild, humid mesothermal climates, D = snowy-forest, microthermal climates, and E = polar climates. Four of the groups, which are given in greater detail in Table 4.10 are defined by temperature and one by precipitation to evaporation ratio. This system of classification has the advantage that it is easy to apply.

The necessary climatic data are available for many places, because only monthly means are needed. The relationships between the climatic zones and vegetation are not, however, exact. However, the groups of the classification do broadly delimit some important boundaries. Group D is related to the poleward limit of tree growth, while group B is too dry for tree growth. Group E is associated with tundra vegetation in high northern latitudes.

Exercise 4.20. Climate and vegetation

Use the data provided in Table 4.11 to plot the

Table 4.9 Continentality data

Place	Latitude		Mean annual temperature range (°C)		
Europe					
Reykavik	64	°N	12·78	1	⎱
Archangel	64		29·61	1,300	
Lisbon	38		11·66	1	
Geneva	47		19·28	580	
Warsaw	52		22·20	1,100	⎬ km from west coast
Palermo	38		13·78	70	
Plymouth	51		7.94	1	
Norwich	53		12·67	380	
Moscow	53		30·00	1,800	⎰
Asia					
Verkhoyansk	68		67·67	5,420	
Aden	12		7·22	200	
Colombo	7		1·89	1	
Singapore	1		1·39	1	
Barnaul	52		38·00	5,200	
Tokyo	36		22·33	330	
Batavia	7	°S	0·83	1	
Australia					
Darwin	12		5·00	1	⎱
Alice Springs	23		19·83	1,800	⎬ to nearest coast
Sydney	34		10·50	1	⎰
Africa					
Lagos	6	°N	3·06	1	
Timbuktu	17		13·44	1,500	
Khartoum	16		11·67	1,800	
In Salah, Sahara	27		25·00	1,500	
Nairobi	1	°S	3·56	600	
Cape Town	34		8·61	1	
North America					
Godthaab, Greenland	64	°N	16·94	1	⎱
Churchill	58		41·67	2,360	
Winnipeg	50		37·94	2,100	
San Diego	33		8·33	1	⎬ to west coast
New York	41		24·17	4,350	
Denver	40		23·22	1,650	
New Orleans	30		15·28	10	⎰
Central and South America					
Dominica	15		3·06	1	⎱
Belem	1	°S	0·83	150	
Quito	0		0·83	220	⎬ to nearest coast
Lima	13		6·94	1	
Bahia	13		4·33	1	
Valparaiso	33		6·11	1	⎰
Buenos Aires	35		13·89	1,250	⎱ to west coast
Punta Arenas	53		9·17	150	⎰

Table 4.10 Koppen–Geiger climate classification

Major groups	

A	Tropical	Average monthly temperature more than 18°C, high rainfall
B	Dry	No water surplus, potential evaporation exceeds precipitation
C	Temperate	Coldest month temperature less than 18°C but more than −3°C. At least one month more than 10°C
D	Cold	Coldest month less than −3°C and warmest month more than 10°C
E	Ice climate	Warmest month less than 10°C

Subgroups defined by second letter

−f Moist, adequate precipitation throughout the year, applies to A, C and D
−w Winter dry season
−s Summer dry season
−m Monsoon type precipitation in A climates

(B)S Steppe climate—semi-arid 38–76 cm precipitation/year
(B)W Desert climate—arid less than 25 cm precipitation/year

ET Tundra climate at least 1 month more than 0°C
EF Perpetual frost climate no month more than 0°C

temperature and precipitation for the 17 stations. Locate the places on the map provided in Figure 4.21. Plot the vegetation by means of symbols on a graph, on which the mean annual temperature is plotted against the annual precipitation. Draw lines on the graph to separate the forest zones from those with long grass and those with short grass, tundra or desert scrub and other desert vegetation. Give the equations for the lines in terms of temperature and precipitation.

Another form of classification is that proposed by Thornethwaite, who has proposed several different versions. His classification is more complex, but it also is based on the relations between climate and vegetation. He devised a precipitation efficiency (P–E) index in which the temperature and evaporation play a part. His first equation was $11·5 (r/t − 10)^{10/9}$, with r = mean monthly rainfall in inches and t = mean monthly temperature in °F, summed for the whole year, giving the P–E index. The relationship between the index and vegetation is given in Table 4.12.

He also devised a thermal efficiency index in the form $(t − 32)/4$ °F for each month summed for the whole year. Zero represents a frost climate and over 127 is tropical. The primary factor in the classification is precipitation rather than temperature.

In his second classification he used a potential evapotranspiration index (PE) calculated from the mean monthly temperature in °C with corrections for day length. PE in cm $= 1·6 (10 \, t/I)^a$, where I is the sum of 12 months of $(t/T)^{1·514}$ and a is a complex function of I. The value of I is used to develop a moisture index I_m, where $I_m = (100 \, S − 60 \, D)/PE$, S is a water surplus and D is a water deficit calculated for each month from an assessment of the water budget. The PE value also incorporates a function of temperature and so can be used for a thermal efficiency index as well. In his final version $I_m = 100 (S − D)/PE$. The resulting groups are given in Table 4.13. This classification does not use vegetation boundaries deliberately, although in some cases they fit the divisions, for example in eastern North America.

Table 4.11 Climatic data and vegetation type. Temperature in °C, precipitation in cm

	1		2		3		4		5	
	Temp.	Precip.	Temp.	Precip.	Temp.	Precip.	Temp.	Precip.	Temp.	Precip.
Jan.	−35·0	0·51	−22·0	0·25	−23·0	3·30	4·0	11·68	0·0	19·81
Feb.	−32·5	0·51	−24·0	0·25	−21·0	3·56	5·0	10·41	1·0	17·02
Mar.	−25·3	0·76	−23·0	0·25	−14·0	4·32	9·0	15·24	2·0	15·49
Apr.	−16·7	0·76	−18·0	0·25	− 6·0	3·81	15·0	10·92	5·0	13·97
May	− 5·5	0·51	− 6·0	0·25	+ 2·0	6·86	20·0	9·65	8·0	10·67
Jun.	+ 2·5	1·78	+ 2·0	0·51	+ 9·0	9·40	24·0	10·41	14·0	8·38
Jul.	+ 9·5	2·29	+ 5·0	1·27	+13·0	10·41	26·0	10·16	13·0	10·92
Aug.	+10·0	2·79	+ 3·0	1·27	+12·0	10·41	26·0	9·14	13·0	18·29
Sep.	+ 2·5	2·03	− 3·0	1·02	+ 7·0	7·87	22·0	8·64	11·0	26·42
Oct.	− 8·6	1·27	− 9·0	0·25	+ 1·0	7·87	19·0	6·60	8·0	32·51
Nov.	−19·7	0·76	−15·0	0·25	− 8·0	6·10	9·0	8·89	4·0	25·91
Dec.	−27·5	0·76	−23·0	0·51	−18·0	4·32	5·0	10·16	2·0	23·11
Year	− 6·2	14·73	−11·1	6·35	− 4·0	78·23	15·0	119·63	7·0	218·5

	6		7		8		9		10	
	Temp.	Precip.	Temp.	Precip.	Temp.	Precip.	Temp.	Precip.	Temp.	Precip.
Jan.	8·0	18·03	17·0	6·60	12·0	10·16	13·0	7·87	− 1·0	3·30
Feb.	8·0	15·24	17·0	7·37	13·0	11·18	13·0	7·62	0·0	4·32
Mar.	9·0	13·21	15·0	13·21	16·0	12·19	14·0	7·11	+ 6·0	6·60
Apr.	10·0	8·38	12·0	23·37	19·0	9·91	16·0	2·54	13·0	8·13
May	11·0	4·57	9·0	36·07	23·0	8·38	17·0	1·02	18·0	12·45
Jun.	13·0	1·78	8·0	44·96	26·0	11·68	19·0	0·25	23·0	11·94
Jul.	13·0	0·25	8·0	39·37	27·0	16·76	21·0	< 0·25	26·0	10·41
Aug.	13·0	0·51	8·0	32·77	27·0	19·30	22·0	< 0·25	25·0	10·41
Sep.	13·0	2·54	9·0	20·83	26·0	14·48	21·0	0·51	21·0	11·68
Oct.	12·0	5·84	12·0	12·70	21·0	10·41	18·0	1·52	14·0	7·11
Nov.	11·0	13·21	13·0	12·45	16·0	9·91	16·0	3·05	7·0	4·83
Dec.	9·0	16·00	15·0	10·41	12·0	11·43	14·0	6·60	1·0	3·30
Year	10·8	100·84	12·0	260·10	20·0	145·80	17·0	38·10	13·0	94·74

	11		12		13		14		15	
	Temp.	Precip.	Temp.	Precip.	Temp.	Precip.	Temp.	Precip.	Temp.	Precip.
Jan.	23·0	7·87	− 1·0	0·76	− 3·0	3·30	18·0	0·51	28·0	26·16
Feb.	23·0	7·11	+ 1·0	1·52	0·0	3·81	18·0	0·76	28·0	19·56
Mar.	19·0	10·92	5·0	1·78	+ 5·0	5·08	14·0	0·76	27·0	25·40
Apr.	17·0	8·89	10·0	4·04	10·0	5·08	11·0	1·02	27·0	26·92
May	13·0	7·62	15·0	4·04	15·0	5·08	7·0	2·03	27·0	30·48
Jun.	9·0	6·10	21·0	3·30	19·0	2·03	3·0	2·03	27·0	23·37
Jul.	9·0	5·59	23·0	5·08	25·0	1·52	3·0	1·52	27·0	22·35
Aug.	11·0	6·10	23·0	4·57	24·0	2·03	6·0	1·27	28·0	18·29
Sep.	13·0	7·87	18·0	2·29	18·0	2·54	8·0	1·02	29·0	12·95
Oct.	15·0	8·64	12·0	1·52	12·0	3·81	12·0	0·76	29·0	17·53
Nov.	19·0	8·38	4·0	1·02	3·0	3·56	14·0	0·51	28·0	18·29
Dec.	22·0	9·91	0·0	1·27	− 1·0	3·56	16·0	0·76	28·0	26·42
Year	16·0	95·00	11·0	31·24	11·0	41·40	11·0	12·95	28·0	267·72

Table 4.11 cont.

	16		17	
	Temp.	Precip.	Temp.	Precip.
Jan.	24·0	2·29	21·0	0·0
Feb.	23·0	3·05	22·0	0·0
Mar.	21·0	2·79	21·0	0·0
Apr.	18·0	1·27	19·0	0·25
May	14·0	1·02	18·0	< 0·25
Jun.	11·0	0·76	17·0	< 0·25
Jul.	11·0	0·51	14·0	0·51
Aug.	13·0	0·76	14·0	0·25
Sep.	15·0	1·27	16·0	< 0·25
Oct.	17·0	1·78	17·0	0·25
Nov.	21·0	1·78	18·0	< 0·25
Dec.	22·0	1·78	19·0	0·00
Year	17·0	19·05	15·0	1·27

Figure 4.21 Map to show the positions of stations for which vegetation and climatic data are given in Table 4.11 in North and South America.

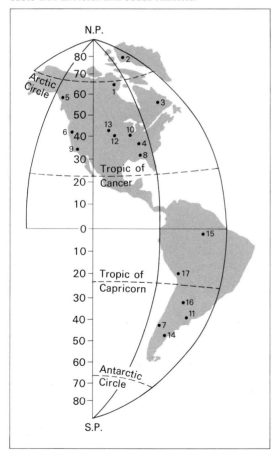

Place	Vegetation	Number
Baker Lake	Tundra	1
Thule	Tundra	2
Nitchequon	Boreal forest	3
Nashville	Deciduous summer forest	4
Sitka	Coast and Lake forest	5
Eureka	Coast and Lake forest	6
Valdivia	Southern hemisphere evergreen mixed forest	7
Pensacola	Broad leaved evergreen rain forest	8
Los Angeles	Sclerophyllous scrub	9
Kansas City	Long-grass steppe	10
Buenos Aires	Long-grass steppe	11
Pueblo	Short-grass steppe	12
Salt Lake City	Semi-desert scrub	13
Sarmiento	Semi-desert scrub	14
Manaos	Tropical rain forest	15
Cordoba	Semi-desert scrub	16
Arica	Desert	17

Exercise 4.21. Climate through the year

Use the data given in Table 4.14 to plot graphs of the type illustrated in Figure 4.22, on which temperature in °C is plotted against precipitation and also against relative humidity for Madras, Nullagine and Winnipeg. Number the months to illustrate the seasonal variation in the weather. Climates with contrasting seasons are more stimulating to live in, but where the contrasts are very extreme, as in the continental interiors of middle latitudes, they can be trying. The climate of Britain has much to recommend it, as has that of New Zealand.

Another aspect of global climate that will probably become more important is the modification of the climate by human intervention, both purposefully and accidentally. Until more is understood of the complex processes operating to maintain the delicate balance of the atmospheric circulation over the earth as a whole it would be foolhardy to interfere with the natural climatic processes. They rely on complex feedback mechanisms that could easily be upset, and a train of events could be initiated that might well be difficult to stop once it got underway. Recent work has

Table 4.12 Thornthwaite's 1948 climatic classification

Humidity index, $I_h = 100\, s/n$ Aridity index, $I_a = 100\, d/n$ Moisture index, $I_m = I_h - 0.65$
s = water surplus, d = water deficiency, n = water need,
n = PE (potential evapotranspiration) $I_m = 400\, s - 60\, d/n$,
I_m is positive in moist climates and negative in dry ones
Climatic types—moisture

A	Perhumid I_m	more than 100
B_4	Humid	80 – 100
B_3	Humid	60 – 79
B_2	Humid	40 – 59
B_1	Humid	20 – 39
C_2	Moist subhumid	0 – 19
C_1	Dry subhumid	−20 – 0
D	Semi-arid	−40 – −19
E	Arid	−60 – −39

Seasonal aspect indicated by second letter, referring to A, B and C_2 (moist climates)

r	little or no water deficiency	I_a	0–16·7
s	moderate summer deficiency		16·7–33·3
w	moderate winter deficiency		16·7–33·3
s_2	large summer deficiency		more than 33·3
w_2	large winter deficiency		more than 33·3

referring to C_1, D and E (dry climates)

d	little or no water surplus	I_h	0–10
s	moderate winter water surplus		10–20
w	moderate summer water surplus		10–20
s_2	large winter surplus		more than 20
w_2	large summer surplus		more than 20

suggested that changes of climate occur throughout the world. For example, positive correlations have been found between temperature in central England and New Zealand, and between central England and north Finland. A strong positive correlation was found between the number of days with a southwest wind in London and snow at the South Pole. Thus any change may have far reaching effects. All the climatic zones appear to move north or south together. Ice from Antarctica spread further north when it also retreated to the north around Iceland in the 1930s, and both retreated south about the 1770s and about 1820.

At present the evidence suggests that natural tendencies dominate the weather changes. There has been a cooling trend which set in about 1950, which is still continuing although it has levelled off, while warming

Table 4.13 Thornthwaite's thermal climatic index

	Temperature	T–E index
A′	Megathermal	114·0 and above
B'_4	Mesothermal	99·7–114·0
B'_3	Mesothermal	88·5– 99·7
B'_2	Mesothermal	71·2– 88·5
B'_1	Mesothermal	57·0– 71·2
C'_2	Microthermal	42·7– 57·0
C'_1	Microthermal	28·5– 42·7
D′	Tundra	14·2– 28·5
E′	Frost	less than 14·2

dominated the first half of the twentieth century. One of the most obvious effects of human activity is to increase the carbon dioxide in the atmosphere, this addition, with heat from other sources, is thought to cause a warming of the climate. It has been estimated

Table 4.14 *Temperature, precipitation and humidity data*

Location	Month	Precipitation (mm)	Temperature (°C)	Location	Month	Precipitation (mm)	Temperature (°C)
Benares	Jan.	20	16	Paris	Jan.	36	3
	Feb.	18	18		Feb.	33	4
	Mar.	13	27		Mar.	31	6
	Apr.	10	30		Apr.	33	9
	May	18	33		May	48	13
	Jun.	122	31		Jun.	53	17
	Jul.	310	29		Jul.	56	18
	Aug.	292	28		Aug.	53	18
	Sep.	178	28		Sep.	48	14
	Oct.	56	26		Oct.	56	10
	Nov.	10	20		Nov.	48	6
	Dec.	10	16		Dec.	51	3
							Relative humidity (%)
Honolulu	Jan.	96	22	Sydney	Jan.	68	18
	Feb.	107	22		Feb.	73	19
	Mar.	96	22		Mar.	76	18
	Apr.	56	23		Apr.	77	16
	May	48	24		May	78	13
	Jun.	23	25		Jun.	76	11
	Jul.	28	26		Jul.	75	10
	Aug.	33	26		Aug.	72	11
	Sep.	38	25		Sep.	68	12
	Oct.	96	25		Oct.	66	14
	Nov.	107	25		Nov.	65	16
	Dec.	107	23		Dec.	66	18

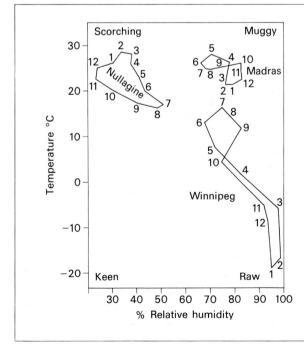

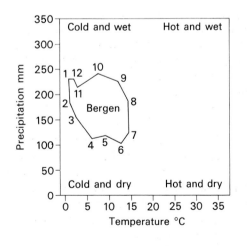

Figure 4.22 Monthly weather graphs based on temperature and precipitation or relative humidity. The figures refer to the months. (After A. A. Miller, 1953)

that man's output of carbon dioxide since the industrial revolution would be sufficient to raise temperatures by about 0·3 to 0·5 °C, over half being within the present century. This has probably moderated the natural cooling that has taken place during the last two and a half decades. It is thus becoming increasingly necessary to monitor all aspects of the global climate and assess the changes that are taking place.

4.6 The hydrological cycle—world relationships

Several aspects of the hydrological cycle that operate on the small and medium scales have already been mentioned. In this section some aspects that need to be considered on a global scale will be commented upon. The hydrological cycle provides the link between the circulations of the atmosphere and oceans, and plays an important part in the geomorphological processes that operate within the cycle of erosion on land.

Properties of water. Water has several unusual properties relating to its atomic structure. It consists of a molecule formed by the bonding of two hydrogen atoms and one oxygen atom. This is called a covalent bond in which the hydrogen has a positive charge and the oxygen a negative one. The charges are opposite at either end. This allows the molecule to separate under some conditions, allowing water to be an active agent in some chemical reactions that affect weathering.

Water can split into two oppositely charged ions H^+ and OH^-. The latter is important in weathering as hydroxyl, and the former determines the acidity of the watery solution, which is its pH value. The pH value is given as the negative logarithm of the free H^+ concentration in grams per litre, pH = 7 has 0·000,000,1 g of free H^+ ions per litre of water. Water has an exceptionally high solvent power in contact with other molecules, es-

pecially those held together by opposite electrical charges, common salt being an example.

Important properties of water, such as surface tension and capillarity, can be explained by its structure, whereby tetrahedral groups of molecules form by hydrogen bonding. The positively charged hydrogen is attracted to the oxygen ion so that clusters of molecules connected by hydrogen bonds form. These clusters are separated by unbound water molecules that rotate freely, thus lubricating the substance and allowing flowage. The individual clusters and molecules frequently interchange. Each intermolecular hydrogen bond breaks and reforms about 10^{12} times each second, the rate depending on the temperature. This results in the viscosity of water varying with the temperature, its absolute viscosity at 20 °C is about half that at 0 °C.

Water achieves its maximum density at 4 °C and below this temperature it begins to expand in volume and the tetrahedral clusters begin to take on a hexagonal structure, which produces ice. These hexagonal structures are important in the ability of ice to flow, this is relevant in glacial geomorphology amongst other things. The cleavage of ice is parallel to the basal plane and its growth rate is at a maximum normal to this plane. The cleavage property allows ice to flow by slippage across the basal plane, while the freezing property is important in weathering processes by mechanical action. Considerable pressures on the freezing surface can be set up as the ice freezes in needle form, where water can be drawn towards the freezing surface from below. Thus frost is an effective weathering agent and ground ice formation is important in many periglacial phenomena.

Water is one of the very few substances that exists naturally on the surface of the earth in all its three states. Water vapour in the atmosphere plays a very important part in many meteorological processes, such as convection, cloud formation and precipitation. The liquid phase of water is best known and

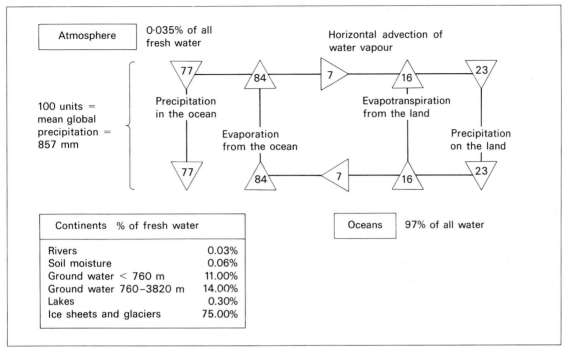

Figure 4.23 Amounts and percentages of water involved in various parts of the global hydrological cycle. (From R. J. Chorley (Ed.), 1969)

most important from many points of view. Its solid form is also significant in many respects, not least as a major form of storage. It also assists in the precipitation process, as well as being an important geomorphological agent.

Water absorbs or releases a large amount of heat in changing from one state to another. Latent heat is given off as condensation takes place, as vapour changes to liquid, and heat is also released as water changes to ice. An increase of pressure raises the temperature at which ice melts, thus facilitating the movement of ice over its bed in a glacier when the ice is near the pressure melting point temperature. The process of pressure melting and subsequent refreezing is called regelation, as the reduction of pressure allows refreezing.

Heat is needed to change ice to liquid water, and liquid water to vapour. This is because the movement of the molecules is more rapid as the water changes from solid, through liquid to vapour. Water in its liquid form can absorb a great deal of heat, which it can then give off

to the surrounding air when it is cooler than the water. This explains the important modifying effect of oceans and large lakes on the surrounding climate.

The amounts of water involved in the different phases of the hydrological cycle are illustrated diagrammatically in Figure 4.23. By far the greatest amount of all water on the earth is found in the oceans, which hold an estimated 97 per cent of all water. Most water is saline and the very large ocean reservoir provides the major storage element of the whole hydrological cycle. Of the 3 per cent of all water that is fresh, the greatest volume is found in ice sheets and glaciers. These solid water masses contain 75 per cent of all fresh water. The great bulk of this water is in the Antarctic and Greenland ice sheets, which are estimated to contain about 27 million km³ and 2·6 million km³ of ice respectively. The water equivalent of all the ice on earth is roughly 26 million km³, indicating the small contribution of the other ice masses and glaciers. The proportion of the hydrological cycle occupied by fresh liquid water is very small, amounting to a little less than 1 per cent of all water, i.e. 25 per cent of 3 per cent.

The next most important source of fresh water is that in the ground. Ground water accounts for 14 per cent of all fresh water, much of it is in deep storage at depths between 760 and 3,620 m. Eleven per cent of fresh water is found in shallower ground storage at depths less than 760 m.

A very small quantity in percentage terms is found in lakes, soil moisture and rivers. Lakes are estimated to hold 0·3 per cent, soil moisture 0·06 per cent and rivers 0·03 per cent. Atmospheric moisture is very small on a percentage scale, being 0·035 per cent of fresh water, or about 12×10^{18} cm^3, although it is of vital significance to the operation of the global hydrological cycle.

The smallest amounts of water are in those parts of the cycle where the water is moving fastest. The water in the atmosphere, which is carried by wind or convection currents, probably moves fastest of all on average, apart from unusual features such as waterfalls. The rivers, soil and lakes also carry water fairly fast. By far the largest volume of water is in storage in one or other form, mainly as sea water, ice sheets or deep ground storage, where on average the movement is very slow. The active part of the cycle in which water is transferred from one store to another holds relatively little of the total supply.

Figure 4.23 shows how the mobile water is distributed in the different processes involved in the cycle. The largest single quantity is the evaporation of water from the ocean surface, which covers 70 per cent of the earth's surface. Units of 100 are used to compare the quantities and on this basis 84 units are evaporated from the oceans, while only 77 fall as precipitation on the sea. The balance of 7 units represents the water moving in the form of vapour or cloud droplets horizontally from the oceans to the land surface. To these 7 units is added a further 16 derived by evaporation and transpiration from the land surface, including lakes

and other fresh water bodies, to form the total of 23 units falling as precipitation on the land. The balance is maintained by the runoff from the land to the sea of 7 units, in the flow of rivers and some small amount may reach the ocean through underground submarine springs and glacier calving. The value of the unit in this diagram is based on mean annual global precipitation, which is 85·7 cm averaged over the whole earth's surface. The total volume of water in the system is about $1·31 \times 10^{24}$ cm^3, while the amount in the atmosphere is only about 12×10^{18} cm^3.

The hydrological cycle is continuous. From the point of view of life on land one of the most essential steps is evaporation from the sea surface. This is the input into the atmospheric part of the cycle. The distribution of this loss from the oceans is not uniform over the globe. The highest rate of evaporation occurs over the subtropical oceans where the high pressure cells are situated. This pattern has already been considered in discussing the surface salinity of the sea. The highest rate of evaporation exceeds 200 cm yr^{-1}. It is very high over the Gulf Stream and Kuro Shio currents in winter, while in the southern hemisphere the trade wind zones provide the highest values, again exceeding 200 cm in the central oceanic area.

By contrast values over land rarely exceed 100 cm, except over small parts of the equatorial forests of the Amazon basin, where transpiration from the tropical forest plays an important part. Over the deserts and higher latitude land area rates are mostly between 30 and 60 cm/yr. They are very much less over the driest areas and highest latitudes, where the cold air can hold very little moisture.

Most of the water in the atmosphere is in the form of vapour. Only about 4 per cent of the total is in the form of clouds, which contain water droplets and ice crystals. The amount of water vapour in the air depends on temperature. The minimum values of between 0·1 and 0·2 cm water equivalent, occur in winter over northern continental interiors at high latitudes, and secondary minima occur over continental deserts with values of 0·5 to 1·0 cm in low latitudes. The maximum values occur over southern Asia in the summer monsoon and over equatorial areas of Africa and South America, where values reach 5 to 6 cm. The average over the world as a whole is about 2·5 cm, which is sufficient for only about 10 days rainfall over the whole earth. This small quantity means that there must be a rapid turnover related to continuous evaporation, condensation and precipitation.

Precipitation cannot fall without moisture in the air, but this is not sufficient for precipitation to occur. The processes that lead to condensation and precipitation must operate effectively to convert the water vapour to moisture on the ground through precipitation. It has been calculated that only about 20 per cent of the water vapour in the air above the Mississippi basin falls on it as precipitation. Precipitation falls in a variety of forms of which the most important are rain, snow and then drizzle. Hail and sleet produce appreciable amounts, while dew, fog and hoar frost are locally and occasionally significant. The world pattern of precipitation reflects the world weather systems, which are affected by the pattern of land and sea. Important features include the equatorial maximum related to the convergence of the trade winds in the ITCZ, and the summer hemisphere monsoon, especially in southeast Asia and West Africa. Amounts of 200 to 250 cm or more can fall in these zones, owing to the high temperatures. The west coasts of the continents in middle latitudes have relatively heavy precipitation, due to the depressions travelling in the westerlies. This precipitation is frontal in character and relatively stable in amount. Variations are often determined by orographic effects.

These two zones contrast with the other two major zones, both of which are relatively dry. The subtropical high pressure areas include the major deserts both on land and sea. Total precipitation may be less than 15 cm, and it is erratic as well as being small, on account of its showery, convectional character. The con-

tinental interiors in higher latitudes are also dry. Precipitation is also low over very high latitudes, largely on account of the small amount of moisture that very cold air can hold. The small amount of precipitation is compensated by very low evaporation rates, so that these tundra areas often have a surfeit of surface water during the short summer season, when water can exist in the liquid form.

The relationship between precipitation and evaporation is very important and can be appreciated from the latitudinal cross section shown in Figure 4.24. The pattern of meridional moisture transfer can be explained by the values of precipitation–evaporation for different latitudes. Moisture is carried from zones of surface divergence, where evaporation exceeds precipitation, to zones of surface convergence, where precipitation exceeds evaporation, in general from the subtropics to the equatorial and middle latitudes.

From the world point of view, the annual totals and patterns of precipitation are most relevant to the global scale dimension. Six types of precipitation are generally recognized. These are: (1) The equatorial regime with rain throughout the year, but often with two maxima at the equinoxes, when the sun is overhead at the equator. Values range between 250 and 300 cm or more. Rain may fall in spells of wet weather when disturbances occur. (2) The second type is the tropical with pronounced wet and dry seasons and a summer maximum. Totals vary between 25 and 100 cm and up to 200 cm in the humid tropics. The monsoon areas are included in this category, and hurricanes can add to the rainfall totals. (3) The third type is the Mediterranean with a winter precipitation maximum and a dry summer, the totals varying from 60 to 75 cm. Westerly depressions bring the rainfall in winter. (4) Temperate continental interiors provide the fourth type, with annual totals of 35 to 50 cm, mainly made up of convectional showers in late spring and summer with light winter snowfall. The amounts vary from year to year.

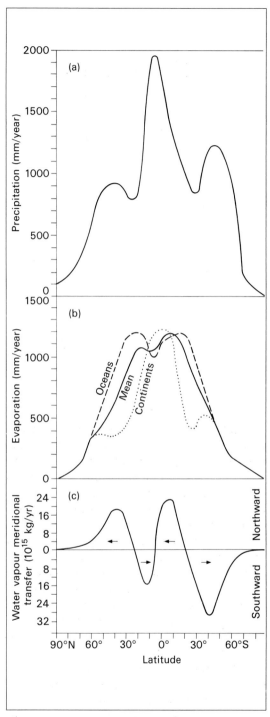

Figure 4.24 Average annual latitudinal distribution of (a) precipitation in mm, and (b) evaporation in mm, and (c) meridional transfer of water vapour. (After W. D. Sellars, 1965)

(5) The fifth type is the west coast temperate oceanic type. Totals are moderately high, between 75 and 100 cm, with a considerable orographic effect increasing amounts to over 200 cm in coastal mountains. The number of precipitation days approaches 200 and the variability is low. Most of the precipitation is frontal and snow falls in winter especially at high elevations. (6) The final type is the polar with a low annual total of 12 to 40 cm, mainly falling as rain in summer in the north. Snowfall is light in winter due to very low temperatures, and most of the precipitation is frontal.

Precipitation can be disposed of in a number of ways. Some is evaporated or transpired, some runs off in rivers, some is stored as snow, which may melt later or turn slowly to ice. Snow accumulation occurs on 14 to 24 per cent of the earth's surface with $\frac{2}{3}$ on land $\frac{1}{3}$ on sea ice, according to the season. The remainder of the precipitation infiltrates to form soil moisture and ground water. There is a maximum retention of water on land during the period March to April, when the extensive northern land masses have large areas of snow-covered ground. Ground water recharge during the winter makes the water table highest at this season. Ground ice is also extensive at this time of year.

The oceans show a reverse pattern, holding their maximum water amount in October, when they have 7.5×10^{18} cm^3 more than in March. This, however, only amounts to a sea level change of 1 to 2 cm. The residence time of the water on land varies mainly between 10 and 100 days, although very much longer periods occur in the major ice masses. Total runoff from the land masses can be calculated from the differences between precipitation and evaporation, if allowance is made for inland drainage areas. This accounts for about 25 per cent of stream runoff, mainly in Asia, Africa and Australia. The average runoff for all continents is 26.7 cm, but values are much higher in South America and the Malayan archipelago, which provide 12 per cent of the runoff from 2 per cent of the area. The proportion provided by the different con-

Table 4.15 Runoff relationships

| | Annual runoff | |
Continent	Million km² land area	% total global runoff
Africa	29·8	17
Asia	42·2	20
Australia and New Zealand	8·0	2
Europe	9·6	7
Greenland	3·8	2
Malayan Archipelago	2·6	12
North and Central America	20·5	18
South America	17·9	22

Runoff relationships for selected river basins in USSR

Zone	Precipitation (cm)	Evaporation (cm)	Surface runoff/ Total runoff (%)
Tundra	45	11	97
Taiga with permafrost	40	21	95
Taiga	50	16	73
Mixed forest	58	37	76
Mixed forest	61	43	60
Forest steppe	41	33	79
Steppe	35	32	87
Semi-desert	20	19	90

tinents is given in Table 4.15, which also includes data concerning the runoff relationships with precipitation and evaporation for different climatic zones in the USSR.

Exercise 4.22. Runoff and precipitation

Calculate the percentage runoff per km² × 10⁶ for the areas given. Account for the differences of percentage total global runoff given in the first part of Table 4.15, by reference to the climatic situation of the continents and islands listed. Calculate for the lower set of values for the USSR climatic regions, the runoff/ precipitation percentage, and explain the results by reference to the type of vegetation and climate. Comment on the values given for surface runoff/total runoff percentages.

Ground water is an important part of the hydrological cycle because it provides the moisture that enables plants to grow and which keeps the rivers flowing when there is no precipitation. Most of the ground water is meteoric in origin, coming from precipitation and infiltrating into the ground. It first forms soil moisture and then penetrates deeper to the zone of permanent saturation below the water table. This is called the phreatic zone to distinguish it from the vadose zone which is not saturated, but through which water passes downwards to the water table. The movement of ground water depends very much on the type of rock. It will travel easily through coarse grained rocks with high permeability. These rocks provide the best aquifers, from which ground water can be extracted most readily, and in which springs will flow copiously. Sandstone that is not highly indurated, cavernous basalts and other volcanic rocks, and calcareous rocks, including chalk and well-jointed limestones provide excellent aquifers. Aquifers are important for water supply purposes, and their recharge is an important aspect of the hydrological cycle. The process can be increased by artificial recharge where surplus runoff exists in a suitable location. The rate of movement of ground water is very variable, depending on the nature of the rocks, the rate of recharge and the head, which is the difference in height between the intake point and the point of outflow.

In many areas ground water is probably of considerable age. In the great Artesian basin of Australia, for example, it has been established that the ground water at a distance of 30 to 80 km from the recharge areas was about 20,000 years old, denoting a travel rate of between 1·5 and 4 m/yr. This value is considerably slower than that in many aquifers, where rates of several m/day have been recorded. In the very porous basalts of Idaho, for instance, which occupy 30,000 km^3 in the Snake River region, the rate of water movement is about 10 to 15 m/day, with residence times of up to several hundred years.

Some of the greatest stores of ground water are inherited from different climatic conditions. There are very large ground water reserves beneath the surface of North Africa in the Nubian sandstone. At present this area receives little or no rain. Rainfall is 0 to 25 cm/yr in most parts and evaporation is high, present recharge rates are, therefore, low, and recharge only takes place when the occasional flash flood occurs. The amount of water in storage in this area is reckoned at 600 10^3 km^3. The water probably entered the aquifer during a wetter period in the Pleistocene between 30,000 and 40,000 years ago. These stores of ground water are known as inherited ground water to differentiate them from the cyclic ground water which is undergoing continual renewal and circulation.

Water is becoming an increasingly short and valuable commodity from the human point of view and as a result the storage of water within the hydrological cycle is of great significance. Some ground water stores have already been mentioned. The method of recharge of ground water will increase in importance, because it has the advantage that the water does not suffer from evaporation during storage, and movement through the aquifer brings about natural purification. The method is being used to maintain the water table in the chalk of southeast England, where the water demand is high in one of the driest parts of Britain. Other areas where ground water recharge is possible are the high plains of Texas and New Mexico. Local storms produce high run off, which flows into playas. These are mainly floored by impermeable material so that much of the water evaporates. Where wells can be dug through the playa floor into permeable strata at depth, the water can penetrate and be withdrawn when needed. The system works well in places where the wells do not become clogged with silt and where the water can move freely underground to where it is needed for extraction.

Storage in glaciers and ice sheets provides another potentially important water source. It has been suggested that large icebergs could be towed from the Antarctic to California or the Middle East where water is very short. Glaciers have the advantage that they melt during the hotter season when the water is required in large amounts for irrigation and other purposes. There is the problem of silt in the water, however, although glacial lakes, which are common, help in this respect as much of the fine sediment settles in them.

The most readily available source of water in many areas are the rivers, which provided the main water supply for a long time. As a total storage, rivers contribute relatively little in total bulk, but the water is moving rapidly and so is replenished fast. The Amazon is the largest river in the world. For the channel length of 800 km downstream of Obidos, with an average width of 2·5 km and a depth of 60 m, the channel storage would be 120 km^3, which would only maintain the flow at the mouth for about 8 days. An estimate of the storage upstream from this point is about 160 km^3, giving a total for the whole river of 280 km^3, which would maintain the flow at the river mouth for only about 19 days. This river system covers 7 per cent of non-arid non-glaciated land area, but it yields much more water than most river systems. The Mississippi provides only a tenth of the runoff.

The average daily discharge of all rivers is about 80 km³, which would provide flow for about 3 weeks without replenishment. It is, therefore, remarkable that rivers continue to flow considering that precipitation is so variable in time and space. The water that sustains the flow is largely ground water, but lakes help to a certain extent. About 80 per cent of the water in lakes is in 40 large ones, the total amounting to 125,000 km³. Lake Baikal has the largest volume of 22,000 km³, because of its great depth. It contains nearly as much water as the five Great Lakes of North America together. Lakes in all amount to 70 per cent of the surface, fresh liquid water. Saline inland seas, such as the Caspian, are nearly as voluminous as the fresh water lakes, having a volume of 105,000 km³.

The water stored in vegetation and available to it as extractable soil moisture is important. Vegetation uses a great deal of water. One tree transpires about 200 litres a day. The soil zone within about 2 m of the surface contains about 10 per cent by volume of water, amounting to about 20 cm of water, except in the arid areas. Over an area of 82 million km² this amounts to 16,500 km³. Moisture in dry areas and unmelted snow increases the total to 25,400 km³. This is about 15 times the water stored in rivers. The actual amount of water stored in vegetation is estimated at 5·3 km³, while the amount passing through plants and organic systems has been estimated to be about 500 km³/yr. It has been suggested that the amount of water needed for photosynthesis is 650 km³. These values are probably too low. The amount of precipitation falling on the earth annually is about 107,000 km³. Evapotranspiration from the land is about 71,000 km³. If only a third of the total is accounted for by transpiration, this would amount to about 24,000 km³, which is probably nearer the right order of magnitude.

There are many constraints on the human use of water resources, some of which are physical and some economic, conservational or aesthetic. There have been several suggestions for major interference with the natural hydrological cycle. Reversing the flow of the Congo, or making more use of the deep water supplies of the desert regions have been suggested. Smaller scale attempts to moderate or modify the weather, by rain-making, hurricane prevention and encouragement of snow fall have all been investigated. They should all be treated with caution until more is known of the complex feedback relationships that exist within the hydrological cycle, because all the different parts are interconnected. For example, if hurricanes were stopped, beneficial rain would be denied to areas that need it badly.

The massive engineering works that have been carried out already and are planned in the USSR provide good examples of major schemes to modify the natural hydrological system. Some of this work has been centred on the dry area of central Asia. The levels of the Caspian and Aral Seas have fallen rapidly as a result of artificial water extraction from the system. The Caspian has fallen 2·3 m since 1930 and the Aral has fallen 3 m since 1961. The level of these inland drainage lakes depends on the balance between evaporation and river inflow. The inflow deficit to the Caspian was 1,012 km³ between 1929 and 1962 or nearly 31 km³/yr, while the figures for the Aral Sea are 186 km³ for the period 1961 to 1973 or 15·5 km³/yr. For comparison the annual discharge of the Thames at Teddington is 2·4 km³ and the Volga at Volvograd between 1879 and 1930 was 269 km³.

The loss of inflow into the Caspian is partly natural and partly due to the construction of reservoirs and irrigation works. The flooding of 20,000 km² by the building of dams has considerably increased evaporation. A total of 14 km³ will eventually be lost in this way. At present 4·55 km³ are lost by this process. Changing agricultural practice also has increased evaporation and transpiration. The Aral Sea has lost water through the tapping of

Table 4.16 Water resources of Central Asia

Period	Flow at mountain edge		Inflow into Aral Sea		Combined inflow into Aral Sea
	Syr Dar'ya at Chadara	Amu Dar'ya at Kerki	Syr Dar'ya	Amu Dar'ya	
1926–50	37·3	70·0	13·0	41·3	54·3
1951–60	43·7	74·5	15·2	40·1	55·3
1961–68	36·2	68·9	8·2	30·2	38·4
1969	65·0	109·5	16·6	56·3	72·9
1970	43·3	73·3	9·3	28·4	37·7

Decrease in runoff of major USSR rivers as affected by proposed industrial and economic agricultural activity

Type of activity	1970		1981–86		1991–2000	
	km³	%	km³	%	km³	%
Irrigation	16	29	60	51	111	61
Additional evaporation	14	26	19	16	21	11
Agricultural practices	10	18	16	13	19	10
Industrial and other water supply	11	20	21	18	27	15
Other	4	7	3	2	5	3
Total	55	100	119	100	183	100

its major source of supply the Syr Dar'ya and the Amu Dar'ya for irrigation. In 1976 100 km³ of water was used to irrigate 6 million ha of arable land and 150 million ha of pasture. This water was distributed through 250,000 km of canal. The Kara Kum canal, started in 1956, alone is planned to reach 1,400 km in length by the 1980s. Table 4.16 shows the pattern of water resources flowing into the Aral Sea up to the present, and future plans associated with the decrease of river runoff.

Before remedial measures are carried further it is necessary to assess the optimum levels of the Caspian and Aral Seas. It will probably be decided that the present lower level of the Caspian Sea be maintained. Although this would have deleterious effects on the productive fishing carried out in the shallow, warm waters of the northern part of the sea, it would save considerably in the expense on engineering works. Needless evaporation would also be reduced from this shallow northern part. 1 km³ of water could irrigate 500,000 ha or raise the Caspian Sea by 0·25 cm. Plans to alleviate water loss are ambitious. Some involve taking water from Lakes Ladoga and Onega in northwest USSR as well as water from the north-flowing Ob and its tributaries.

If Lake Ladoga were raised 50 cm, 9 km³ of extra water would be available for diversion via the Volga–Baltic canal. Lakes Ladoga and Onega, which have a volume of 1,203 km³, could not, however, provide sufficient water to halt the fall of the Caspian. The diversion of west Siberian rivers could, however, provide the necessary water. Several schemes have been suggested. The west Siberian rivers carry 1,283 km³/yr to the Arctic, which is three times the present inflow to the Caspian and Aral Seas and 10 times the predicted water deficit in AD 2000. The watershed between the Ob and Central Asia is only 124 m high at the Turgay Gates, and water could flow from

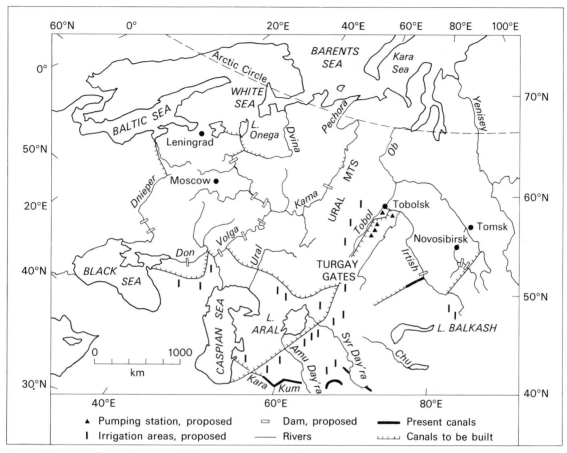

Figure 4.25 Map of part of western USSR to show hydrological developments.

here to the Aral Sea and thence to the Caspian, which is 80 m lower. Figure 4.25 illustrates some of the schemes.

The most favoured scheme at present could be developed in three stages, starting in 1990. The first stage involves dams on the Irtysh at its confluence with the Tobol, and pumping water along the Ob–Caspian canal through the Turgay Gates for distribution to the south. The second stage involves taking the water from the Novosibirsk and Kamen reservoirs on the Ob to the headquarters of the Irtysh. Finally the Ob would be further regulated. This whole scheme is based on canal construction and pumping, rather than on massive flooding to form huge inland seas as in earlier schemes. It will assist in the drainage of swamps on the lower Ob. Eventually 300 km³/yr could be

transferred south.

Possible repercussions of this scheme and others that have been suggested are difficult to foresee. The extra evaporation from the large irrigated area could increase precipitation over central Asia by 100 mm and in the Volga basin by 10 to 30 mm (5–30 per cent) in the summer. A reduction in the northerly flowing Ob could make the Arctic more saline and lead to loss of ice formation, reducing the albedo and causing further warming. On the other hand it has been argued that the reduction in relatively warm water from the Ob could result in cooling of the swamps, extension of the permafrost on the taiga and tundra areas and further ice formation in the Arctic, which would increase the albedo and cause further cooling; another positive feedback situation (Hollis, 1978).

Table 4.17 *Caspian and Aral Seas level and river flow into them*

	Caspian Sea Level below sea level (m)	Aral Sea Level above sea level (m)	Deviation from average natural discharge (m³ s⁻¹)				
				Volga (1887–1930)		Amu Dar'ya (1930–1950)	Syr Dar'ya
1850	25·9		1931	−1,600	1952	+200	+200
1855	26·1		1933	−2,000	1953	+200	+100
1860	26·2		1935	−1,900	1954	+200	+200
1865	26·0		1937	−2,700	1955	−100	0
1870	25·5		1939	−2,400	1956	0	0
1875	25·6		1941	− 800	1957	−600	−300
1880	25·6		1943	−1,300	1958	+100	+100
1885	25·4		1945	−1,000	1959	0	+100
1890	25·3		1947	+ 200	1960	−100	+200
1895	25·3		1949	−1,300	1961	−500	−200
1900	25·4		1951	−1,200	1962	−600	−400
1905	25·5		1953	− 400	1963	−400	−100
1910	25·8		1955	+ 200	1964	−200	−100
1915	25·6		1957	0	1965	−700	−400
1920	25·8	53·0	1959	−1,500	1966	−300	−100
1925	26·2	53·1	1961	− 200	1967	−500	−200
1930	25·7	52·6	1963	− 300	1968	−300	−300
1935	26·0	53·2	1965	−1,800			
1940	27·4	52·7	1967	−2,800			
1945	27·6	52·7					
1950	27·8	52·8					
1955	27·9	53·2					
1960	27·9	53·3					
1965	28·0	52·0					
1970	28·1	51·0					
1973		50·3					

Exercise 4.23. Caspian and Aral Seas' hydrology

Calculate the trend for the discharge of the Volga from 1931 to 1967, and the Amu Dar'ya and Syr Dar'ya from 1952 to 1968, and for the level of the Caspian Sea firstly for 1850 to 1930 and, secondly, for 1930 to 1965, and the Aral Sea for 1920 to 1960 and 1965 to 1975 from the data given in Table 4.17. Plot the curves and relate the cumulative discharge decline to the level of the Caspian Sea and Aral Sea.

4.7 Conclusion—the interaction between the major circulations and cycles

The hydrological cycle provides a link between many aspects of physical and human geography. It is especially important in the strong link between the two major global circulating systems. The circulations of the atmosphere and oceans come into contact at their mutual interface. Water is withdrawn from the oceanic reservoir and is carried to the land to water the earth and make life possible. The two circulations are linked with the global energy budget. There is a positive energy balance in low latitudes but a negative one in middle and high latitudes, the point of balance being about 35 degrees latitude. The imbalance requires heat transfer from low to high latitude, a process achieved by atmospheric transport of sensible heat and latent heat, as well as the transport of warm water by ocean currents. About 80 per cent of the transport takes place in the atmosphere, mainly by the transfer of sensible heat. The global winds carry latent heat both poleward and equatorward from the subtropical high pressure systems. Sensible heat and heat carried by ocean currents is continuously taken away from the equator as shown in Figure 4.17.

The global air circulation is caused by the pattern of solar energy reaching the earth's surface. In turn the air circulation affects the pattern of moving water. There is a close similarity between the global wind system and the ocean currents, as a result. The air circulation does not, however, show the asymmetry of the ocean current system, because the air is free to move over the whole earth. The water, on the other hand, is confined by the land masses, and is thus influenced by friction at the basin margins. The Coriolis effect varies in strength with latitude, which helps to account for the asymmetry of ocean currents.

The oceans in turn exert a strong modifying effect on the climate of the land around them. The western side of the landmasses benefit most from the warmth carried poleward by the ocean currents. The wind also plays other important roles with respect to the oceans in addition to initiating the major current system. It aids upwelling, which gives fertility, and is also responsible for generating the waves that form such a conspicuous feature of the ocean surface. Wave action is the most important process modifying the world's coastlines.

There are also many ways in which the oceans affect the atmosphere. It has been suggested that the temperature of the north Atlantic water may influence the development of blocking anticyclones, and thus the weather over large parts of the world. The feedback relationships between the two major circulation systems are highly complex and not yet fully understood. They include variations in sea level related to the increase and decrease of solid water storage as ice in glaciers and ice sheets, which depend in part at least on climatic variations.

On a longer time scale there are also connections between the major circulations discussed earlier. The cycle of movement within the earth's interior by convection activity results in the wandering of the crustal plates over the earth's surface and changes in the pattern of land and sea. In the very long term, oceans are created and closed, plates collide to form new mountain ranges and separate to form new oceans. Plate movement influences the climate and when the land and sea patterns are favourable can help to initiate ice ages. During the Tertiary, the initiation of the ice age may have been triggered by the drifting of Antarctica to the south polar area, while the north polar ocean became nearly land-locked. All three major circulations are intimately linked to give the earth its essential physical character.

There appears to be a cyclic element in many of these circulations, and the systems are delicately self-regulating. Energy remains evenly distributed, and the earth maintains its remarkable temperate character, varying only within a narrow range. The earth's surface is finite and the circulations must fit this basic framework. Plates may move, but as crust is created in one place, so it must be destroyed elsewhere in equal amount. Excess heat must be transferred to areas of deficit to maintain the balance. The laws of conservation must be respected in all these complex interactions. The earth is a place in which life can flourish in great variety, providing species to fit into the very wide range of available habitats. All life depends on the life-giving energy from the sun.

References and further reading

Barry, R. G. and Chorley, R. J. (1976) *Atmosphere, weather and climate*. 3rd Edn, Methuen, London.

*Bridges, E. M. (1970) *World Soils*. Cambridge University Press, Cambridge.

Chorley, R. J. (Ed.) *Water, earth and man*. Methuen, London.

Churchill, D. M. (1965) The displacement of deposits formed at sea level 6,500 years ago in southern Britain. *Quaternaria* **7**, 239–49.

*Dewey, J. F. (1972) Plate tectonics. *Scientific American*.

Dietz, R. S. and Holden, J. C. (1970) The breakup of Pangea. *Scientific American*.

Doodson, A. T. and Warburg, H. D. (1941) *Admiralty manual of tides*. HMSO, London.

*Eyre, S. R. (1963) *Vegetation and soils: a world picture*. Arnold, London.

*Gass, I. G., Smith, P. J. and Wilson, R. C. L. (1971) *Understanding the earth*. Open University, Artemis Press, Horsham, Sussex.

Haltiner, G. L. et al. (1957) *Dynamical and Physical Meteorology*. McGraw-Hill, New York.

Hollis, G. E. (1978) The falling level of the Caspian and Aral Seas. *Geographical Journal* **144**, 62–80.

Isacks, B., Oliver, J. and Sykes, L. R. (1968) Seismology and the new global tectonics. *Journal of Geophysical Research* **73**, 5855–99.

*King, L. C. (1967) *The morphology of the earth*. 2nd Edn, Oliver and Boyd, Edinburgh.

Koppen, W. (1936) Das Geographische system de Klimate. Vol. I Part C in *Handbuch der Klimatologies*. Eds W. Koppen and R. Geiger, Borntraeger, Berlin.

Lamb, H. H. (1966) *The changing climate*. Methuen, London.

*Macmillan, D. H. (1966) *Tides*. CR Books, London.

Miller, A. A. (1953) *The skin of the earth*. Methuen, London.

Peltier, L. C. (1950) The geographical cycle in periglacial regions as it is related to climatic geomorphology. *Annals of the American Association of Geographers* **40**, 214–36.

*Press, F. and Siever, R. (1974) *Earth*. Freeman, San Francisco. 2nd Edn, 1978.

Sellars, W. D. (1965) *Physical Climatology*. University of Chicago Press, Chicago.

*Strahler, A. N. and Strahler, A. H. (1976) *Elements of physical geography*. Wiley, New York.

Sverdrup, H. U., Johnson, M. W. and Fleming, R. H. (1946) *The oceans*. Prentice-Hall, Englewood Cliffs.

*Tarling, D. and Tarling, M. (1971) *Continental drift*. Doubleday, New York.

Taylor, R. L. S. and Smalley, I. J. (1969) Why Britain tilts. *Science Journal* **5**, 54–9.

Thornthwaite, C. W. (1948) An approach towards a rational classification of climate. *Geographical Review* **38**, 55–94.

Umbgrove, J. H. F. (1947) *The pulse of the earth*. M. Nijhoff, The Hague.

Umbgrove, J. H. F. (1950) *Symphony of the earth*. M. Nijhoff, The Hague.

*Wilson, J. Tuzo (compiler) (1970) Continents adrift. Readings from *Scientific American*.

CHAPTER 5

Conclusion

5.1 Scales of investigation

Physical geography covers a very wide field, and in order to study some of the most important aspects at a reasonable level of detail, the material has been arranged according to the scale of activity. This arrangement also has the advantage that it leads from the more familiar to the less easily grasped phenomena that operate over a very large area.

Small scale phenomena can be experienced at first hand in the field, where physical geography comes to life. Field-work and field techniques, which enable the observations to be based on a more precise and quantitative foundation, are very significant in local scale investigations. At this scale it is possible to measure changes over relatively short periods of time, so that the dynamic element, the processes, can be investigated directly.

On the medium or regional scale a considerable area is covered, for example the whole British Isles or a natural region, such as the equatorial forest zone, the arctic tundra or tropical desert. The emphasis at this scale is on a clearly defined unit or natural region, whether it be based on a land area, such as Britain, or a natural division, based on one or other of the different branches of physical geography. The natural regions can be extended from those more familiar on land to the oceans, where natural boundaries separating areas of different characteristics can also be distinguished.

The next scale of study is the global scale covering the whole earth. At this scale it is essential to appreciate the nature of the earth's surface as a finite, closed and spherical surface, which operates as a single unit in many important respects. At this scale one is concerned with the major circulations and cycles that keep the life of the earth as well as its surface features running in a reasonably uniform and stable way. The establishment and maintenance of equilibrium through the constant recirculation of all types of matter is very significant at this scale. It could be called the scale of the cycle.

The earth is a much smaller proportion of the whole in the next and largest scale, in which the earth is examined as a member of the solar system. At this scale those elements of physical geography that depend on the earth as a planet are considered, particularly the relationships between the earth and the sun and moon.

5.2 Interrelationships and interactions between the different aspects of physical geography

Throughout the book the importance of interrelationships between the different aspects of physical geography has been stressed, and a few of the numerous feedback interactions have been commented upon. The major branches of the subject include the study of the land surface, the field of geomorphology, and the processes that are altering and shaping it.

The atmospheric envelope of the earth gives the weather and climate that influences all other aspects of the physical scene. They are also affected by the nature of the surface, and on the large scale especially, by the distribution of land and sea. The interactions are so numerous and complex that they are difficult to unravel. They operate at all the scales of study, from the diurnal changes of wind and temperature in a small valley, to the worldwide global pattern of air movement, and the effect of the earth's movements relative to the sun in the pattern of solar radiation receipt.

Weather and climate, together with the nature of the surface material determine the formation of the soil. The soil in turn is necessary to the vegetation, which also affects geomorphological processes and is affected by them. The vegetation provides the basic sustenance for the animals. Biogeography usually includes the study of soils, and thus provides the link between geomorphology, meteorology and climatology, and the organic world.

The oceans play an important part in all aspects of physical geography. They also produce plants which sustain animals. In the oceans climatology and meteorology also play a part in providing the nutrients, which are the oceanic equivalent of the soil. They make plant life possible in the sea in the presence of sunlight for photosynthesis. This process is essential both in the water and on land in creating organic matter from inorganic nutrients. The earth's place in the solar system influences the incidence of light on the earth's surface, and hence the production of organic matter through photosynthesis. The relationship is an important link between the small scale and global scale studies.

Water is the element that is common to all

the scales of study and all aspects of physical geography, and it is essential to the operation of many of the processes in all fields. The subject of hydrology is increasing in importance, and water provides the link in many interactions. Hydrology is primarily concerned with the hydrological cycle, the circulation of water which impinges on all aspects of physical geography. The reservoir of the ocean plays a large role in the cycle. It should, therefore, play an important part in physical geography. The ocean basins hold 97 per cent of all water on earth. These large water receptacles are the low areas in the earth's crust. They are younger than the water they contain, as has been discovered in the development of the theory of plate tectonics and sea floor spreading. Their development forms part of another even longer cycle, that of internal convection which causes the drift of the plates and the creation of new oceans as others close.

The hydrological cycle, however, is of more direct relevance on the time scale of most events of physical geography. The circulation of water in the ocean basins is slow compared to that in the atmosphere. The evaporation of water from the oceans plays a vital part in the atmospheric circulation, and thus is often taken as the first step in the hydrological cycle. The moisture in the atmosphere is mainly in the form of vapour, but its condensation to the liquid form in clouds and rain is necessary to the next important stage of the hydrological cycle, that of precipitation on land, without which terrestrial life would be impossible. Precipitation is related to the third state in which water can exist, the solid state. Ice particles are important in the process of precipitation. They can fall as snow, transferring solid water to land, where some of it builds up the great ice sheets and smaller glaciers. These ice masses provide 75 per cent of fresh water on land.

The much smaller amount of liquid water in the land stages of the hydrological cycle are most immediately involved in physical geography and life in general. Part of the rainfall is used directly by the vegetation and animals, some is stored as ground water to keep the rivers flowing under dry conditions, and to replenish the underground storage. Runoff and infiltration are, therefore, the next stages of the hydrological cycle to be considered. There are many paths through the cycle, all of which are of direct concern in physical geography, in all branches of the subject. Water is the common element at all the scales of study.

5.3 The complexity of the natural environment—ecosystem

Life cannot exist without water, whether it be in the sea or on land, but it also requires other elements, including nutrients for plants and food for animals, and a means of reproduction. Thus animals depend upon plants and on each other, both animals of the same species and other species. The complex environment within which organisms live is the ecosystem. Each natural environment has its own particular ecosystem, which under natural conditions reaches a state of equilibrium, in that numbers of the different species become adapted to their surroundings.

The conditions of the earth's surface vary greatly, from the bleak heights of the great mountain ranges, through the plainlands, the deserts and tropical forests, to the coastal zone and the ocean depths. In each of these very different environments plants and animals have become adapted to live together in ecosystems. Some provide food for the others. There are the vegetarians, feeding directly on the plants, and the carnivores, feeding on other animals, so that a complex food web is set up, in which each individual has its niche. The system is not very efficient in that at each step in the food web there is a great loss of energy in the creation of living matter. Nevertheless the system is adjusted very precisely and delicately to the environment, with every organism depending on many others for its livelihood. All are adjusted to the type of surface, the soil, the weather and climate. All branches of physical geography must be involved in the study of the ecosystems.

The ecosystem can be studied in great detail within a very small area, on the local scale, in which each plant and animal can be considered. Alternatively it can be studied as a particular environmental type that will cover a fairly large area, the regional scale of study. On the global scale the study includes the classification of the different ecosystems of the regional scale. Natural regions can be recognized from the point of view of their natural vegetation and the associated animal life, as well as the climate on which they depend. Their geomorphological setting must also be considered, including mountain ecosystems, plains and coastal ones, for example.

As with other aspects of physical geography, the time element is very important, and is usually associated with the areal and spatial elements. Small areas can be studied over short time spans, whereas larger areas, particularly the global aspects, can be studied over much longer time spans. Evolution is involved at the longest time span, covering geological time, and climatic change at all time spans, including fairly short ones.

5.4 The significance of physical geography to geography and other disciplines—man-made problems

Man is playing an increasingly important and active part in the operation of natural ecosystems, which is the field of physical geography. Geography as a whole includes the study of man and his activity. As man plays an increasingly effective part in altering the natural ecosystems into a man-dominated world, it is becoming more and more necessary for physical geography to form the foundation of geography as a whole. There is as yet far too little known of the effects of man's interference in the natural ecosystem, and until this knowledge is available there are bound to be mistakes, such as those that have occurred in the past.

The problem of soil erosion is only one of the most obvious effects of man's interference with the natural system. It is now often referred to as accelerated erosion, in that soil erosion is a natural process, but when man interferes the process is made much too rapid. The natural production of new soil is out-stripped, and a serious loss of fertility results, with considerable changes to the surface morphology by the creation of gullies and the increase of sediment yield to streams. Activity that modifies climate and weather, both accidental and purposeful, could set up chain reactions that might be difficult to control. In all these operations the economic aspect is an important one.

The overuse of the resources of the sea cannot make sense from the environmental point of view or even the long term economic point of view. It is only the short term economic greed that allows such gross in-terference with the well balanced natural system. The eventual result is usually to the disadvantage of man as well, as some desirable species may be made extinct. The loss of any species must be deplored because it is an irreversible process. Man cannot recreate lost species, although it is possible to breed new species of both plants and animals.

Physical geography, especially with its human implications and interferences, touches on many other disciplines. The natural sciences of geology, geophysics, pedo-logy, biology, meteorology and oceano-graphy are all involved. Aspects of the social sciences are also relevant in the contact between physical and human geography, including economics, sociology and others. Physical geographers should be able to ap-preciate the interrelationships between all the aspects of their own subject, as well as its connections with human geography and related subjects. It is a very wide field, but very important in many aspects of modern living, as it is directly relevant to an under-standing of the world we live in.

Appendix: solutions of and suggestions concerning the exercises

Exercise 1.1

As an example of the calculation of a length of latitude the length of 53 degrees is given.

Take $R = 6{,}378$ km. The length is given by

$$2\pi R \cos 53 = 2 \times 3{\cdot}14159 \times 6{,}378 \times 0{\cdot}60181 = 12{,}058{\cdot}6 \text{ km}$$

The distance along the whole meridian = $2\pi R = 40{,}074$ km

The distance to the equator = $53/360 \times 40{,}074$ km = $5{,}900$ km

The distance to the pole = $37/360 \times 40{,}074$ km = $4{,}118$ km

The distance along $1°$ of latitude = $111{\cdot}316$ km

The distance along $1'$ of latitude = $1{\cdot}855$ km

The distance along $1''$ of latitude = $30{\cdot}9$ m

Exercise 1.2

(1) The pole star moves around the pole within 1 degree of its position. At its maximum altitude it could be seen at a distance of 21 m from the wall. Its elevation is, therefore, equal to the tangent given by $16{\cdot}8$ m $- 1{\cdot}5$ m$/21$ m $= 15{\cdot}3/21 = 0{\cdot}73$. Tan $0{\cdot}73 = 36°$. At its minimum elevation it could be seen from $22{\cdot}6$ m. The elevation at this time is similarly tan $15{\cdot}3/22{\cdot}6 = 0{\cdot}68 = 34°$. The latitude must be the mean of these two values and is, therefore, $35°$N.

(2) The observation was made at the equinox when the sun is overhead at the equator and its declination is zero. The altitude of the sun must have been at $45°$ to cast the shadow half way across the water surface, therefore the latitude must have been $45°$N as France is in the northern hemisphere.

(3) Nick goes south along $1°$ of a meridian, which is $2\pi R/360$ from $48°$N to $47°$N and then west along latitude $47°$N for $1°$, which is $2\pi R \cos 47/360$. If R, the radius of the earth is taken as $6{,}367{,}244$ m, then the east–west distance covered by Nick is $75{,}790$ m. Steve goes north along $1°$ of a meridian from $47°$ to $48°$ and then east along latitude $48°$N for $1°$. He covers an east–west distance of $2\pi R \cos 48/360$. This is $74{,}356$ m, so that his journey is about $1{,}434$ m shorter. This is because the meridians converge towards the poles and a degree along a parallel of latitude decreases in proportion to the cosine of the latitude polewards. Note that 1 m is approximately $1/10{,}000{,}000$ of a quadrant of the globe through Paris, or 1 nautical mile = $6{,}080$ feet = 1 minute along a meridian.

265

(1) The difference of longitude is 36°, which is 2 hours 24 minutes in time. The plane must cover the distance in 24 hours less the difference in time resulting from the longitude difference, and this is 21 hours 36 minutes. The speed of the plane is thus 3,840/21·6 km/h, which is 177·78 km/h. The date is the same as the plane crosses the date line during the flight from Wellington to Tahiti.

(2) The length of the shadow indicates that the maximum elevation to the north at the summer solstice is 82°. The latitude is equal to $23\frac{1}{2}° - (90 - 82) = 15\frac{1}{2}°$N. The maximum elevation to the south occurs at the winter solstice of the northern hemisphere and is 51°. The latitude is, therefore, $90 - 51 - 23\frac{1}{2} = 15\frac{1}{2}°$N. These angles are illustrated in Figure A1.1. The angles of the sun's elevation are derived from the tangent of the length of the flag-pole divided by the length of the shadow. The time of the eclipse indicates that the island is 61°W of Greenwich. The island he was on is Maria Galante Island in the West Indies.

(3) The boy scout used apparent or solar time and not mean time by using the shadow, for this reason the midway position is not due south. On 4 November the equation of time reaches a maximum of +16·22 minutes so that apparent time is later than mean time by this amount. His mark is, therefore, 4° to the west of north, which is the amount of his error.

Figure A1.2a shows the pattern of the sun's declination throughout the year, with the equinoxes in March and September and the solstices in June and December when the sun is overhead at the northern and southern tropics respectively. Figure A1.2c shows the details of the declinational changes and the equation of time over the northern winter solstice and the times of sunset and sunrise for mean time and solar time. The time of sunrise is progressively later until early January

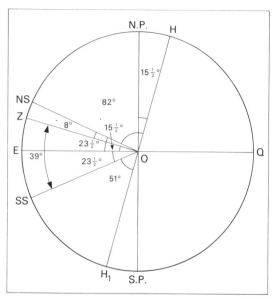

Figure A1.1 Diagram to illustrate angles in the meridian plane. Maximum elevation to N at summer solstice 82°. Latitude $= 23\frac{1}{2} - (90 - 82) = 15\frac{1}{2}°$. Maximum elevation to S at the winter solstice is 51°. Latitude $= 90 - 51 - 23\frac{1}{2} = 15\frac{1}{2}°$. Length and direction of shadow at the equinox is $51 + 23\frac{1}{2}°S = 74\frac{1}{2}°$, tangent $74\frac{1}{2}° = 3·6058$, therefore shadow $= 40/3·6058 = 11·1$ feet.

because at this time the equation of time becomes negative from its high positive value through late November and December. It is this fact that makes the darkest evening or earliest sunset occur in mid-December.

The pattern is not so asymmetrical at the northern summer solstice as the equation of time is smaller at this season. Nevertheless the latest sunset is delayed by a few days after the solstice and the earliest sunrise occurs a few days before the solstice on 21 June. This pattern is shown in Figure A1.2d.

The totals for the different latitudes listed are as follows:

90°N	1,906	30°N	3,828	50°S	3,074
70°N	2,106	0°	4,374	70°S	2,263
50°N	2,986	30°S	3,885	90°S	2,019

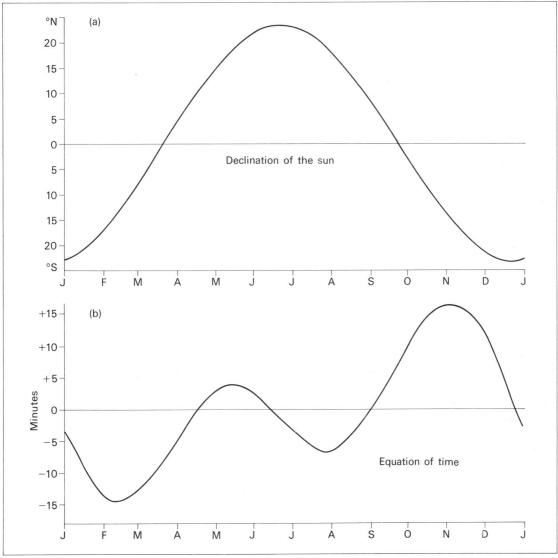

Figure A1.2 (a) Graph of the declination of the sun, (b) graph of the equation of time, (c) details of declination of the sun and equation of time at the northern winter solstice. Times of sunrise and sunset are given. (d) Details of declination of the sun and equation of time at the northern summer solstice. Times of sunrise and sunset are given.

The variation at different latitudes is due to the varying distance between the sun and the earth as it moves around the sun. The earth receives more energy when it is nearest to the sun at perihelion on 3 January, amounting to 7 per cent excess over the aphelion value received on 4 July. January temperatures in theory should be about 4 °C higher than those in July, but other factors, such as the distribution of land and sea, mask this effect.

Exercise 2.1

Figure A2.1 shows the pattern of Thiessen polygons derived from the rainfall data shown

267

Figure A1.2 cont.

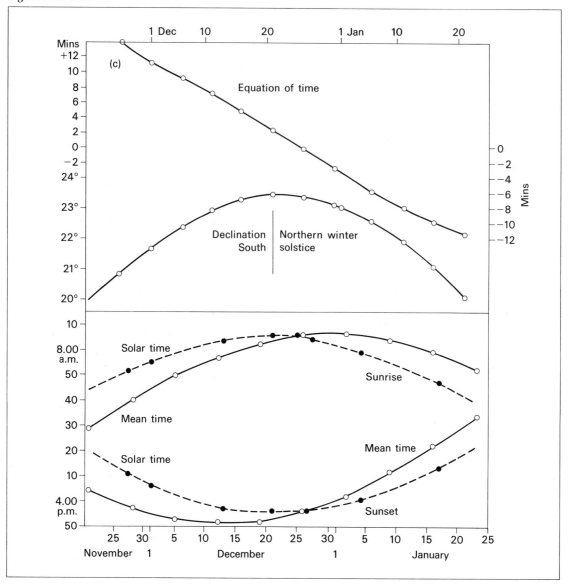

in Figure 2.2 for a small catchment. The total rainfall falling over the catchment is found by summing the individual values for the area of the polygons.

$$0·5 \text{ cm} \times 55 = 27·5$$
$$0·4 \text{ cm} \times 48 = 19·2$$
$$0·3 \text{ cm} \times 52 = 16·6$$
$$0·2 \text{ cm} \times 52 = 10·4$$
$$0·1 \text{ cm} \times 30 = 3·0$$
$$\text{Total} = 76·7 \text{ cm}$$

Exercise 2.2

The *t*-test provides a suitable statistical method of assessing the significance of the difference between the sets of figures. The method using the differences between the values provides the simplest means of applying the test. The test for the maxima given in Table 2.1, for Ushaw and Houghall, is given in full and the results of the other tests are given. The difference values cannot be used as the

Figure A1.2 cont.

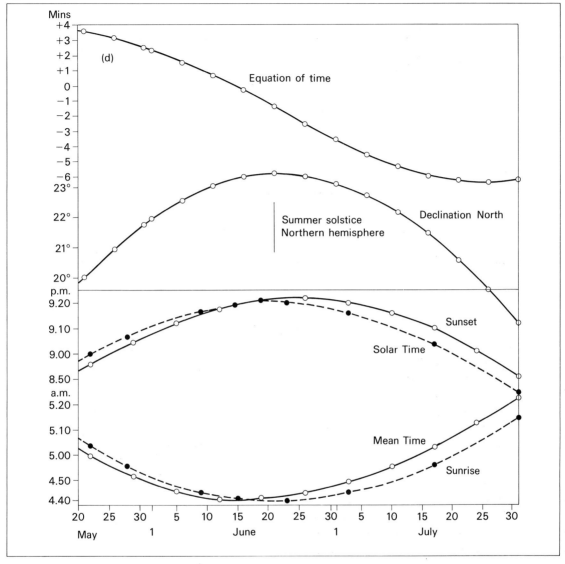

differences between the means are not available, but the latter part of Table 2.1 provides values that can be used to assess the effect of soil type on temperature.

Maxima values differences

d	$d - \bar{d}$	$(d - \bar{d}^2)$
0·90	0·162	0·0262
1·18	0·118	0·0139
1·11	0·048	0·0023
0·95	0·112	0·0125
0·78	0·282	0·0795
0·95	0·112	0·0125
1·11	0·048	0·0023
0·88	0·182	0·0331
1·05	0·012	0·0001
1·88	0·818	0·6691
1·11	0·048	0·0023
0·84	0·222	0·0493

$\Sigma d = 12{\cdot}74 \qquad \Sigma(d - \bar{d}^2) = 0{\cdot}9031$

$\bar{d} = 1{\cdot}062$

$$\frac{\Sigma(d - \bar{d}^2)}{n - 1} = \frac{0{\cdot}9031}{11}$$

$$= 0{\cdot}0821$$

$$s_{\bar{d}} = \sqrt{\frac{0{\cdot}0821}{12}}$$

$$= \sqrt{0{\cdot}00684}$$

$$= 0{\cdot}0827$$

Mean value for Ushaw	$= 11{\cdot}8125$
Mean value for Houghall	$= 12{\cdot}8321$
Difference between means	$= 1{\cdot}02$

$$t = \frac{1{\cdot}02}{0{\cdot}0827} = 12{\cdot}3337$$

d.f. $= 11$, t for $0{\cdot}001$ significance level and 11 d.f. $= 4{\cdot}44$ for 6 d.f. $= 5{\cdot}96$
The result is, therefore, significantly different at a high level of significance.
t values for the minima values for Ushaw and Houghall $= 11{\cdot}739$
t values for the extreme minima values for Ushaw and Houghall $= 35{\cdot}116$
t values for the minima for Lynford and Cambridge $= 14{\cdot}0$ with 6 d.f.
t values for the maxima for Lynford and Cambridge $= 2{\cdot}155$, which is significant at 95%.

Exercise 2.3

The rainfall values for Dale Fort and Glyntawe in Wales can be analysed in the same way, using the difference values for the t-test. The results for January give a t value of $2{\cdot}6635$, which is significant at 95% level for 10 d.f., and for July the t value is $3{\cdot}1153$, which is significant at 98% level for 10 d.f. Figure A2.2 shows the pattern of precipitation at the two stations relative to wind direction for January. The figure shows the dominance of the southwesterly quarter and the comparative shelter from the northeast. The coastal site has very considerably less precipitation than the more exposed higher site inland.

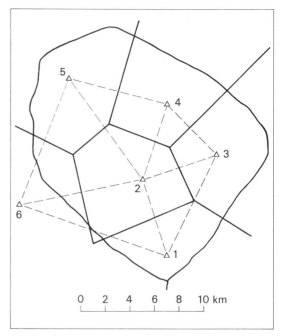

Figure A2.1 Thiessen polygons, figures give precipitation in mm. The polygons are outlined in solid lines.

Exercise 2.4

The regression lines and their equations are given on Figure A2.3 and the values of the correlation coefficient, r. The graphs show that the storm of 1 November 1967, produced considerably greater peak runoff values than that of 4 November. The rainfall during the first storm was approximately $1{\cdot}625$ cm, falling in four hours, while during the second storm $0{\cdot}915$ cm of rain fell in 3 hours. The graphs show that the correlation between the discharge and woodland percentage is negative, while that between enclosed land and discharge is positive. The basin with the largest area in woodland had the smallest discharge, while the most enclosed and cultivated basin had the highest. These results reveal clearly the importance of surface cover in determining local runoff following short, heavy rainfalls in a small catchment basin. The degree of correlation also indicates, particularly in the woodland results, that other factors must also be considered, such as the shape of the basin,

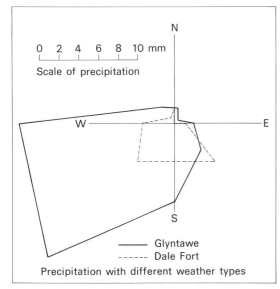

Figure A2.2 The pattern of precipitation relative to wind direction for two stations in south Wales. The distance from the centre is proportional to the amount of precipitation.

the relief and the character of the soil and rocks.

Exercise 2.5

The figure A2.4 shows the much greater variability of the rainfall compared with the evaporation, which has a regular maximum in summer and minimum in winter, the values being very nearly the same in the north and south of England. The precipitation, on the other hand, varies greatly both from month to month and from year to year. There is also much greater differences between the two stations. Over the year as a whole there is nearly always a water surplus at these two stations, in marked contrast to the situation of San Antonio in Texas.

Figure A2.3 Graphs showing discharge plotted against land use for 2 storms in small catchments in south Devon, with regression lines and equations.

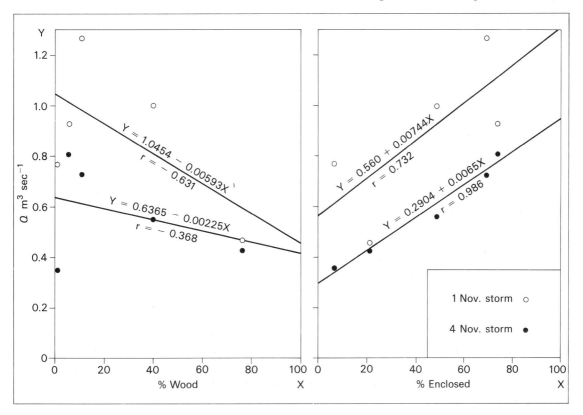

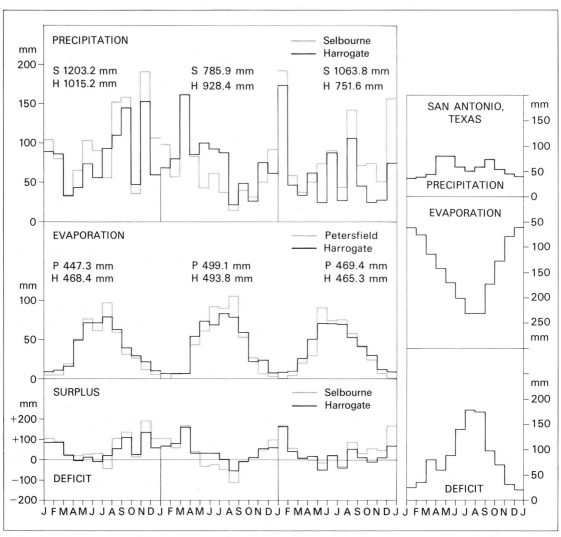

Figure A2.4 Patterns of precipitation, evaporation and water balance for each month for 1946, 1947 and 1948, for Selbourne/Petersfield, Harrogate and San Antonio, Texas.

Exercise 2.6

The percentage rainfall to runoff values are as follows:

Month	1934 %	1936 %	Four year mean %
Jan.	18	71	57
Feb.	108	94	50
Mar.	20	79	70
Apr.	18	68	39
May	35	86	45
Jun.	11	16	15
Jul.	8	12·5	15
Aug.	6	65	16
Sep.	7	12	8
Oct.	10	21	17
Nov.	11	31	31
Dec.	18	46	32

Annual values 1934–5 Oct.–Sep.   9
1936–7 Oct.–Sep. 49

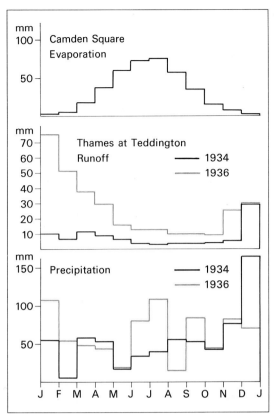

Figure A2.5 Precipitation and runoff for the Thames at Teddington for 1934 and 1936, and mean evaporation at Camden Square, London.

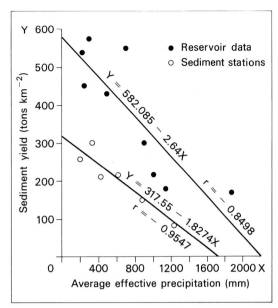

Figure A2.6 Sediment yield plotted against average effective precipitation for reservoir data and sediment stations, with regression lines and equations.

The effect of very variable precipitation accounts for the marked contrasts in the percentage of runoff to rainfall in the two years given. The mean values for four years shows the general pattern of high values in winter and lower ones in summer, when evaporation is much higher. This is also clearly demonstrated in the plot of evaporation against the percentage of runoff to rainfall. (See Figure A2.5.)

Exercise 2.7

The plot of the data in Table 2.7 is shown in Figure A2.6 on which sediment yield is plotted against average effective precipitation. For the reservoir data the regression equation is $Y = 582·085 − 2·64 X$, where X is the precipitation and Y the sediment yield, which is the dependent variable. The value of r, the correlation coefficient is $−0·850$ and n, the number in the sample, is 9. The result is, therefore, highly significant at the 99% level of confidence. The correlation is negative showing that as precipitation increases so the sediment yield decreases because of the greater protective effect of more prolific vegetation. The results for the sediment stations are similar. The regression equation is $Y = 317·55 − 1·8274 X$, and $r = −0·951$, which for $n = 6$, is also significant at the 99% level of confidence.

Exercise 2.10

Analysis of slope form from simple profiles.
(1) Constant slope—this could be caused by rapid and steady river incision.
(2) Convex-constant slope—a slope typical of permeable rocks, such as chalk.
(3) Constant-concave—a result of inefficient basal removal of debris.
(4) Convex-concave—a mature slope with a stable base level.

273

(5) Constant—freeface, constant—concave—hard rock outcrops on a slope with inefficient basal removal.

(6) Constant — convex — accelerating rapid rejuvenation, or more efficient removal of basal debris.

Exercise 2.11

The sign test is a simple and useful non-parametric test to establish whether there is a significant difference between two sets of values according to their relative magnitude. In the example given the zeros are ignored, leaving 19 values, of which 16 are pluses and 3 minuses. It is assumed that there is an equal probability of pluses and minuses, thus $p = q = 0.5$ and $n = 19$. The value of $m = np = 9.5$, the expected value. The value of s is given by

$$s = \sqrt{(npq)} = \sqrt{(19/2 \times 2)}$$

$$= \sqrt{4.75} = 2.179.$$

The value of Z is then found by

$$Z = \frac{(\text{no. of pluses} - m) - p}{s}$$

$$= \frac{(16 - 9.5) - 0.5}{2.179} = \frac{6}{2.179}$$

$$= 2.753$$

The value of $Z = 2.753$ is found as an area under the normal curve, $A = 0.4970$.

The probability is then given by

$$2 \times [(- \text{value of } 2.753 \text{ in } Z \text{ tables}) - 0.5000] = 2 \times 0.003 = 0.006$$

This result is highly significant.

Exercise 2.12

The two-way analysis of variance table can be simplified by deducting the same amount from each value. In the example given it is con-

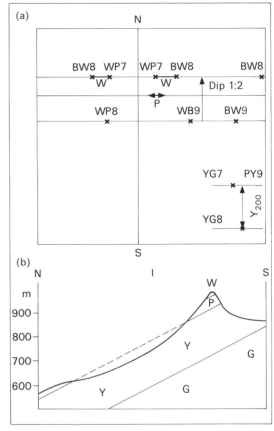

Figure A2.7 Geological map 1, showing strike, dip and measurement of bed thickness (a), north–south section across geological map 1, (b).

venient to deduct 8 from all values, giving the following adjusted table:

Aspect stream	N	E	S	W	Total
No stream	−3	−2	0	1	−4
Stream	0	−1	4	3	+6
Totals	−3	−3	4	4	+2

The correction factor is given by the grand total squared and divided by $N = 8$, i.e. $4/8 = 0.5$.

Between aspect sum of squares is given by $1/4 (16 + 36) - 0.5 = 12.5$.

Between streams sum of squares is given by $1/2 (9 + 9 + 16 + 16) - 0.5 = 24.5$.

Total sum of squares is given by $40 - 0.5 = 39.5$. The analysis of variance table is given

	Sum of squares	Degrees of freedom	*V*	*F*	
Between aspect	12·5	3	4·167	5·00	Not significant
Between streams	24·5	1	24·5	29·40	Significant 95%
Total	39·5	7			
Residual	2·5	3	0·833		

in Table A1.

Exercise 2.13

The sandstone is porous and permeable and so has little surface drainage. The shale is impermeable and its upper outcrop is marked by the appearance of streams at springs, and its lower outcrop by the sink holes where the streams go underground on reaching the outcrop of limestone. The limestone is permeable, mainly through its enlarged joints and bedding planes, so streams appear at its lower boundary, where it rests on impermeable rocks as strong springs. This type of geology and drainage can be studied in the field in the area around Ingleborough and Malham Tarn in North Yorkshire.

Exercise 2.14

Geological map 1. Figures A2.7a and b show that the strike is east–west, and the beds dip uniformly north at 1 in 2 as the strike lines are straight and parallel. The section shows that bed G is the oldest followed by Y, P and W. The isolated outcrop of Y is an inlier as it is surrounded by younger beds. The thickness of Y is 200 m; it can be measured as indicated on the figure in the southeast corner of the map where G outcrops. Bed W is 100 m thick and P is 50 m, bed B is at least 200 m thick.

Geological map 2. There are two sets of beds that are unconformable. Beds Y, W and R are the oldest and dip southwest at 1 in 1·5. They were faulted by a normal fault running northwest to southeast with a 100 m throw down to the northeast. Bed Y is the oldest, bed W is next and is 100 m thick, followed by bed

R, which is at least 250 m thick. The beds were planed off before the deposition of beds B and G, which have a north–northwest to south–southwest strike, dipping east–southeast at 1 in 4·5. The isolated outcrop of bed G in the northeast is an outlier, as it is surrounded by older beds, G being the youngest on the map. Bed B is about 120 m thick. The section illustrates the structure and is drawn along the apparent dip of both sets of beds, the dip of R, W and Y being 1 in 1·6 and of B and G being 1 in 12. Apparent dips are always less than true dips, which are perpendicular to the strike (Figures A2.8a and b).

Figure A2.8 Geological map 2, showing strikes, dips, fault and unconformity (a), northeast–southwest section across geological map 2, (b).

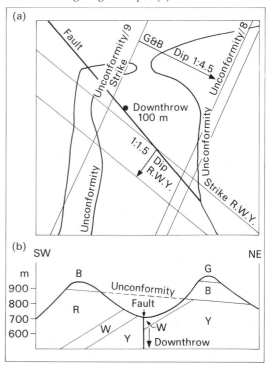

275

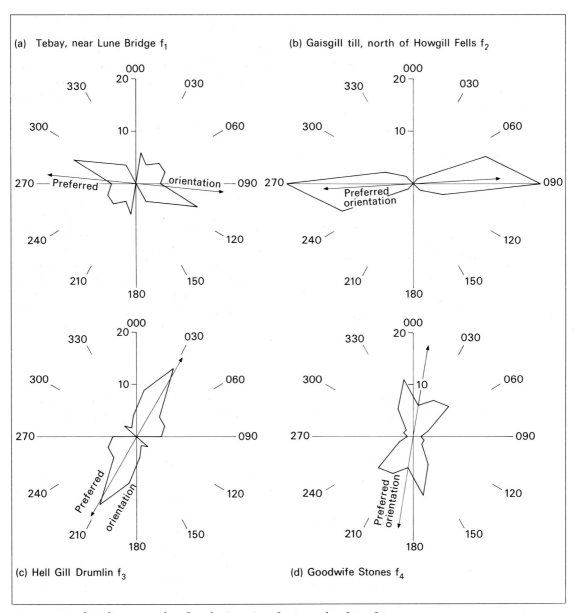

(a) Tebay, near Lune Bridge f₁

(b) Gaisgill till, north of Howgill Fells f₂

(c) Hell Gill Drumlin f₃

(d) Goodwife Stones f₄

Figure A2.9 Fabric diagrams and preferred orientations for 4 samples, f-1 to f-4.

Geological map 3. The strike of the rocks is northeast to southwest at 1 in 7·5. The coal at B is at a depth of 200 m O.D. The ground here is at 500 m so the shaft must be 300 m deep. Measure 750 m on the scale and draw a second strike line, which goes through A. The coal here must be at 100 m O.D., the ground is at 600 m so the shaft must be 500 m deep. Join A to C, in the distance between A and the strike line the coal seam rises from 100 m O.D. to 200 m O.D., measure this distance again until C is reached, i.e. twice more; at C, therefore, the coal seam is at 400 m and the ground is at 700 m, so the shaft must be 300 m deep. At D the seam has fallen to 0 m O.D., as D is 1,500 m from the strike line in a direction at right angles to it. The ground at D is 400 m, so the coal mine shaft must be 400 m deep at D.

Exercise 2.15

As an example of the type of results obtained from roundness tests the following values may be considered: Borrowdale volcanic stones from Blea Water moraine were compared with a sample of 50 obtained from Haweswater moraine a few km further down the valley, the mean and standard deviations were respectively 108 and 153 means, and 49 and 56 standard deviations. The t-test gave a result of 4·3, which is significant at 99 %. Silurian grits were sampled in the upper reaches of Uldale Beck and several km downvalley, giving results of 140 and 202 means respectively and 58 and 74 standard deviations, giving a t-test value of 4·65, again significant at 99 % level. Both results indicate the effectiveness of erosion in rounding stones over a short distance by glacial and fluvial erosion.

Exercise 2.16

f1 is a fabric taken from an exposure of till in the Lune Gorge near Tebay during the construction of M6. The preferred orientation is 95° 45′ in the second quadrant as the sine is + and the cosine −. This easterly orientation probably represents a transverse pattern induced as the ice was constrained to flow south through the Lune Gorge. The L value is 24.5, significant at 0·04.

f2 is a fabric taken from an exposure of till on the banks of the Upper Lune just east of Tebay. The preferred orientation is 86° 18′ and the L value is 82 which is highly significant showing a strongly oriented fabric. The pattern is again transverse for the same reasons as f1, as ice was being forced to rise over the Howgill Fells through Ellergill valley near the sampling point.

f3 is taken from a drumlin near the upper Eden and Ure watershed. The preferred orientation is 31° 30′, a direction that suggests ice movement down the upper Eden valley. The L value is 43·2, which is again significant at 99 %.

f4 is taken from a sample of the orientation of 50 stones in a deposit that is probably the result of landsliding under periglacial conditions in the upper Eden valley high on the hillside. The preferred orientation is 10° 7′, but its strength is not significant. This result tends to confirm the deposit is not a moraine, but the result of slumping or landsliding. (See Figure A2.9a–d.)

Exercise 2.18

The meanders shown in Figure 2.18b (ii) have a mean stream width of 1·5 mm, a mean meander amplitude of 15 mm, a mean meander wavelength of 19·5 mm, giving the ratio of l/a (length/amplitude) as 1·3. The stream length is 38·5 cm and the valley length 19·5 cm, giving a sinuosity index of 1·974. The amplitude of the meander belt is 10 times the width of the stream, which is a commonly observed relationship.

Exercise 2.19

The reduced levels and cumulative distances derived from the levelling data in Table 2.13 are as follows:

Height of instrument	Reduced level	Distance	Remarks
25·17	24·45		Bench mark
29·82	22·08		
37·75	27·44		
32·34	28·06		
22·80	19·68		
17·38	13·43	0	Station 1
	14·10	12	Station 2
	10·21	150	Station 3
12·16	9·35	208	Station 4
	7·89	301	Station 5
	7·26	418	Station 6
	6·68	466	Station 7
9·90	6·74	502	Station 8
	3·43	725	Station 9
11·52	6·47		Return levelling
23·77	9·88		
33·31	21·39		
28·15	26·20		
	24·45		Bench mark

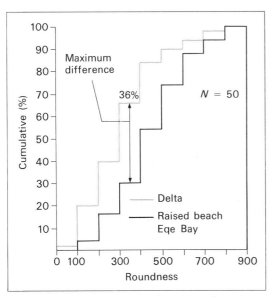

Figure A2.10 Cumulative frequency graph for two samples of stone roundness values.

Exercise 2.20

The cumulative frequency graph shown in Figure A2.10 shows the pattern of stone roundness for two samples of 50 stones taken from a delta and a raised beach respectively. The maximum difference between the two samples is 36%. The Kolmogorov–Smirnov method can be used to test the significance of the difference between samples of stones measured in Exercise 2.6. The data can also be analysed by means of the t-test if the roundness pattern is normally distributed. In this instance the mean and variance of the samples must be calculated.

Exercise 2.21

The data provide an example of the value of transformation of the observed results to provide the best correlation. The beach gradient increases as the material size becomes coarser. The data can be plotted as a gradient, the cotangent of the slope (Figure A2.11a) as a logarithmic value of the cotangent (b), or in degrees against the sand size in mm (c), given

as the median diameter. The results of the three correlation analyses are as follows:

Cotangent values
$$r = -0.7718, Y = 86.45 - 1.2056\,X$$

Logarithmic cotangent values
$$r = -0.867, Y = -2.159 - 0.02\,X$$

Degrees values
$$r = +0.9638, Y = -2.577 + 0.1747\,X$$

X is the independent variable, the sand median diameter in mm, and Y is the gradient of the beach swash slope, the dependent variable. Figure A2.11 also shows the data plotted on semi-logarithmic graph paper (d), which produces a result similar to that where the logarithmic values are plotted on ordinary graph paper. The conclusion reached is that the slope in degrees provides the closest relationship between beach slope and sand size. The correlation is positive in this instance as the smaller values for the sand provide the flatter beaches. The flat beaches have high values when the cotangent is used so the correlation is negative.

Exercise 2.22

Figure A2.12 shows the orthogonals drawn for the two different spit patterns. In (a) the short waves are little refracted, meeting the coast at a considerable angle. Material is, therefore, drifted along the coast to form the spit, which diverts the river mouth downwave in the direction of beach drifting. In (b) the long waves are refracted to a large extent. The spit has formed with its seaward margin parallel to the wave crests, with the orthogonals reaching the shore at right angles. Beach drifting is at a minimum and the river mouth is controlled by the pattern of the refracted swell. Wave energy is less intense on the spit owing to the greater wave refraction indicated by the wider spacing of the orthogonals. The stream exit is where the energy is lowest and the wave height and the spit height least.

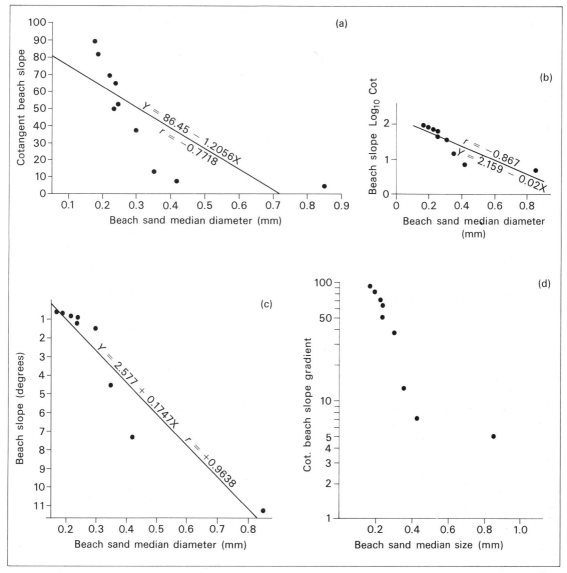

Figure A2.11 (a) Linear cotangent beach slope plotted against sand median diameter in mm. (b) Logarithmic cotangent beach slope plotted against sand median diameter in mm. (c) Beach slope in degrees plotted against sand median diameter in mm. (d) Cotangent beach slope plotted on semi-logarithmic paper against sand median diameter in mm.

Exercise 2.23

The chi square table must include the observed and expected results of the observations. For the first set of values it is as in Table A2. chi square = 11·0057; for one degree of freedom the value of chi square in the table is 10·83 at the 0·001 or 99·9% level of significance.

The second set of values are as shown in Table A3.

chi square = 10·9956; the result is also significant at the 99·9% level of confidence.

279

Table A2

| | Onshore winds | | Offshore winds | | |
	Observed	Expected	Observed	Expected	Totals
Fill	4	8·76	13	8·24	17
Cut	13	8·24	3	7·76	16
Totals	17		16		33

Observed (o)	Expected (e)	(o − e)	(o − e)^2	(o − e)^2/e
4	8·76	4·76	22·6576	2·5865
13	8·24	4·76	22·6576	2·7497
13	8·24	4·76	22·6576	2·7497
3	7·76	4·76	22·6576	2·9198
				$\Sigma = 11\cdot0057$

Table A3

| | Upper beach | | Lower beach | | |
	Observed	Expected	Observed	Expected	Total observed
Fill	4	8·5	11	6·5	15
Cut	13	8·5	2	6·5	15
Total	17		13		30

Observed	Expected	(o − e)	(o − e)^2	(o − e)^2/e
4	8·5	4·5	20·25	2·3824
13	8·5	4·5	20·25	2·3824
11	6·5	4·5	20·25	3·1154
2	6·5	4·5	20·25	3·1154
				$\Sigma = 10\cdot9956$

These results illustrate the importance of the wind on the movement of beach material and show that on this beach material moves mainly from the upper to the lower beach, and vice versa rather than moving throughout the beach profile.

Exercise 3.1

Curve C starts with a steep lapse rate to 900 mbar, indicating warming of the lower layers. This causes initial instability as the path curve lies to the right of the environment curve. Above 900 mbar the lapse rate decreases and the path curve follows the SALR. Clouds would, therefore, form with their base at 900 mbar. The reduction in lapse rate causes the environment curve to pass to the right of the path curve, making the air stable above this point, which is at a height of about 800 mbar. This level would mark the top of the clouds, which would be thin strato-cumulus or fair weather cumulus, as their thickness would be only about 100 mbar.

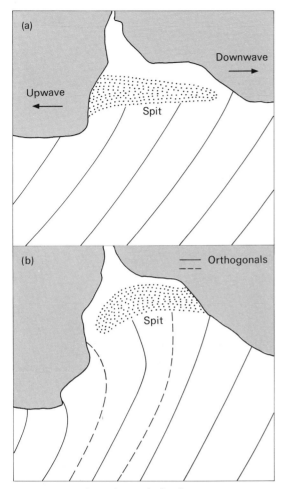

(a)

Downwave →

← Upwave

Spit

(b)

—— Orthogonals
- - -

Spit

Figure A2.12 The orthogonals for the wave patterns shown in Figure 2.24.

Curve A as it stands represents a stable air mass, owing to the inversion between 1,000 mbar and 950 mbar. With diurnal warming, however, this air could become unstable; if the temperature were raised to 20 °C and the water vapour content remained at 8 g/kg, the path curve would run from 20 °C along the DALR to a level of 850 mbar, the point at which the water vapour curve cuts the DALR. Above this level the air would be saturated and the path curve follow the SALR. The path curve would then lie to the right of the environment curve and convective activity leading to summer afternoon showers or thunderstorms could be forecast, with the formation of cumulus and/or cumulo-nimbus clouds, as indicated in Figure A3.1.

Exercise 3.2

Map (a): Arctic maritime air reached Britain from the north on 24 January 1976 giving very cold conditions over the country, temperatures being below freezing in many areas and snow showers common in the north and east, as the air mass is unstable after passing south across the sea.

Map (b): Arctic continental air reached Britain from the northeast around a high pressure centre over Scandinavia, bringing very cold conditions with temperatures well below zero in many places, although the short sea passage renders temperatures higher than on the continent.

Map (c): Polar maritime air is one of the most common to reach Britain, and in autumn provides cool, clear showery conditions, owing to instability in the air as it passes south from the northwest over the Atlantic Ocean to reach Britain. Visibility is good and cumulus clouds are common.

Map (d): Polar continental air comes from the east or northeast around high pressure over northwest Europe. Conditions in spring are cold, and relatively dry owing to the high pressure, but the short sea crossing brings cloud and slightly higher temperatures than further east.

Map (e): Tropical maritime air is typical of the warm sectors of depressions passing across Britain before they become occluded. In winter the conditions are mild and moist, with much cloud, although the air is relatively stable as it moves to higher latitudes.

Map (f): Tropical continental air brings exceptionally hot conditions in summer, and the map illustrates conditions during the hottest part of the heat wave of 1976 in late June, with temperatures exceeding 30 °C at 1800 on 28 June 1976. The air comes from the east and brings hot, dry weather, especially to

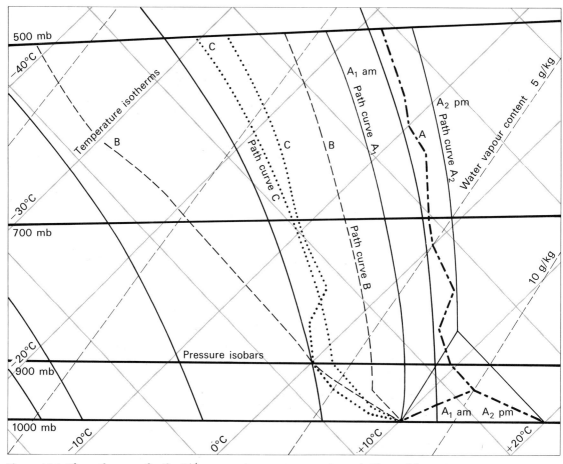

Figure A3.1 The path curves for the T/ϕ gram environment curves shown in Figure 3.1.

the south and central parts of the country, with relatively little cloud and no precipitation as long as the high pressure persists.

Exercise 3.3

The map, Figure A3.2a, shows the position of the warm front, behind which there is a sudden increase of temperature. The cold front separates a very mild tropical maritime air mass from one only a little colder so that the occlusion is a warm one. The cold air ahead of the warm front is much colder, and snow is falling over the Midlands and eastern England, turning to rain behind the front. The isallobars show that the depression is still deepening, as it moves northeast, because pressure is still falling fast ahead of the front and in the warm sector.

The upper air data shown in Figure A3.2b indicate the contrasts of the air masses meeting to form this vigorous front. At Stornoway it occurs in the form of an inversion as the warm air lies above the cold. The front is at 525 mbar at Stornoway, a height equivalent to about 4,500 m at 0900. By this time the front at ground level would be about 400 km away, giving a slope of 1 in 89. At Liverpool the front is much nearer the ground. Valentia is situated in the warm air which extends vertically upwards throughout the (T/ϕ)gram, while Lerwick, on the other hand, is situated entirely in the cold air ahead of the front. From

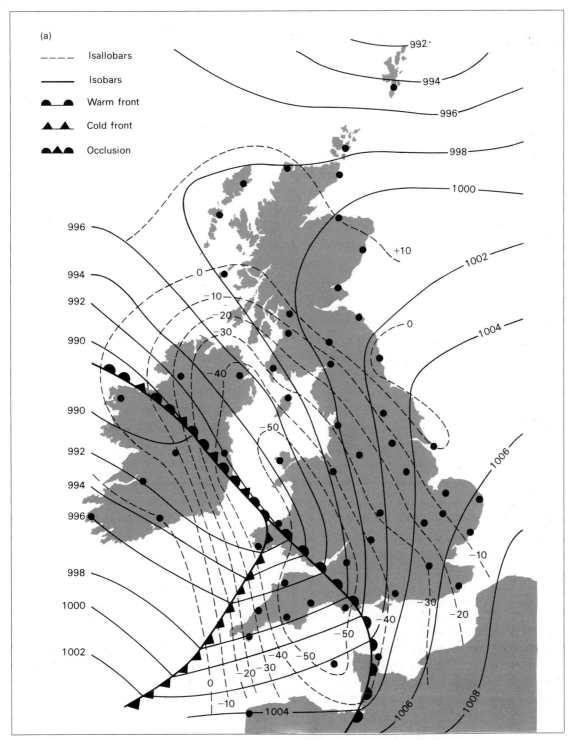

Figure A3.2 (a) Weather map showing isobars, isallobars and fronts for 4 January, 1949, 0600. (b) Upper air data for four stations for 4 January, 1949.

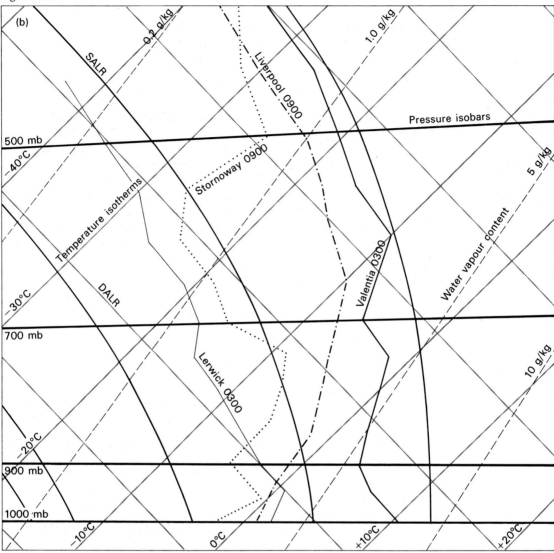

the (T/ϕ)gram it is possible to note the difference in temperature at each level of the two air masses involved in this particular situation, which is of a vigorous frontal system in midwinter, on 4 January 1949. The two air masses can be identified by their characteristics, and the vigour of the system stems from the great contrast between them. The pattern of the wind indicates a vigorous cyclonic circulation, the air moving in an anticlockwise direction around the central low pressure. This pattern accounts for the abrupt change from a southeast wind ahead of the front to a west wind behind it. The high strength of the wind also reflects the steep barometric pressure gradient near the frontal surface on the ground, while weak winds occur in the shallow ridge running over Scotland ahead of the frontal system.

Exercise 3.4

Regression analysis results for runoff and

River Dove	$r = +0{\cdot}6666$	$Y = 1{\cdot}296 + 0{\cdot}398\,X$
River Stour	$r = +0{\cdot}093$	$Y = 0{\cdot}684 + 0{\cdot}018\,X$
River Beauly	$r = +0{\cdot}9174$	$Y = 1{\cdot}340 + 0{\cdot}651\,X$

rainfall for the three rivers is shown in Table A4.

Figure A3.3 shows the regression lines for the three equations, the scatter is very wide for the River Stour, confirming the non-significant result. This is due to the very low runoff in the summer compared with the winter, owing to the high proportion of rain lost by evaporation in the summer compared with the winter. The heavier precipitation and lower evaporation of the Beauly basin ensure a closer and highly significant correlation. The Dove is intermediate in all respects. The figure also shows the close relationship between discharge and runoff for the rivers Dove and Beauly. The slope of the regression lines indicates that a greater proportion of the rainfall runs off as the climate becomes wetter

Figure A3.3 Graphs to show monthly runoff, precipitation and discharge/km² for Rivers Beauly, Dove and Stour. Runoff is also plotted against discharge and precipitation.

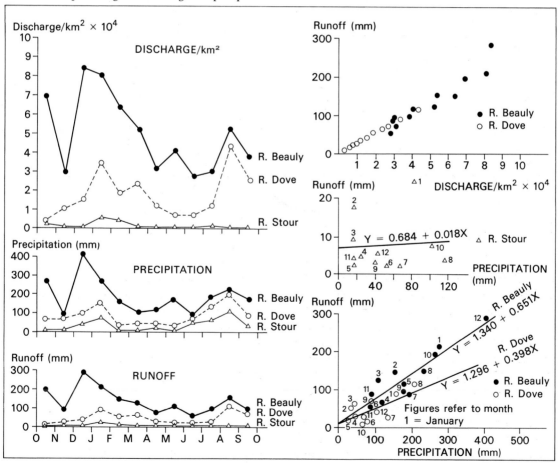

and cooler. Thus the regression line is steeper for the river Beauly and least steep for the driest, warmest area in the southeast of the country. The monthly values show that on the whole the summer months tend to have a lower proportion of runoff to rainfall, the values of 7, 8, 9 and 10 generally lying nearer to the rainfall axis than the runoff one, due to increased summer evaporation. The months of February and March have the highest percentages of runoff to rainfall, when the evaporation is at a minimum and the ground water storage is at a maximum. The precipitation pattern throughout the year shows the characteristically dry spring, during the months of February to May, with high values for the winter months of December and January. In this particular year November was notably dry, while August was exceptionally wet, although in the southeast even the heavy August rains did not increase the runoff very greatly. In the wetter part of the country, however, the runoff for August was exceptionally heavy.

Exercise 3.5

The river regimes are as follows:
  1 = Simple glacial regime
  2 = Simple glacial regime
  3 = Simple glacial regime
  4 = Simple pluvial oceanic regime
  5 = Simple pluvial oceanic regime
  6 = Simple tropical pluvial regime
  7 = Simple tropical pluvial regime
  8 = Simple mountain snow regime
  9 = Simple plains snow regime
 10 = First degree complex snow-rain regime
 11 = Mediterranean regime
 12 = Central continental regime
 13 = Second degree complex snow-glacial—transitional regime
 14 = Second degree complex regime—the Rhine
 15 = Second degree complex regime—the Rhine further downstream
 16 = Second degree complex regime—the Rhine near its mouth

Figure A3.4 Graph to show height distribution, net, summer and winter balances relative to elevation, and net balance related to area for 50 m intervals for Engabreen, Norway.

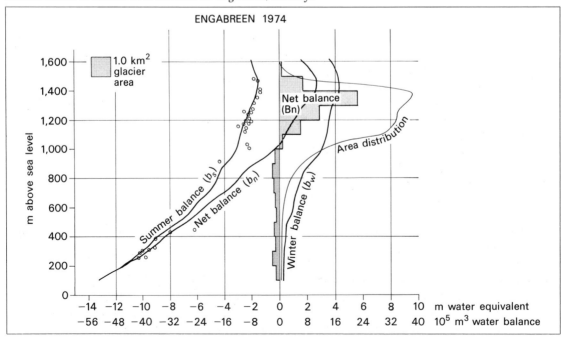

The simple regimes have only one peak while the first degree complex ones have two peaks and the second degree complex several peaks. In the latter rivers the different tributaries provide a more even flow throughout the year with their different sources.

Exercise 3.6

Figure A3.4 shows the pattern of height distribution and net balance for Engabreen. In contrast to Trollbergdalsbreen it has a much higher proportion of its area at a high altitude so that its net mass balance is positive, despite very high ablation rates at low levels in summer compared with the other glacier. The value for the winter mass balance is $129{,}030 \times 10^6$ m³, while the summer balance is $98{\cdot}710 \times 10^6$ m³, giving a total net mass balance of plus $30{,}320 \times 10^6$ m³, which is a large positive mass balance compared with a small negative one for Trollbergdalsbreen. These measurements indicate some of the complications in deducing climatic change from the behaviour of glaciers.

Exercise 3.7

Figure A3.5 shows the relationship visually between the precipitation and the weight of vegetation. The first graph (a) is plotted on semi-logarithmic paper and the second (b) shows the logarithmic values of vegetation weight against the precipitation on the linear scale. The correlation coefficient indicates that there is a close relationship between the density of vegetation production and the precipitation, as the value of $r = 0{\cdot}8248$. The regression equation is $Y = 3{\cdot}0176 + 0{\cdot}0121\,X$, where X is the independent variable, precipitation and Y is the dependent one, density of vegetation. The graph of the relationship shows that coniferous forests produce proportionately more vegetation mass for a given precipitation than the tropical forest, when the data are plotted logarithmi-

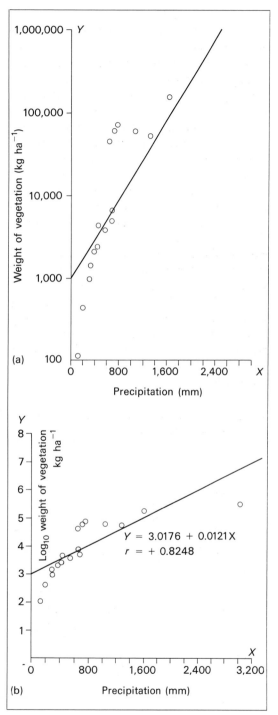

Figure A3.5 (a) Semi-logarithmic plot of vegetation weight against precipitation. (b) Linear plot of logarithmic vegetation weight against precipitation with regression line.

287

cally. This is probably due to the greater effectiveness of the precipitation in the cooler conditions characteristic of the coniferous forest zone compared with the tropical forests. The sparse vegetation of the desert zones produces less vegetable matter in proportion, as these points lie below the regression line.

Exercise 3.8

The data given in Table 3.4 are in the form suitable for analysis by the chi square test to assess the significance of the change in vegetation since the introduction of European species of plants and animals. The expected frequencies are as follows:

	1	2	3	4	5 and 6
Natural cover	11·5	5·5	11	11	11
Present cover	11·5	5·5	11	11	11

$$\frac{\Sigma(o - e)^2}{e} = \text{chi square} = 30 \cdot 31,$$

with $(r - 1)(k - 1) = 4$ degrees of freedom. The value of chi square with 4 d.f. $= 18 \cdot 46$ at the $0 \cdot 001$ significance level. Thus it can be established with a high degree of confidence that the introduction of alien species has resulted in a serious deterioration of the natural vegetation, with a great increase of shingle and rock and a reduction of beech forest and subalpine vegetation.

Exercise 3.9

Figure A3.6 shows the pattern of diatom growth and the factors on which it depends.

Exercise 3.10

Figure A3.7 shows the position concerning

Figure A3.6 Pattern of diatom production through the year in the North Sea with the variables on which it depends. (After A. J. Lee, 1958)

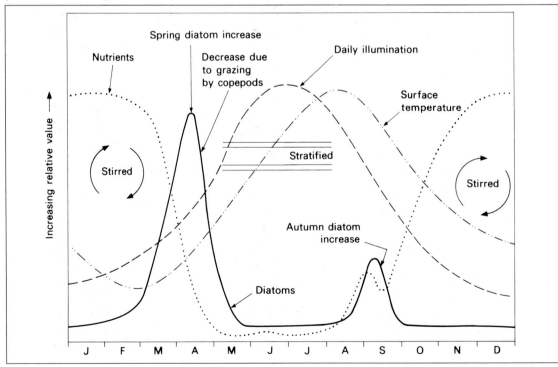

upwelling in the southern hemisphere. Upwelling takes place when the wind blows with the land on the right of its direction, and sinking occurs where the land lies to the left of the wind direction. When upwelling occurs water is transported offshore and the sea surface slopes up in this direction. The reverse occurs when the water tends to sink, with an offshore downslope.

Exercise 3.11

The method of calculating the trends is indicated in the table below:

$Y = a + bX$
$a = 2{,}515/15 = 167{\cdot}7$
$b = -8{,}534/280 = -30{\cdot}48$
$Y = 167{\cdot}7 - 30{\cdot}48\,X$

	Y	X	XY
1950–51	571	−7	−3,997
1951–52	420	−6	−2,520
1952–53	305	−5	−1,525
1953–54	220	−4	− 880
1954–55	173	−3	− 519
1955–56	127	−2	− 254
1956–57	117	−1	− 117
1957–58	131	0	0
1958–59	99	+1	+ 99
1959–60	93	+2	+ 186
1960–61	120	+3	+ 360
1961–62	70	+4	+ 280
1962–63	62	+5	+ 310
1963–64	6	+6	+ 36
1964–65	1	+7	+ 7
ΣY	2,515	ΣXY	−9,812
			+1,278
			−8,534

The data refer to blue whales

Figure A3.7 Diagram to illustrate upwelling in the southern hemisphere. T = transport of water, W = wind direction, D = level of water surface.

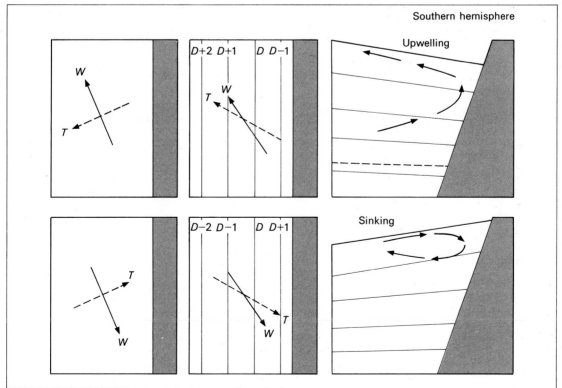

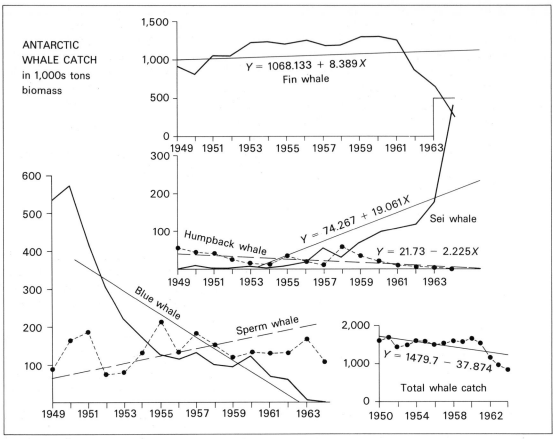

Figure A3.8a Graphs to show trend lines for different species of whales, numbers caught from 1949–64, and total whale catch.

Figure A3.8 shows the data plotted and the trend line calculated for the different whale species and the total whale catch. The trend equations are as follows:

Fin whales $\quad Y = 1,068·13 + \ 8·389 \, X$
Sperm
  whales $\quad Y = \ 143·70 + \ 9·811 \, X$
Humpback
  whales $\quad Y = \ 21·73 - \ 2·225 \, X$
Sei whales $\quad Y = \ 74·27 + 19·061 \, X$
Total whale
  catch $\quad Y = 1,479·70 - 37·854 \, X$

The equations show how the more valuable species, the blue whale, and later the fin whale were hunted until numbers fell drastically and then the less valuable species were increasingly taken, especially the Sei whales.

Exercise 3.12

The values for Spearman's ρ are found by ranking the data and finding the differences between the ranks of each pair of variables, then

$$\rho = 1 - \frac{6 \Sigma D^2}{n^3 - n},$$

where D^2 is the square of the differences and n is the number in the sample ($= 21$). For rock type and rock age $\rho = +0·747$, ranking old rocks low and hard rocks low. For rock type and relief $\rho = +0·815$, with high relief ranked low. The ρ value for rock age and relief is $+0·775$. All three values are significant at the 99% level of confidence because for

290

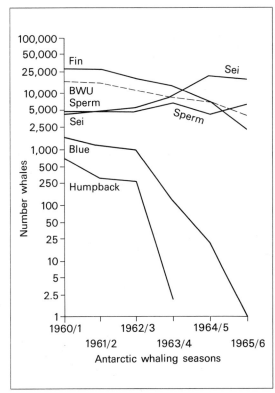

Figure A3.8b Number of whales caught in the Antarctic from 1960 to 1966. BWU = blue whale unit.

$n = 21$ the value of p in tables is 0.534. The test confirms the significance of the relationship between rock type and relief, and between the other pairs of variables. The highest correlation is between rock type and relief.

Exercise 3.13

The grouped means for the three squares are respectively from west to east 511 feet, 409 feet, and 271 feet. The surface represented by the figures slopes down to the east at 1 in 320 at first and then at 1 in 238, or 11′ and 15′ respectively. This method of analysis indicates the warping of a surface cut across the chalk, which is probably the eastern continuation of the warped surface of the Central Pennines, cut across Carboniferous strata.

As an example of the calculation of the grouped mean the eastern square is as follows:

100 × 3	=	300
150 × 13	=	1,950
200 × 20	=	4,000
250 × 32	=	8,000
300 × 21	=	6,300
350 × 7	=	2,450
400 × 4	=	1,600
	=	24,600

24,600/100 = 246, add 25 for the group interval to give 271.

Exercise 3.14

The data for Snowdonia are first simplified by deducting 17 from all values after dividing them all by 100. The correction factor is 5.44 and the analysis of variance table can be set out as follows:

Source of variation	Sum of squares	Degrees of freedom	V	F
Between aspect	21.56	2	10.8	3.92
Between group	74.89	2	37.45	13.46
Total	107.56	8		
Residual	11.1	4	2.78	

The variation between groups is significant but that between aspect is not in Snowdonia. The data for the Lake District can be analysed similarly, using the same initial corrections. The correction factor is 6.75 and the analysis of variance table is as follows:

Source of variation	Sum of squares	Degrees of freedom	V	F
Between aspect	2.92	3	0.97	0.19
Between group	6.50	2	3.25	0.63
Total	40.25	11		
Residual	30.83	6	5.14	

Direction	Northeast	Southeast and southwest	Northwest	Total
Groups 1 and 2	17/19	4/8	15/9	36
Groups 3, 4 and 5	21/19	13/9	3/9	37
Totals	38	17	18	73

Table A5 shows the chi square analysis of corrie frequencies in the Lake District.

The chi square value is 12·22, which is significant at the 99% (0·01) confidence level for 2 d.f. Note that some amalgamation of the groups has been necessary to make the expected values at least 5, a condition necessary for the application of the chi square test. The observed values are given first followed by the expected ones. The dominance of the northeast orientation is apparent in the data.

Exercise 3.15

The sum of the sine values is $\Sigma L \sin \alpha = +42 \cdot 0329$ and the cosine values is $\Sigma L \cos \alpha = +33 \cdot 7878$, the tangent value is $1 \cdot 2440$ which is $51 \cdot 205°$ or $51° \, 9'$. The L value is $53 \cdot 8$, which is significant. The orientation checks with that shown in Figure 3.15.

Exercise 3.16

The normal erosion rates are as follows:
Warm

equatorial	300
intertropical	225
total	360
Temperate	1,050

Cold

extra-polar	550
polar	350
Total unglaciated	2,835
Glaciated	4,000

Total unglaciated all precipitation groups = 3,823·7

The results show that the intermediate precipitation provides the most effective system, as in arid areas water is lacking and the humid areas support thick vegetation which protects the land surface. Glacial processes are the most effective, apart from humid areas which are warm and, therefore, support little ice cover.

Exercise 3.17

The results of the Spearman rank correlation test are as shown in Table A6.

The most significant correlation of the six that can be calculated is that between drainage basin area and suspended sediment load/km², which is negative. Next is that between average discharge and drainage basin area, which is positive. The total suspended load does not correlate significantly with any of the three variables. The average discharge correlates positively with basin area, which is what would be expected provided the precipitation values are not too dissimilar. It is also reasonable that the suspended sediment load/km² should correlate negatively with the discharge and the drainage basin area, as these two variables correlate strongly with each other. The smallest basins have the highest suspended sediment yield per unit area, while the larger ones, the Amazon and Mississippi, have low suspended sediment loads. The large basins have relatively large areas at low gradient compared with the smaller ones in general, so that erosion would be less effective over wide areas of the larger basins. The reason for the lack of correlation between the total sediment load and the other variables

1 = suspended sediment load/km²			3 = average discharge	
2 = suspended sediment load			4 = drainage basin area	

1 and 2	$\rho =$	0·3239	not significant
1 and 3	$\rho =$	−0·5147	significant at 95% negative relation
1 and 4	$\rho =$	−0·8176	significant at 99% negative relation
2 and 3	$\rho =$	0·2971	not significant
2 and 4	$\rho =$	0·1743	not significant
3 and 4	$\rho =$	0·6621	significant at 99%

indicates that the variables that control sediment load are not related to discharge or basin area, but are more likely related to relief, land-use and surface material.

Exercise 3.18

Figure A3.9 shows some of the processes involved in the different climatic zones. The nine morphogenetic types recognized are subject to the following processes:

Glacial Nivation, glacial erosion and deposition, wind action, strong frost action and moderate mechanical weathering.

Periglacial Strong frost action and mechanical weathering, strong nivation and mass movement, moderate to strong wind action and weak running water action.

Boreal Moderate frost action, moderate to slight wind action, moderate running water action and weak chemical weathering.

Maritime Strong mass movement and moderate to strong running water, strong to moderate chemical weathering, moderate mechanical weathering.

Selva Strong mass movement, no wind action, maximum chemical weathering, weak running water action.

Moderate Maximum running water, moderate mass movement, slight frost action, little wind action, moderate chemical and mechanical weathering.

Savanna Strong to weak running water, moderate wind and mechanical weathering, moderate to strong chemical weathering.

Semi-arid Strong wind action, moderate to strong running water, moderate to strong mass movement, moderate to weak mechanical and chemical weathering.

Arid Strong wind action, slight mass movement and running water action, weak chemical and mechanical weathering.

Exercise 3.19

Curve (a) shows eight major warm phases during the last 400,000 years, curve (b) also has eight oscillations, two of which are complex. Curve (c) (iv) only indicates four major warm interglacials. This result could be accounted for if feedback processes prevent glacial melting unless conditions are all favourable together. The absence of data on precipitation variation could be significant.

Exercise 3.20

Figure A3.10 shows the summation of the three Milankovich cycles, which are shown separately in Figure 3.19. A peak that reaches almost the maximum possible occurs after 50,000 years and a trough close to the minimum after 80,000 years. The actual

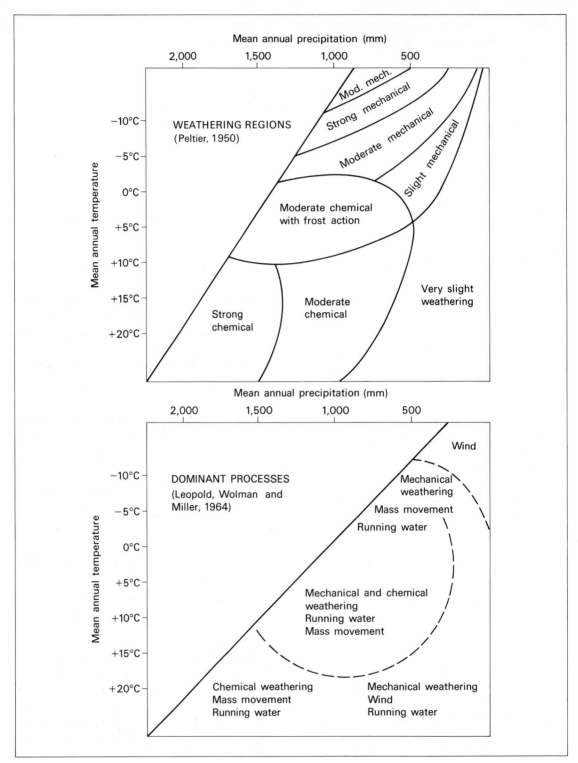

Figure A3.9 Weathering regions and dominant processes in relation to temperature and precipitation.

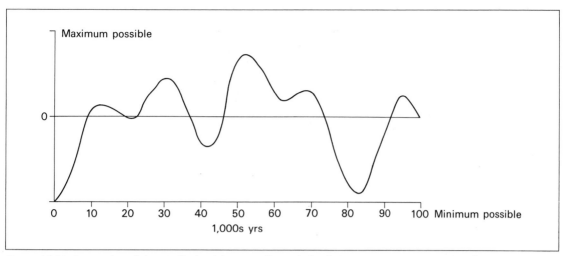

Figure A3.10 Summation of three Milankovich curves for periods of 40,000 years, 21,000 years and 96,000 years.

situation is much more complex than this simple example in which all the curves are of equal magnitude and start at their minima together. The fluctuations also affect every latitude differently.

Exercise 3.21

Figure A3.11 shows the decadal trends of rainfall, mildness/severity winter index for the period from 1730 to 1950, and the west wind frequency from 1880 to 1950. The trend equations are plotted on the graph and they show a rising trend in both variables over the period from 1730 to 1950, as the climate gradually improved after the Little Ice Age. The trend equations are as follows:

$$Y = +0.174 + 0.1937 \ X$$

for the winter mildness/severity index

$$Y = +95.087 + 0.437 \ X$$

for the rainfall in mm

The close relationship between the west wind frequency and the winter index over the period from 1880 to 1950 is clearly shown by comparing the graphs for these two variables. Both rise to a peak in the period 1920 to 1930. This relationship is confirmed by the correlation analysis.

Before calculating the correlation coefficients and regression equations add 10 to the values for the winter severity/mildness index to eliminate the negative values. There are 8 values in these data sets and the results of the analysis are as follows:

West wind days correlated with rainfall, taking west wind days as X and rainfall as Y_1

$$r = +0.4766 \quad Y_1 = 82.157 + 0.1337 \ X$$

West wind days correlated with winter index, west wind days $= X$, winter index $= Y_2$

$$r = +0.9649 \quad Y_2 = -11.613 + 0.2014 \ X$$

Rainfall correlated with winter index, rainfall $= Y_1$, winter index $= Y_2$

$$r = +0.6095 \quad Y_2 = -32.44 + 0.45335 \ Y_1$$

These data are illustrated in Figure A3.11 in graphical form. The closest correlation is between the winter index and the number of westerly winds. This result confirms the importance of the zonal index, related to the character of the upper westerly wind system, to the winter weather. A high zonal index is associated with strong westerly winds and milder winters.

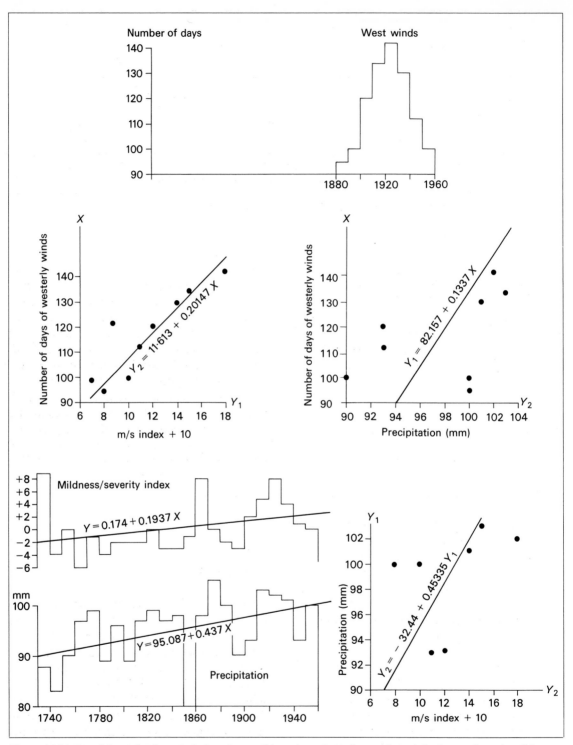

Figure A3.11 Decadal trends of west winds, winter mildness/severity index, and precipitation, and graphs of the relationships between these variables.

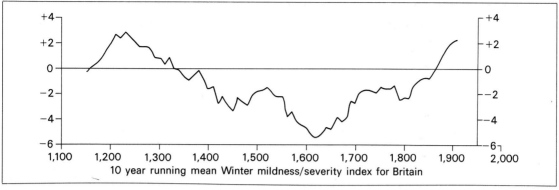

Figure A3.12 10-year running mean for winter mildness/severity index for Britain.

Exercise 3.22

Figure A3.12 shows the variation of the winter mildness/severity index plotted both as decadal values and as a 10-year running mean. The running mean shows clearly the gradual deterioration over the period from 1230 to 1610, the time of the Little Ice Age, and the gradual improvement up to the beginning of the twentieth century.

Exercise 3.23

The correlation of the time series given in Table 3.15 gives the following results:

Britain and Russia, 38 pluses, 39 minuses, 7 zeros. The chi square value is 0·013, which is not significant because there are nearly as many pluses as minuses.

Britain and Germany, 48 pluses, 17 minuses, 19 zeros. The chi square value is 7·392, which is significant at the 99% confidence level with 1 d.f.

Germany and Russia, 35 pluses, 44 minuses, 5 zeros. The chi square value is 0·513, which is not significant.

The results suggest that the winters of Britain and Germany vary in the same way. The fluctuations in Russia are on the whole more extreme and do not correlate with either of the other series. There was a very severe period of hard winters in Russia in the seventeenth century.

Exercise 3.24

The rise of sea level is shown in Figure A3.13 as a much straighter line on the semi-logarithmic scale, although the correction is rather overdone as the curve slopes the opposite way to the linear plot. The pattern of the rise suggests that the melting of the land-based northern hemisphere ice sheets, which was mainly responsible for the sea level rise, was very rapid at first and then slowed down markedly as their bulk was substantially reduced and they retreated to the colder, higher and more northerly locations. Small remnants still survive in these areas, including the Barnes ice cap of Baffin Island and the Jostedalsbreen of Norway.

Exercise 3.25

Figure A3.14 shows the uplift and emergence curve. The uplift curve is calculated by taking the total uplift of 59 m as 100% at 10,000 years since deglaciation. The emergence curve is plotted by allowing for the eustatic effect. Thus at 10,000 years it is 59 − 31 m = 28 m, which is the marine limit. As uplift slows down the eustatic effect becomes smaller so the emergence curve flattens. Note that the two time scales go in opposite directions for the two curves.

297

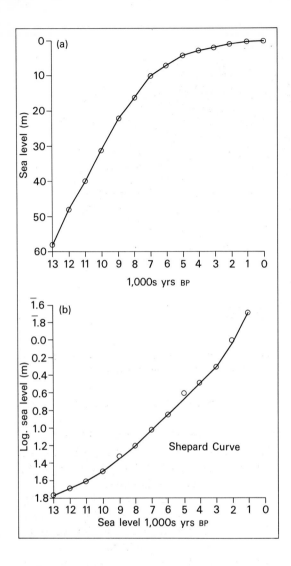

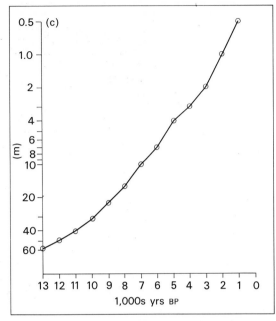

Figure A3.13 (a) Shepard's sea level curve for the last 13,000 years – linear plot. (b) Shepard's sea level curve for the last 13,000 years – logarithmic values. (c) Shepard's sea level curves for the last 13,000 years – semi-logarithmic paper plot.

Figure A3.14 Uplift and emergence curves for Cape Henry Kater, east Baffin Island where deglaciation took place 10,000 years ago.

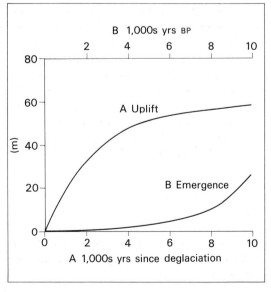

Exercise 4.1

The area–height curve illustrates the important contrast between the crust of the oceanic plates and that of the continental part of the plates. The two main levels are those of the continental shield areas and the deep ocean basins, representing the continental and oceanic crustal types respectively. The isostatic balance is maintained by the compensation between density and thickness of the two types of crust. The total mass of 40 km of oceanic crust and mantle is 114·0 and of 40 km of continental crust and mantle is 114·4.

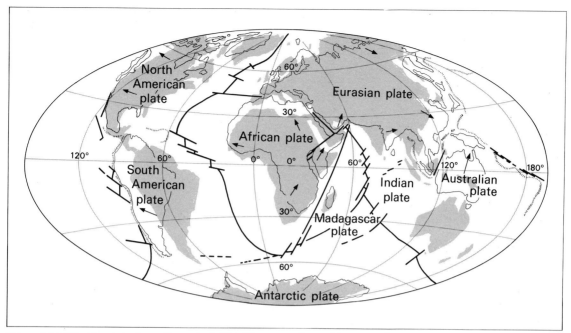

Figure A4.1 World geography as it may look 50 million years hence if present plate movements continue. (From 'The breakup of Pangea', R. S. Dietz and J. C. Holden. © 1970 Scientific American, Inc. All rights reserved)

The greater density of the former crust compensates for the greater thickness of the latter crust.

Exercise 4.2

Figure A4.1 shows the projected pattern of the plates 50 million years hence according to Dietz and Holden (1970). The Atlantic Ocean would be considerably wider and the Pacific Ocean smaller. Britain would have moved eastwards, Africa northeastward and Australia would lie across the equator, while North and South America would be further west. Active sea floor spreading would account for the movements.

Exercise 4.3

The radius of the earth is used to convert angular distance along the great circle of the earth into km:

$$\frac{2\pi R}{360} = \text{distance of } 1° = 111 \cdot 1 \text{ km.}$$

(1) Therefore for $2 \cdot 1°$ the value is $233 \cdot 31$ km in 10 million years, or $2 \cdot 33$ cm/yr. (2) $1 \cdot 8$ cm/yr, (3) $4 \cdot 95$ cm/yr, (4) $4 \cdot 4$ cm/yr, (5) $2 \cdot 5$ cm/yr, (6) 340 km maximum displacement.

The Atlantic Ocean is $50°$ wide at $30°$N, which is 5,555 km. If the spreading rate were $2 \cdot 3$ cm/yr it would take 242 million years to form, at $2 \cdot 5$ cm/yr it would take 222 million years, and at $3 \cdot 0$ cm/yr it would take 185 million years.

Exercise 4.4

The three types of margin are distinguished on Figure A4.2. The association of deep sea trenches and island arcs with destructive margins is clearly shown, particularly in the Pacific Ocean, while the constructive margins are associated with the central oceanic ridges. The conservative margins are more restricted in extent.

299

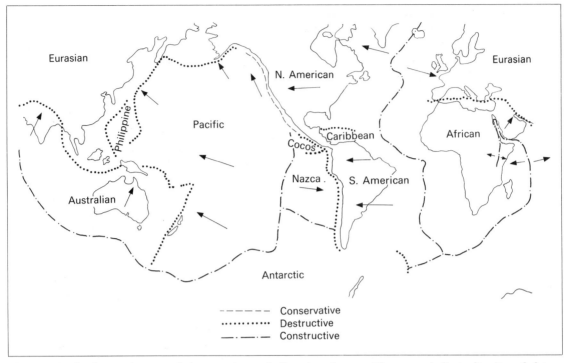

Figure A4.2 The three main types of plate margins are indicated on the map. The arrows indicate direction of plate movement.

Exercise 4.5

The correlation coefficient is $r = +0\cdot7743$, which is significant at $0\cdot01$ confidence level in a two-tailed test for $n = 19$. The regression equations are shown in Figure A4.3 against the appropriate line. The dependent variable, Y, is best taken to be the distance from the ridge with age being the independent variable, X. This equation gives the slowest rate of expansion of $1\cdot8$ cm/yr. The other regression equation gives a rate of $3\cdot0$ cm/yr, while the mean of the two gives a value of $2\cdot4$ cm/yr, which is a very reasonable value.

Figure A4.3 Graph to show the relationship between the age of volcanoes and their distance from the mid-oceanic ridge in the Atlantic and Indian oceans. The equations for the regression lines are given.

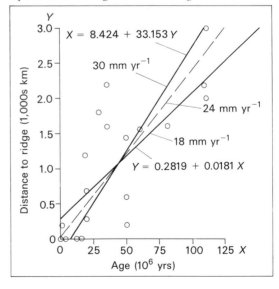

Exercise 4.6

The orogenic belts around the Siberian shield get progressively younger away from the shield. A similar pattern applies on the east of the Australian shield and on the southeast of the Canadian shield. The Cainozoic or Alpine orogenic belt is formed where plates are colliding with continental crust at one or both margins. The Andes and Rockies illustrate the former and the Alps and Himalayas the latter. The Urals mark a zone along which the European and Asian plates collided and became welded into one plate.

Exercise 4.7

The shield areas are generally of low relief, being higher in Africa, which was central Pangea before the splitting up of the plates and the creation of the Atlantic and Indian Oceans. It is in these areas that L. C. King has identified old, extensive erosion surfaces, referred to as the Gondwana surface and African, etc. surfaces. The shields have not been involved in orogenic processes since at least the pre-Cambrian period. Approximate altitudes are as follows:

North Canadian shield	about	200 m to 1,000 m
Scandinavian shield	about	200 m
Siberian shield	up to	1,000 m
African shield	about	1,000 m to 2,000 m
Indian shield	up to	1,000 m
Chinese shield	about	1,000 m
Australian shield	up to	600 m
Brazilian shield	about	1,000 m

Exercise 4.8

Figure A4.4 illustrates the subsidiary currents, which have a velocity of about $\frac{1}{5}$ those of the main current. The divergences occur at the equator, where the Coriolis effect reverses, and at about 10 °N. Convergences occur at about 4 °N and about 25 °N and S. The pattern is due to the thermal equator lying north of the geographical equator so that the southern

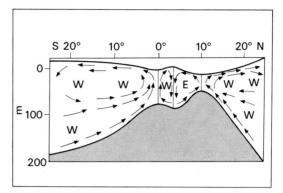

Figure A4.4 The subsidiary currents caused by the rotation of the earth in the equatorial zone. Convergences occur at 4 °N and 25 °N and S, divergences occur at 0 ° and 10 °N.

hemisphere west-flowing equatorial current straddles the equator. The vertical exaggeration is 11,110 times. This shows that the surface currents take place in a very shallow layer of water compared with the depth of the ocean, which is about 5,000 to 6,000 m, or about $\frac{1}{60}$th of the depth.

Exercise 4.9

The correlation coefficients for the two sets of data and the regression equations are as follows:

Atlantic Ocean
$$r = +0.850,$$
$$Y = 34.996 + 0.0149 \, X$$

Pacific Ocean
$$r = +0.489,$$
$$Y = 34.580 + 0.0098 \, X$$

X is the difference between evaporation and precipitation, and Y is salinity. The correlation shows a significant relationship between the two variables, significant at the 99·9% level for the Atlantic and at the 95% level for the Pacific. As evaporation increases relative to precipitation so the surface salinity increases. The values fall to the south of the thermal equator (including 5 °N) in the Pacific Ocean,

301

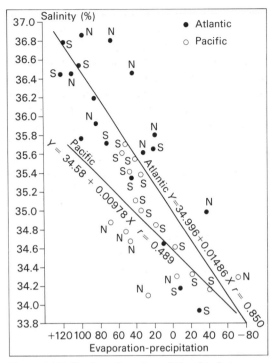

Figure A4.5 Graph to show the relationship between salinity and evaporation-precipitation for the Atlantic and Pacific oceans with regression lines.

and are all more saline than those in the northern hemisphere. The same pattern applies, but rather less consistently in the Atlantic Ocean. This is accounted for by the higher precipitation values that are commonly recorded in the northern hemisphere. For the Atlantic Ocean from latitudes 10° to 40° the total precipitation in the north is 439 cm/yr and 283 cm/yr in the south, while in the Pacific the values 563 cm/yr and 524 cm/yr for the north and south respectively. These results can be explained by the higher temperatures in the northern hemisphere, giving a greater water-holding capacity to the air and hence the greater precipitation. Figure A4.5 shows the data plotted in graphical form with the regression lines plotted.

Exercise 4.10

The T/S curve is shown in Figure A4.6. It may

be compared with that shown in Figure 4.9. The two curves show a similar pattern, with salinity first increasing with depth and then decreasing. The two curves represent the same water mass which occurs at different levels at the two stations recorded. The water has the same character at 277 m depth and 590 m depth, and at 461 m and 790 m in the two samples respectively, the second sample thus lies deeper than the first.

Exercise 4.11

The salinities of water masses at 15 °C of the central and equatorial surface water masses are given in order of decreasing salinity as follows:

Mediterranean water	more than 36·5‰
North Atlantic central water	36·0‰
South Atlantic central water	35·5‰
Indian Ocean central water	35·5‰
West Pacific south central water	35·45‰
Indian equatorial water	35·2‰
North Pacific equatorial water	35·15‰
South Pacific equatorial water	35·1‰
East Pacific south central water	35·0‰
West Pacific north central water	34·6‰
East Pacific north central water	34·4‰

The relative densities may be assessed by noting the salinity values for one temperature from the T/S curves, as indicated above. The Atlantic and north Indian Oceans are the most land-locked and hence have a higher salinity for a given temperature. The north Atlantic has the highest values for salinity owing to its narrowness and large adjacent land masses. The Pacific has two closed circulations and hence four different central water masses. The least dense are those of the north, which are warmer relative to their salinity, owing to the cooling effect of the Antarctic. All the equatorial waters are similar and in general have a smaller range of salinity than the central waters, owing to lesser seasonal variations in temperature.

302

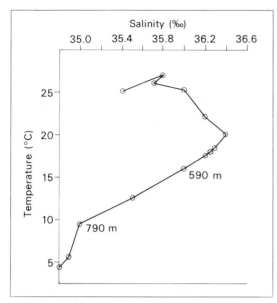

Figure A4.6 *T/S* curve on which salinity is plotted against temperature. Depths are given in m on curve.

the surface, and 9 million m³ s⁻¹ of Antarctic intermediate water moving immediately below the surface water. At the bottom 3 million m³ s⁻¹ of Antarctic bottom water is moving north. At the equator the amount of north-moving Antarctic intermediate water has been reduced by 7 million m³ s⁻¹ to 2 million m³ s⁻¹. Thus 8 million m³ s⁻¹ of water is moving north in the upper layers with 9 million m³ s⁻¹ moving south below it composed of north Atlantic deep water and Mediterranean water. To balance the flows 1 million m³ s⁻¹ must move north on the bottom in the Antarctic bottom water mass; this means that 2 million m³ s⁻¹ must turn back southwards in this lowest and densest water mass between 30 °S and the equator.

Exercise 4.12

Since the amount of water moving north at 30 °S amounts to 18 million m³ s⁻¹ this amount must also move south. The north-moving water is made up of 6 million m³ s⁻¹ moving at

Exercise 4.13

Figure A4.7 shows the northerly component split into its semi-diurnal and diurnal elements with the observed tide force corrected for the constant term. The pattern for the easterly components is also shown. The calculations are as follows in Table A7.

Table A7

	Northerly component						Easterly component				
Hour	0–11	Hour	12–23	0–11 Semi-diurnal	0–11 Diurnal	Hour	0–11	Hour	12–23	0–11 Semi-diurnal	0–11 Diurnal
0	−0·65	12	−0·15	−0·40	−0·25	0	0·00	12	0·00	0·00	0·00
1	−0·59	13	−0·11	−0·35	−0·24	1	−0·34	13	−0·47	−0·40	+0·06
2	−0·42	14	+0·02	−0·20	−0·22	2	−0·58	14	−0·83	−0·70	+0·12
3	−0·17	15	+0·18	0·00	−0·18	3	−0·63	15	−0·99	−0·81	+0·18
4	+0·08	16	+0·33	+0·20	−0·13	4	−0·48	16	−0·92	−0·70	+0·22
5	+0·29	17	+0·42	+0·36	−0·07	5	−0·16	17	−0·65	−0·40	+0·24
6	+0·41	18	+0·41	+0·41	0·00	6	+0·25	18	−0·25	0·00	+0·25
7	+0·42	19	+0·29	+0·36	+0·07	7	+0·65	19	+0·16	+0·40	+0·24
8	+0·33	20	+0·08	+0·20	+0·13	8	+0·92	20	+0·48	+0·70	+0·22
9	+0·18	21	−0·17	0·00	+0·18	9	+0·99	21	+0·63	+0·81	+0·18
10	+0·02	22	−0·42	−0·20	+0·22	10	+0·83	22	+0·58	+0·70	+0·12
11	−0·11	23	−0·59	−0·35	+0·24	11	+0·47	23	+0·34	+0·40	+0·06

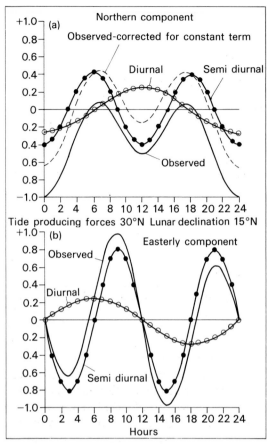

Figure A4.7 Graphs to illustrate (a) northerly and (b) easterly components of the observed, and diurnal and semidiurnal components of the lunar tidal tractive force at 30 °N for a lunar declination of 15 °N.

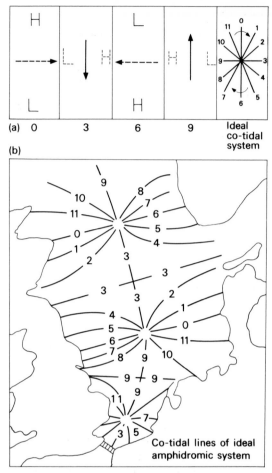

Figure A4.8 (a) Ideal amphidromic system in the southern hemisphere. (b) Ideal amphidromic system in the North Sea.

Exercise 4.14

Figure A.4.8 shows the ideal co-tidal chart for the North Sea. The tide rotates around the amphidromic points in an anticlockwise direction. The pattern would be reversed in the southern hemisphere.

Exercise 4.15

The values for the different elements of the atmospheric energy systems are as follows:

A	3 + 17 =	20
B	=	23
C	=	4
D	24 + 21 =	45
E	=	4
F	=	113
G	=	107
H	=	97
I	=	63
J	=	6
K	=	23
L	=	6
M	=	4
N	=	31
O	=	69
P	=	100

Total space	$N = (B + C + E)$,	$O = (I + J)$

Gain	Loss
P = 100	N + O = 31 + 69 = 100

Atmosphere

Gain from				Loss to		
Short wave	M + A	=	4 + 20	Space	I	= 63
Long wave	G	=	107	Surface	H	= 97
Convection	K	=	23			
Conduction	L	=	6			
			160			160

Earth

Gain from			Loss to		
Short wave	D	= 45	Atmosphere (long wave)	F	= 113
Long wave	H	= 97	Convection	K	= 23
			Conduction	L	= 6
		142			142

The completed table is as above (Table A8).

Exercise 4.16

Figure A4.9 shows the total transfer of heat as the sum of the elements given in Figure 4.17. The curve reaches a peak at 40 °N of just under 4×10^{19} kcal/yr. The peak in the southern hemisphere is at about 37 °S and is rather larger than that in the northern hemisphere, giving steeper gradients, as both curves fall to zero at the poles and equator. The curves for sensible heat and latent heat are different in the two hemispheres owing mainly to the distribution of land and sea, and hence variations in the pattern of evaporation and surface heating. The steeper gradients in the southern hemisphere are related to the stronger pressure gradients and wind velocities are higher in the south.

Exercise 4.17

Figure A4.10a illustrates the trade wind system with air moving from the subtropical high pressure centre towards the equatorial low pressure zone, and deflected to the west to form the northeast trade winds in the northern hemisphere. Figure A4.10b illustrates the westerly wind system with air moving in general from subtropical high pressure centres

Figure A4.9 Total transfer of energy in the earth-atmosphere system. (After R. G. Barry and R. J. Chorley, 1976)

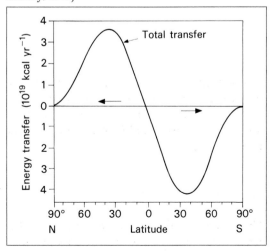

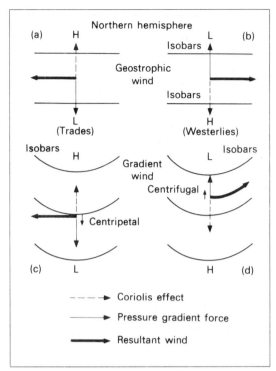

Figure A4.10 Diagram to illustrate the influence of the Coriolis effect and the isobar curvature on the geostrophic and gradient winds for the northern hemisphere.

centre. Figure A4.10d illustrates the circulation around a low pressure centre with curving isobars, causing outblowing (c) and inblowing (d) winds respectively.

Exercise 4.18

Figure A4.11 shows the data plotted for January, giving the pattern of wind frequencies from 1500 to 1950. The chi square test shows that there is a very significant difference in the frequencies of the winds during the period from 1550 to 1690 compared with the period from 1700 to 1850. The chi square value is 38·601 which is significant at the 99·9% level of confidence. The calculations are set out as shown in Table A9.

The result indicates that there were substantially more northern and eastern winds in the early period while westerly winds were also more frequent, but southerly winds were very greatly reduced in frequency. This would relate to the cold weather of the Little Ice Age during the early period. In the later period southerly winds become more frequent and the climate is generally warmer as a result. The weather reflects changes in the winter frequencies rather than the summer, as the chi square test applied to the July figures gives a non-significant result, with chi square equal to only 3.15. The data are shown in Table A10.

towards the low pressure zone of the polar front to form southwesterly winds in the northern hemisphere. Figure A4.10c illustrates the circulation around a high pressure

Table A9

Frequencies for 15 decades	Observed/Expected				
	North	East	South	West	Total
January 1559–1690	49/41·2	51/46·15	41/71·4	137/119·2	278
January 1700–1850	26/33·8	33/37·85	89/58·6	80/97·8	228
	75	84	130	217	506

Table A10

	North	East	South	West	Total
July 1550–1690	49/45·4	6/9·2	52/55·5	120/117	227
July 1700–1850	50/53·6	14/10·8	69/65·5	135/138	268
	99	20	121	255	495

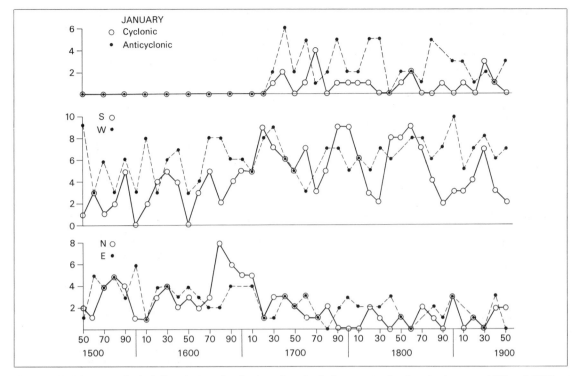

Figure A4.11 Pattern of wind direction, anticyclonic and cyclonic frequencies for decades between 1550 and 1950 for January. Data from H. H. Lamb, 1966.

The tests do, however, show that there is a very significant difference in wind frequencies at the two seasons for both the early and late decades. In the early period easterly winds are much less common in summer, when they would bring warmer conditions than in winter, when they would bring cold air from the east to western Europe. The data are as in Table A11.

The same pattern continues to a lesser degree in the later period, for which the chi square value is 20·90 compared with a value of 68·03 for the earlier period. The data are as in Table A12.

Table A11

		North	East	South	West	Total
July	1550–1690	49/44	6/25·6	52/41·8	120/115·5	227
January	1550–1690	49/54	51/31·4	41/51·2	137/141·5	278
		98	57	93	257	505

Table A12

		North	East	South	West	Total
July	1700–1850	50/41	14/25·4	69/85·4	135/116·2	268
January	1700–1850	26/35	33/21·6	89/72·6	80/98·8	228
		76	47	158	215	496

The Spearman correlation coefficient gives a value of $\rho = 0.9161$, which is highly significant. The details are set out as follows:

Place	Rank— west coast distance	Rank— temper- ature	d	d^2
Reykavik	2	4	−2	4
Archangel	7	7	0	0
Warsaw	6	6	0	0
Plymouth	2	1	+1	1
Norwich	5	3	+2	4
Moscow	8	8	0	0
Barnaul	11	10	+1	1
Verkhoyansk	12	12	0	0
Godthaab	2	5	−3	9
Churchill	10	11	−1	1
Winnipeg	9	9	0	0
Punta Arenas	4	2	+2	4
				$\Sigma d^2 = 24$

$$\rho = 1 - \frac{24 \times 6}{1,716} = 1 - 0.0839$$

$$= 0.9161$$

Exercise 4.20

The pattern of precipitation and temperature variation throughout the year at the stations given range from the tundra of North America through the various forest and desert zones of both North and South America. The relationship between vegetation and climate is shown by plotting the vegetation type against the precipitation and temperature, using the mean annual values. The range of climates illustrated is wide, ranging from the great temperature range of Baker Lake with its very cold winter to the perpetual heat of Manaos, with its high and well distributed rainfall, and associated tropical forest vegetation. The very low values of precipitation at Arica in north Chile with its moderate temperature provides another striking contrast.

The line separating forest from grassland runs through a precipitation value of 25 cm

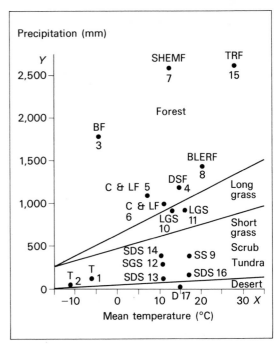

Figure A4.12 Graphs to show the relationship between precipitation and temperature, and vegetation type. (See Table 4.11 for type of vegetation.) The lines separate forest from long grass (upper) and long grass from short grass, scrub and tundra (lower).

with a temperature of −15 °C to a precipitation of 125 cm with a temperature of 25 °C, while that separating long grass from short grass and desert and tundra vegetation runs from the same lower point to a precipitation of 75 cm with a temperature of 20 °C. The equations for these lines are $Y = 62.5 + 2.5 X$ for the first and $Y = 47.5 + 1.43 X$ for the second, which has a lower gradient. Y is the precipitation and X is the temperature. They are shown on Figure A4.12.

Exercise 4.21

Figure A4.13 shows the data for Sydney, Benares, Paris and Honolulu. Sydney is notable for its relatively equable climate, and Paris also shows relatively little annual variation, with a greater temperature range than precipitation range, while Sydney has a

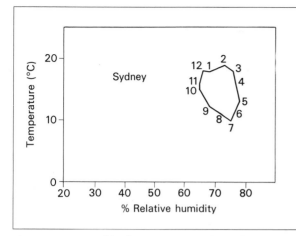

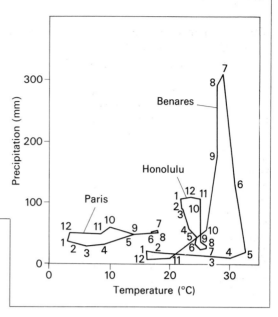

Figure A4.13 Monthly climatic variation for Sydney, Honolulu, Paris, and Benares.

moderate range of both variables. Honolulu has a small temperature range, with high values throughout the year and more rainfall variation. Benares, on the other hand, shows extreme rainfall variations owing to the summer monsoon, and its temperatures are continuously high except for the moderate winter values.

Exercise 4.22

The following figures give the % runoff per $km^2 \times 10^6$.

Africa	0·57
Asia	0·47
Australia and NZ	0·25
Europe	0·78
Greenland	0·53
Malay Archipelago	4·61
North and Central America	0·88
South America	1·25

These figures illustrate the importance of climate and vegetation, which in turn is dependent on climate and relief, in accounting for the great variation in the proportion of total global runoff. The dry areas, Australia and Asia provide the least, while Africa provides an average amount, as its large desert area is compensated by its wet equatorial belt, and Greenland counteracts low precipitation with low temperatures, thus reducing evaporation. Europe has a temperate climate with fairly heavy precipitation and moderate temperatures. The really high values are provided by the countries with large areas in the equatorial rain forest zone, the Amazon basin of South America, and the area of heavy monsoon rain in southeast Asia, the Malaysian area.

The following figures give the percentage runoff in the different zones of the USSR.

Tundra	76
Taiga and permafrost	49
Taiga	48
Mixed forest	36
Mixed forest	30
Forest steppe	19
Steppe	9
Semi-desert	5

The largest percentage of runoff occurs in the tundra and taiga zones, where precipitation is moderate and evaporation fairly low. At the other extreme is the high evaporation of the steppes and semi-desert, which accounts for

Table A13

Volga	1931–1967,	$n = 19,$	$Y = 12 \cdot 0 + 0 \cdot 509\ X$
Amu Dar'ya	1952–1968,	$n = 17,$	$Y = -2 \cdot 118 - 0 \cdot 417\ X$
Syr Dar'ya	1952–1968,	$n = 17,$	$Y = -0 \cdot 706 - 0 \cdot 270\ X$
Caspian level	1850–1930,	$n = 17,$	$Y = 25 \cdot 7 - 0 \cdot 011\ X^*$
Caspian level	1930–1970,	$n = 9,$	$Y = 27 \cdot 4 + 0 \cdot 282\ X^*$
Aral level	1920–1960,	$n = 9,$	$Y = 53 \cdot 0 + 0 \cdot 023\ X$
Aral level	1960–1970,	$n = 3,$	$Y = 52 \cdot 1 - 1 \cdot 150\ X$

*Note that the sign for the trend of the Caspian level is positive when the water level is falling and negative when it is rising because the level is recorded in metres below sea level, hence values increase as the level falls.

Figure A4.14 Trend lines for the time series of variations in discharge of the Volga, Syr Dar'ya, Amu Dar'ya, and levels of the Caspian and Aral Seas.

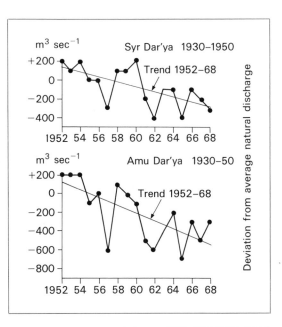

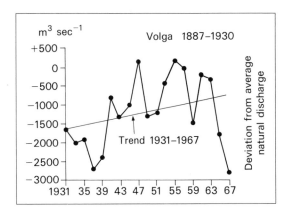

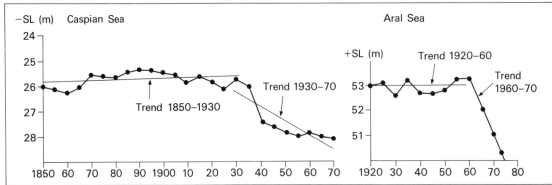

310

almost all the precipitation which is in any case low. The permafrost is important in accounting for the high surface runoff of the tundra and northern taiga, while the heavy, episodic but rare precipitation of the dry areas accounts for the high percentage of surface runoff in these zones.

Exercise 4.23

The trend equations are as shown in Table A13. The trend lines are plotted on Figure A4.14. The rapid fall in level of the Caspian took place around 1935 to 1940, when the Volga showed large negative anomalies in its discharge. The fall in the level of the Aral Sea took place mainly after 1960, since this time both the Amu Dar'ya and the Syr Dar'ya have shown strong negative anomalies in their discharge. All three rivers show very erratic discharges, due in part to climatic variability.

Glossary

Ablation: All the processes that lead to a reduction of mass in a glacier, including melting, calving, evaporation, wind erosion, avalanching.

Accumulation: All the processes that lead to an increase of mass in a glacier, including snow fall, rain if it freezes on impact, sublimation (direct condensation from vapour), avalanching onto the glacier, refreezing of water, wind-blown snow.

Active layer: The upper layer of soil or rock in a periglacial climate that melts annually.

Adiabatic: A change of air temperature brought about by vertical displacement without addition or subtraction of heat. The vertical displacement causes a change of pressure, which results in a change of volume of the air. The change in volume results in a change of temperature.

Air mass: A large body of air that usually obtains its characteristic temperature and humidity in a zone of high pressure. It also has a characteristic lapse rate that may be altered as it moves away from its generating air, causing it to become either more or less stable.

Albedo: The proportion of incident radiation that is reflected from a surface, or reflection coefficient.

Altimetric analysis: The analysis of the distribution of height in geomorphology. It may be done by taking all the summit heights over a given area, using closed contours, or a random or regular series of elevations may be analysed.

Amphidromic system: The basic pattern of tidal motion, consisting of a wave moving anti-clockwise around the amphidromic point in the northern hemisphere and clockwise in the southern. The tidal range increases away from the amphidromic point. The associated tidal streams are also rotatory.

Anabatic: A local upslope wind generated as a result of heating in the valley.

Angular momentum: The tendency of the earth's atmosphere to move around the axis of rotation with the earth. It is proportional to the rate of spin, the angular velocity and the square of the distance from the axis of rotation.

Antarctic convergence: The zone around the southern ocean in about 50°S where there is a rapid increase of temperature northwards. The Antarctic Intermediate water sinks to the north of it and the Antarctic Circumpolar Water rises to the south of it.

Anticyclone, blocking: A high pressure system that prevents the normal west to east movement of fronts in the temperate zone near the polar front. It results in long spells of weather in any one area, and is related to the pattern of Rossby waves in the upper westerlies.

Aquifer: A permeable stratum from which

water can be readily abstracted and into which ground water can penetrate easily. Chalk and uncemented sandstone form good aquifers, or well-jointed limestone and volcanic rocks.

Arête: A sharp ridge separating two cirques and formed by glacial erosion.

Artesian: Ground water that rises naturally to the surface from an aquifer beneath impermeable strata where the intake for precipitation is higher than the ground level at the artesian spring.

Aseismic: A zone in which earthquake activity is unlikely to occur. Used to refer to oceanic ridges not actively developing at present.

Asthenosphere: A zone within the earth's mantle at a depth of about 100 to 200 km where the viscosity is low. It lies below the lithosphere, which is more rigid, and is sometimes referred to as the Low Velocity Zone, owing to the slower speed with which earthquake waves travel through it.

Bankfull discharge: That discharge that fills a river to the top of its banks but does not overflow.

Base flow: The normal dry weather flow in a river that is maintained by the outflow of ground water into the stream.

Benioff zone: The steeply dipping (about 45°) plane along which subduction of the crust takes places at the destructive plate margins. It is the zone along which deep focus earthquakes take place to a depth of 700 km, owing to the low temperature and high rigidity of the descending crust.

Benthos: Organisms that live on the bottom of the sea, including sea-weeds and a wide variety of animals.

Cailleux roundness: A measure of the degree to which a stone has been rounded by erosion. It is given by $2r/l \times 1000$, where r is the minimum radius of curvature in the principal plane and l is the length of the longest axis.

Catena: The variation of soil type along a slope, reflecting changes in drainage and other properties down the slope.

Chernozem: Dark soils characteristic of steppe grassland, and referred to as black earths. A fertile soil.

Climatic optimum: The period of most favourable weather in the post-glacial time lasting from about 4000 to 2000 B.C. when temperatures were 2–3 °C higher than now. There was a secondary optimum with a peak from 800–1000 A.D.

Conduction: The passage of heat by means of transfer of adjacent molecular motions. Air is a poor conductor, but earth is a good one.

Confidence limit: A statistical term for the degree of likelihood of an event occurring. A 95 per cent confidence level indicates that the event could occur by chance one time in twenty; a 99 per cent level once in 100 times.

Convection: The chief method of atmospheric heat transfer. Free thermal convection results from surface heating. Forced convection can occur when eddies develop as a result of air flow over uneven surfaces.

Coriolis effect: The deflection of moving objects owing to the rotation of the earth. The deflection is to the right in the northern hemisphere and to the left in the southern. It varies with the sine of the latitude, decreasing from the poles towards the equator.

Covalent bond: A bond between atoms in which they share outer electrons.

Cyclothem: A cycle of sedimentary sequence well exemplified in the Yoredale rocks of the Carboniferous, in which limestone is overlain by shale, sandstone, a thin seat earth and sometimes a narrow coal seam. It is repeated about 8 times and indicates a rhythmic change of conditions of gradually shallowing sea to become estuarine, with increasingly coarse sediment becoming available.

Date Line: The meridian which is approximately antipodal to the Greenwich meridian across which time changes by 24 hours.

Declination: The tilt of the earth's axis of rotation to the plane along which it rotates around the sun. It is now $23\frac{1}{2}°$, but varies cyclically. At times of maximum declination the sun is overhead at the tropics at the solstices, while at the equinoxes the sun is overhead at the equator and the declination is zero.

Demersal fish: Fish that live near the bottom of the sea. Some are adapted for life near the bottom by becoming flat fish, such as plaice, others are round, such as cod.

Dendrochronology: The use of tree rings for dating. The trees develop an annual pattern in areas where there is a strong contrast in the seasons.

Desertification: The process whereby the over-exploitation of the land leads to a deterioration of the soil and vegetation, giving rise to desert conditions.

Diatom: A minute oceanic plant that has a case of silica and which reproduces by splitting, thus increasing in geometric proportion when conditions are suitable. It can synthesize organic matter from the inorganic nutrients in sunlight.

Dip: The angle of inclination of bedded strata. The maximum slope of the strata. The angle is measured on a plane perpendicular to the strike.

Earthquake, deep focus: Deep focus earthquakes can take place along the Benioff plane down to a depth of about 700 km, where the crust is subducted into the mantle in the northwest and southeast marginal zone of the Pacific Ocean.

Ecosystem: The interaction of plants and animals with their environment produces an ecosystem.

El Nino: The abnormal situation in which the cold water upwelling along the coast of Peru and Chile is temporarily prevented and warm water extends southwards along the coast to the detriment of the marine life. It occurs approximately every 7 years, there having been a particularly serious occurrence in 1972–3.

Endemic: Plants and animals that have developed over a long time in an area.

Endogenetic: Processes that operate within the earth.

Equation of time: The difference between

mean time and solar time; it is due to the obliquity of the ecliptic and the ellipticity of the earth's orbit, and reaches a maximum of 16 minutes.

Equilibrium line: The line across a glacier which separates the accumulation zone above from the ablation zone below. The line of no net gain or loss of ice.

Ergodic principle: The use of space for a surrogate for time.

Eustatic: Changes of sea level that are world wide and brought about by changes in the amount of water in the oceans or the capacity of the ocean basins. Glacioeustatism is the change due to the variations in the volumes of ice on land, and these changes are the most significant at present.

Evapotranspiration: The combination of evaporation and transpiration in producing the total loss of moisture from the land surface and vegetation.

Exogenetic: The processes that operate on the surface of the earth, including subaerial processes, subglacial and submarine processes.

Exponential: An exponent is the power to which a mathematical expression is raised. The rate increases or decreases at an accelerating speed until the inflection point is reached, when it reverses sign.

Feedback: The interaction between variables whereby a change in one is the cause of a change in the one that first caused the change. Negative feedback is self regulating, leading to dynamic equilibrium, while positive feedback is self generating, leading often to a vicious spiral.

Flandrian transgression: The rapid postglacial rise of sea level resulting from the rapid melting of the northern hemisphere ice sheets between about 15000 and 5000 years ago.

Fourier analysis: The analysis of time series to establish the presence of cycles within the data and the relative amount of energy in each cycle.

Fournier ratio: The relationship between the

maximum monthly precipitation and the mean annual precipitation, p^2/P. It is significant in assessing the rate of erosion as determined by sediment yield in streams.

Front: The plane along which two different air masses come into contact. The polar front usually separates polar air masses from tropical ones, and at a warm front tropical air replaces polar, while at the cold front polar air undercuts tropical air. An occlusion or occluded front is one where the warm air has been lifted off the surface and the two colder air masses come into contact.

Geomorphological map: A map on which specific landforms are shown. Often forms are shown symbolically, and colour may be used to differentiate different processes, such as glacial erosion or fluvial deposition.

Geostrophic flow: In the free air above the zone of friction the pressure gradient force is modified by the Coriolis effect to produce the geostrophic wind. The wind instead of blowing from high to low pressure blows parallel to the isobars, with low pressure to the left and high to the right in the northern hemisphere and the reverse in the southern. The speed varies with the pressure gradient and also to a lesser degree with the latitude.

Geosyncline: A major downwarp of elongated form in the earth's crust in which great thicknesses of sediment accumulate in the early stages of orogeny over a long period often of several 100 million years. Sediments can reach thicknesses of over 10 km, and can eventually form folded mountain ranges, as the Caledonian mountains.

Glacier activity index: The ratio of the change in mass balance over vertical position at the equilibrium line of the glacier, given in mm/m. A low activity index indicates a slow change of mass balance and is characteristic of cold, polar glaciers with low accumulation and ablation rates. These glaciers are sluggish compared with those with a high activity index, often in temperate glaciers with large accumulation and ablation rates and rapid flow.

Gley: Soils associated with conditions of poor drainage with periodic waterlogging. The pH is low and a process of reduction occurs, especially affecting iron. The soils are often mottled as a result of the reduction of iron.

Gradient wind: The geostrophic wind is modified when the isobars are curved by the addition of the centripetal force, causing the air to move on a curved path in towards the centre of low pressure and out from a centre of high pressure.

Greywacke: A rock often formed in a developing geosyncline and often deposited by turbidity currents in deep water. The sediment is poorly mixed and often contains much felspar.

Hadley cell: The cell named after G. Hadley, who in 1735 put forward the idea of a thermally direct cell of vertical atmospheric circulation, with air rising at the equatorial low pressure zone moving poleward aloft to subside in the subtropical high pressure zone and then move equatorward as the trade winds on the surface. A somewhat modified version of the idea is still accepted.

Helicoidal: A type of flow that has a spiral element. It occurs when a river is flowing around a meander, with the flow having an upward element at the outside of the bend and a downward one at the inside, leading to the formation of a point bar.

Humus: The organic element in the soil. Acid humus is called 'mor' and basic humus is called 'mull'.

Hydrograph: The relationship between precipitation and the flow of a river. The flood hydrograph shows the effect of a single precipitation event of large dimension on the flood flow of a stream, which is divided into a rising limb and a recession limb, when the flow returns to normal base flow level.

Hypsometric: The relationship between area and elevation. The world hypsometric curve shows the two main levels of the continental and oceanic sectors of the earth's crust.

Inflection point: The point on a curve where

the direction of curvature changes. It may separate a negative from a positive exponential element of a double exponential curve or the convex from the concave element of a slope profile.

Inherited groundwater: Water that is not being replenished in the ground under present hydrological conditions, but which accumulated during a wetter phase.

Inlier: An outcrop of rock that is entirely surrounded by younger strata.

Intertropical Convergence Zone: The zone where the northeastern and southeastern trade winds converge. A zone of calms, the doldrums, with strong upward convectional activity.

Interstadial: A warmer period within a glaciation that does not last long enough to be called an interglacial, but which is marked by minor glacial retreat.

Isallobar: Lines joining points of equal barometric tendency, which is the rate of rise or fall of pressure, usually over a period of three hours.

Isobar: Lines joining places with equal barometric pressure, usually drawn in millibars. Sea level pressure is approximately 1000 mb.

Isohyet: Lines joining places of equal precipitation.

Isostatic: Isostatic adjustment is brought about by changes in mass. Isostatic depression takes place as mass is added, such as the build up of ice sheets on land, and recovery and uplift take place as the ice melts and mass is reduced. Other processes including sedimentation, erosion and addition or subtraction of water can lead to isostatic adjustment. Isostatic anomalies refer to areas where the balance is not in adjustment, as along the deep sea trenches.

Isotherm: Lines joining places of equal temperature.

Jet stream: Relatively narrow bands of high velocity air flow in the upper atmosphere close to the tropopause. These rapid air flows move from west to east above the polar front and the subtropical jet flows eastwards above the subtropical high pressure centre at a rather higher altitude than the polar front jet. Velocities can reach 45 to 70 m-sec^{-1} or even up to 135 m-sec^{-1}.

Katabatic: Down slope winds that blow down the lee side of mountain ranges. The winds are usually warm and dry as the rising air cools at the S.A.L.R. and warms on the lee side at the D.A.L.R. The Föhn wind of the Alps, and the Chinook of the Rockies are examples of katabatic winds. The term 'katabatic' is also used to describe the drainage of cold air down a mountain side at night.

Kinetic energy: The energy provided by motion. Heat energy can be converted into kinetic energy by the rising of air that is warm or the sinking of cold air.

Krill: The Norwegian term for *Euphausia superba*, a small shrimp-like member of the zooplankton, which is the main food of the large Antarctic whalebone whales.

Krumholz: Literally 'crooked tree', a term used to describe trees distorted by strong winds near the upper limit of tree growth in exposed places.

Lapse rate: The rate of change of temperature with height. The dry adiabatic lapse rate is 9·8 °C-km^{-1} and occurs when nonsaturated air rises. The saturated lapse rate is less than the dry and varies with temperature, which in turn controls the amount of moisture in the air. The difference is caused by the release of latent heat of condensation as the air rises. For high temperatures the S.A.L.R. can be as low as 4 °C-km^{-1}.

Latent heat: The change in state from solid to liquid and from liquid to gas requires energy. The energy is provided by the removal of heat causing a heat loss called latent heat. The reverse process of condensation or freezing releases latent heat. The latent heat of vaporization to evaporate 1 g of water at 0°C is 600 calories.

Lichenometry: The use of lichen measurement to establish the date at which the substrate became available for lichen growth. The method can be used for absolute dating when a lichen growth rate curve has been established. The method extends over several hundred years and can date moraines, for example.

Little Ice Age: The period lasting from about 1500 to 1850 A.D. when the climate was considerably colder and glaciers advanced to their historic maxima.

Mass balance: The relationship between ablation and accumulation on a glacier or ice sheet. The mass balance is positive when accumulation is greater than ablation and negative when ablation is greater.

Meridional: A distribution or movement in a north-south direction, as opposed to a zonal pattern, which is an east-west one.

Milankovich cycles: The regular variations caused by the movement of the earth around the sun that are thought to influence world climate. Three cycles are involved, the variation in the tilt of the earth's axis with a period of 40,000 years, the precession of the equinoxes with a period of 21,000 years and the ellipticity of the earth's orbit with a period of 96,000 years.

Morphogenetic region: A region that can be differentiated on the basis of its landforms that are a result of the effect of different climates, such as the glacial, periglacial or semiarid landform regions.

Morphological map: A map that represents the form of the earth's surface in terms of changes and breaks of slope and slope units.

Morphometry: The analysis of landforms by quantitative measurements from maps of the dimensions of particular features, for example the area of drainage basins or the length of streams.

Munsell colour chart: Three variables are considered in assessing the colour of soil. The 'hue' indicates relationship to spectral colours, the 'value' indicates its lightness or darkness, and the 'chroma' represents the strength relative to a neutral colour of the same value. The soil to be tested is matched against sample colours using variables listed.

Nappe: A large scale overfold involved in the orogenic activity that has created the Alpine mountain chain. The material involved has been intensely folded and metamorphosed, indicating high temperatures and great pressures deep in the earth's crust when deformation took place.

Neogene: The later part of the Tertiary period, including the Miocene and Pliocene.

Niche: A term implying a particular environment to which an individual species is adapted.

Nivation: Processes associated with the activity of snow, giving rise, for example, to nivation hollows.

Non-parametric: Applies to statistical tests where values are given in the ordinal or nominal scale or in terms of frequencies. The requirements of normality of the data distribution are no longer necessary.

Ooze: Deep sea deposit formed of the remains of minute organisms that have accumulated slowly over a long time period. Siliceous oozes include diatom ooze formed of minute plants, and radiolarian ooze. Calcareous oozes include Globigerina and pteropod oozes.

Orogenesis: The process of fold mountain formation, usually beginning with a geosyncline in which thick sediment collects, and then folding occurs before the emergence of the mountain chain as a relief feature, the whole process taking hundreds of million years.

Orthogonal: A line at right angles to another, used to refer to a line at right angles to wave crests in wave refraction close to the shore.

Outlier: An isolated outcrop of rock entirely surrounded by older strata.

Ozone: Ozone is a recombination of oxygen (O_3) and is concentrated mainly between 15 and 35 km up in the atmosphere. It absorbs solar and terrestrial radiation and thus affects the heat budget of the earth. Its creation is caused by ultraviolet radiation from the sun mainly between 30 and 60 km in the upper atmosphere.

Palaeomagnetism: The reconstruction of the earth's ancient magnetic field from observations of remanent magnetization in ancient rocks. It can be used to deduce former latitudinal positions of the continents and to measure the rate of sea floor spreading from change in magnetic polarity in the ocean crust alongside the spreading ridges. Dating of deep sea cores can also be achieved by reference to magnetic reversals recorded in the cores.

Parametric: Statistical tests that use data on the interval or ratio scale of measurement, which are actual measurements, are called parametric. These tests assume that the distribution of values in the sample tested is normal, and that the sample is a random one.

pH: A measure of acidity or alkalinity in soil or water. $pH = \log (1)/(H^+)$. The hydrogen ion concentration for a neutral solution is $0 \cdot 0000001$ or 1×10^{-7} gram of H^+ per litre of solution, the $pH = \log (1)/0 \cdot 0000001) = 7$, higher values are alkaline and lower ones acid.

Palynology: The study of pollen grains to establish the vegetation at the time that they collected. The method can be used also for relative dating and correlating different deposits. Peat is a good preservative of pollen grains.

Pediplanation: The process of erosion based on the formation of pediments and the back-wearing of mountain scarps. The process operates most effectively in a semiarid or savanna type climate.

Pelagic fish: Fish that swim in the open sea, often near the surface. The herring is an example and it comes to the surface at night, descending deeper during the day. Tuna and mackerel are other pelagic fishes.

Peneplanation: The process of erosion that leads eventually to the formation of a peneplane mainly by subaerial slope processes and fluvial erosion.

Periglacial: An environment that is marginal to glacial areas, occurring in high latitudes and at high altitudes. Processes associated with frost and ground ice are important in this environment.

Permafrost: Perennially frozen ground. Permafrost covers a more restricted area than the periglacial environment, and it can be either continuous, as in very high latitudes, discontinuous, towards the margins, or sporadic, where it only occurs in the most favourable localities for freezing. The term 'talik' refers to small unfrozen areas within the permafrost.

Permeability: The ability of a rock or sediment to transmit water, either through pores or joints and bedding planes.

Photosynthesis: The process whereby living matter is created from nutrient elements in the presence of sunlight by plants, either on land or in water. Organic matter can only be produced where inorganic nutrients and light occur together.

Plankton: The drifting community of the ocean organisms. Phytoplankton is the plant element of the community, consisting of organisms such as diatoms. Zooplankton is the animal element of the plankton, and includes the larval and egg stages of various fishes and other creatures, and many other small animals, such as foraminifera, tiny shrimp-like animals, and many others.

Plate tectonics: The theory that considers that the earth's crust is divided into 6 large and several small plates that move about across the earth's surface at speeds about 2–5 cm-year^{-1}. The plates are about 100 to 150 km thick and therefore very thin relative to their areal extent. They may consist of both oceanic and continental crust.

Polycylic: A landscape that has developed under the influence of several different base levels, usually due to periodic falls in sea level, which may be due to sea level changes or changes in the level of the land.

Polygenetic: Landscapes that have developed under the control of different climatic regimes. Glacial conditions may, for example, have been followed by periglacial and/or temperate conditions, as in many parts of Britain.

Porosity: The percentage of a sediment or rock that is pore space. If the pores are relatively large the material is also permeable, but if they are very small, as in clay, it will be impermeable.

Pressure melting point: The temperature at which a glacier will melt at its base. The temperature of the melting point falls as the pressure increases. Thus the thicker the ice, the higher the pressure, and the lower the melting point is. Even the Antarctic ice sheet is at pressure melting point in its deeper basins, despite low temperatures.

Progressive wave: A type of wave that advances with a current flowing in the direction of advance at its crest and in the opposite direction at its trough. Its speed of advance depends only on its length and period in deep water ($d/L > 0.5$), but in very shallow water its velocity is $\sqrt{gd}$, $d = $ depth.

Protruberance: Undulations formed along long sandy beaches as a result of the development of a cellular pattern of circulation in the nearshore zone, often involving the formation of seaward directed rip currents.

Quadrat: A method of sampling vegetation by means of randomly located square sampling areas in which the vegetation types are counted.

Radiation: Solar radiation is the source of energy on which the earth's dynamic circulations depend and its life. Electromagnetic waves can transfer solar energy in the form of heat and light through space to the earth. The earth's atmosphere restricts the transfer of solar radiation to some specific wave lengths. The atmosphere receives most of its heat by re-radiation from the earth's surface.

Radiocarbon: The use of radiocarbon is an important method of dating that can be applied to organic remains and covers a period of about 40,000 years. The half life of Carbon 14 is 5730 ± 40 years. The radiocarbon trapped in the tissues of a living organism when it is alive starts to decay when it dies, by measuring the remaining radio carbon it can be dated.

Rating curve: The relationship between discharge and the height of a river. Once the rating curve has been established only the height need be recorded to obtain the discharge.

Resonance: A water body will respond to the tide producing force if its natural period of oscillation is close to that of the tide producing force. It depends on the length and depth of the water body.

Riffle: A shoal in a gravel river where the water is shallower. Riffles tend to be spaced 5 to 7 river widths apart and may be an early stage in meander development as they form in relatively straight reaches of the river.

Rocks: Rocks are divided into three main groups. *Igneous* rocks are those formed directly from molten material either deep within the earth to form plutonic rocks or on the surface as volcanic rocks. *Hypabyssal* rocks are formed at moderate depth often as sills or dykes. *Plutonic* rocks are normally coarsely crystalline, while volcanic ones are fine grained or glassy often. *Metamorphic* rocks form the second group. They are rocks altered by heat, pressure or both, and include quartzite, gneiss, schist, slate and marble. The third major group is the *sedimentary* rock group, consisting of consolidated sediments worn by erosion from earlier land masses. They include sandstones, siltstone, clay, limestone, chalk, conglomerates, greywackes and several other types.

Rossby waves: The long waves that develop in the upper westerlies above the polar front. They exert an important effect on the surface weather conditions. Three to six Rossby waves form around the earth and are influenced by the major relief features, including the Rocky Mountains.

Running mean: The calculation of the running mean smooths a time series. A number of years observations are averaged and placed opposite the central year. For the next value the earliest is dropped and the next added.

Salinity: The salinity of the sea is largely composed of sodium chloride, common salt.

Salinity is given in parts per thousand, ‰, and a common value is about 35‰, of which 30‰ are sodium chloride. Sulphate and magnesium are the next most common elements, but traces of very many elements are present. 75 per cent of all ocean water lies between 34·5 and 35‰ salinity.

Schist lustré: Highly metamorphosed rocks found in the central nappes of the Swiss Alps, mainly south of the Rhone valley. They indicate great pressure and temperature in their metamorphism.

Sea floor spreading: The theory that new ocean crust is created at the mid-oceanic ridges where plates are moving apart at their constructive boundaries. Thus much of the ocean floor is relatively young, and much younger than the water.

Shield: Those parts of the continental crust that have not undergone orogenic upheaval since at least the beginning of the Cambrian, although they do show evidence of earlier orogenic activity. Shields form the rigid core of all the continental plates, including the Canadian shield, the Baltic shield much of Africa, Brazil and western Australia.

Sinuosity: The degree of curvature of a river. It can be measured as the ratio of the length of the river to the valley within which it is flowing.

Skeletal soil: Soil that has not had time to develop a full profile, for example on rocks recently exposed by glacial retreat.

Skewness: A measure of the asymmetry of a distribution curve. It is a measure of the difference between the mean and the median values. If the ϕ notation is used, where $\phi = -\log_2$ mm, then a positive skewness implies a tail of fine particles and a negative skewness implies a tail of coarse particles. Skewness can be referred to as the third moment measure, where sorting is the second and the mean is the first.

Soil structure: The arrangement of the soil particles to form aggregates or peds provides a measure of the structure. It is the degree of aggregate development. Terms used to describe structure include structureless – with no observable peds, weak, moderate or strong structure according to the development of peds. Platy – vertical axis much shorter than horizontal. Prismatic – vertical axis longer than horizontal. Blocky – peds of similar dimensions. Crumb – soft, porous, granular peds, like bread crumbs.

Soil texture: The distribution of particle size less than 2 mm. Sand is 2·0–0·5 mm. Silt is 0·5 to 0·002 mm, and clay is less than 0·002 mm. The proportion of these constituents indicates the texture of the soil.

Sorting: The second measure used to describe the size distribution of a sediment sample. It is the equivalent of the standard deviation, indicating the spread of the data around the mean.

Spectral analysis: The analysis of time series to isolate the pattern of cyclic frequencies and the relative amount of energy that they contain.

Spherical excess: The amount by which the angles of a triangle exceed 180° when the sides of the triangle form parts of the great circles on a sphere. For example a triangle having two sides of 90° along meridians 90° apart, the angles add up to 270°.

Stability and instability in the atmosphere: The degree of stability can be defined as the relationship between the environment curve giving the change of temperature with height and the path curve, which gives the temperature that air under given surface conditions of temperature and humidity would have if it were raised from the surface. In stable air the path curve lies to the left of the environment curve so that rising air is colder than its surrounding and rises no further. In unstable air the path curve lies to the right of the environment curve, so that rising air is warmer than the air around and continues to rise, being less dense.

Stationary wave: When two progressive waves move in opposite directions, for example on reflection from a barrier, then the combined wave forms a stationary or standing wave. The wave pattern can be divided into oscillating zones every half wave length as no streams occur at points every half wave length. Within one oscillating system it is high water at one end and low at the opposite end, when the surface is flat streams reach their maximum and the pattern reverses.

Stereographic projection: A projection drawn by projecting the lines of latitude and longitude onto a plane tangent at one pole from the opposite pole. It has the property of preserving shape and angles correctly, and thus is useful in tidal force analysis and geological mapping problems.

Strike: A horizontal line at right angles to the dip of a rock stratum. It is the equivalent of a contour line on the surface of a rock stratum.

Surge: The term is used in two senses. In one it is applied to a local and temporary disturbance of sea level due to extreme meteorological conditions. The North Sea is liable to surges of this type, one serious one occurring in 1953 and another in 1978. The term is also used to describe a sudden glacier advance caused by instability between the accumulation and ablation zones. The upper part of the glacier falls in level while the lower part is raised and the front advances rapidly down valley for a short period, usually a few months.

Tacheometry: The measuring of distance by means of a level while surveying profiles. The upper and lower cross wires in the level eye piece are used to read the staff and the difference between these readings multiplied by 100 gives the distance between the level and the staff.

Taiga: The zone that lies between the tundra and the boreal forest, forming a broad belt in northern Eurasia and Canada. It is mainly forested by trees tolerant of harsh conditions.

Tephrochronology: A method of dating using layers of volcanic material ejected by known volcanic events. It has been used mainly in Iceland, New Zealand, Japan and the western United States. When the eruptions can be dated the deposits can be correlated by reference to the ash layers.

Threshold: A point at which there is an abrupt change in the variables that control any particular process. For example a threshold may be reached in the stability of a slope so that it suddenly collapses and a new set of controls are imposed.

Till fabric: The pattern of orientation and dip of elongated stones in a deposit of glacial till. It is generally true that the preferred orientation indicates the direction of ice movement, although there are exceptions.

Tombolo: A deposit of marine sediments that links an island to the mainland or one island to another.

Tractive tidal force: The force induced by the gravitational attraction between the moon and sun on the waters of the ocean that has an element parallel to the surface, thus impelling the water over the surface and setting up tidal currents and changes of level.

Transform faults: A fault that cuts across a spreading oceanic ridge to enable the moving plates to fit the earth's spherical surface. The opposite sides of the fault move in opposite directions both due to spreading and movement along the fault, giving an active earthquake zone between the two displaced sections of ridge.

Tropopause: The tropopause separates the lower part of the atmosphere, the troposphere, from the upper part, the stratosphere. It is about 16 km high above the equator and about 10 km above the poles. At the tropopause there is an inversion, making the troposphere more or less self-contained. The official definition of the tropopause is the lowest level at which the lapse rate decreases to less than or equal to $2\,°C–km^{-1}$.

T/S curve: The temperature-salinity curve that is used to define a water mass. It is obtained by plotting temperature against salinity for observations made at different depths within the water mass.

Tundra: The vegetation characteristic of the permafrost zone. It is a more or less treeless area, the only trees being dwarf birch and arctic willow, which rarely exceed about 30 cm in height.

T/φ gram: A diagram on which upper air data are plotted. Temperature is plotted against pressure. The curve showing the conditions in a vertical section is called the environment curve and it can be related to the path curve, which indicates the conditions of rising air under given surface conditions. On it pressure, temperature, DALR and SALR and water vapour content lines are shown.

Unconformity: The plane on a geological map between two sets of strata of differing ages. Usually one set has been tilted and eroded before the deposition of the younger set.

Vector: A line that has both direction and magnitude. The term is also used for a one dimensional array of data.

Vegetation: See Table 2.17 for a list of different types of vegetation.

Water mass: A large mass of water that can be recognized by its temperature-salinity relationships. Surface water masses occupy the uppermost 200 m of the ocean approximately, while intermediate, deep and bottom water masses fill the deeper basins. The only sources of deep and bottom water are found in high latitudes in the Atlantic Ocean.

Wave refraction: The bending of wave fronts as they approach the shore owing to the influence of depth on wave velocity and wave length. The wave crests tend to turn more nearly parallel to the depth contours. Orthogonals, drawn at right angles to the wave crests, can indicate zones of convergence of wave energy and zones of divergence, where waves will be lower.

Wave spectrum: The combination of waves of varying length and direction of movement that is normal in the open sea is called the wave spectrum. It can be resolved into the individual wave trains that make up the spectrum by spectral or Fourier analysis.

Western intensification: The asymmetry characteristic of ocean currents in which those flowing away from the equator on the western side of the ocean on the surface are much stronger and faster than those flowing towards the equator on the eastern side of the ocean on the surface. This asymmetry is the result of the variation of the Coriolis effect with latitude.

Woodhead seabed drifters: Plastic saucers weighted to move near the bottom of the sea. They are released at a known point and become stranded on the shore where they may be found and returned, thus giving information concerning the movement of water in the nearshore zone, normally dominated by tidal streams.

Zenith: The point immediately overhead. The zenith is 90° from the horizon.

Zonal index: The magnitude of the curvature of the Rossby waves in the upper westerlies blowing in the upper troposphere is described by the zonal index. A high zonal index implies a strong west to east component and minimum curvature, while a low zonal index implies maximum curvature with a well marked north-south component.

Index